Palaeobiogeography and Biodiversity Change: the Ordovician and Mesozoic–Cenozoic Radiations

Reviewing procedures

The Society makes every effort to ensure that the scientific and production quality of its books matches that of its journals. Since 1997, all book proposals have been refereed by specialist reviewers as well as by the Society's Books Editorial Committee. If the referees identify weaknesses in the proposal, these must be addressed before the proposal is accepted.

Once the book is accepted, the Society has a team of Book Editors (listed above) who ensure that the volume editors follow strict guidelines on refereeing and quality control. We insist that individual papers can only be accepted after satisfactory review by two independent referees. The questions on the review forms are similar to those for *Journal of the Geological Society.* The referees' forms and comments must be available to the Society's Book Editors on request.

Although many of the books result from meetings, the editors are expected to commission papers that were not presented at the meeting to ensure that the book provides a balanced coverage of the subject. Being accepted for presentation at the meeting does not guarantee inclusion in the book.

Geological Society Special Publications are included in the ISI Science Citation Index, but they do not have an impact factor, the latter being applicable only to journals.

More information about submitting a proposal and producing a Special Publication can be found on the Society's web site: www.geolsoc.org.uk.

It is recommended that reference to all or part of this book should be made in one of the following ways:

CRAME, J. A. & OWEN, A. W. (eds) 2002. *Palaeobiogeography and Biodiversity Change: the Ordovician and Mesozoic–Cenozoic Radiations.* Geological Society, London, Special Publications, **194.**

COPE, J. C. W. 2002. Diversification and biogeography of bivalves during the Ordovician Period. *In*: CRAME, J. A. & OWEN, A. W. (eds) *Palaeobiogeography and Biodiversity Change: the Ordovician and Mesozoic–Cenozoic Radiations.* Geological Society, London, Special Publications, **194,** 35–52.

GEOLOGICAL SOCIETY SPECIAL PUBLICATIONS NO. 194

Palaeobiogeography and Biodiversity Change: the Ordovician and Mesozoic–Cenozoic Radiations

EDITED BY

J. A. CRAME
British Antarctic Survey, Cambridge, UK

and

A. W. OWEN
University of Glasgow, UK

2002
Published by
The Geological Society
London

THE GEOLOGICAL SOCIETY

The Geological Society of London (GSL) was founded in 1807. It is the oldest national geological society in the world and the largest in Europe. It was incorporated under Royal Charter in 1825 and is Registered Charity 210161.

The Society is the UK national learned and professional society for geology with a worldwide Fellowship (FGS) of 9000. The Society has the power to confer Chartered status on suitably qualified Fellows, and about 2000 of the Fellowship carry the title (CGeol). Chartered Geologists may also obtain the equivalent European title, European Geologist (EurGeol). One fifth of the Society's fellowship resides outside the UK. To find out more about the Society, log on to www.geolsoc.org.uk.

The Geological Society Publishing House (Bath, UK) produces the Society's international journals and books, and acts as European distributor for selected publications of the American Association of Petroleum Geologists (AAPG), the American Geological Institute (AGI), the Indonesian Petroleum Association (IPA), the Geological Society of America (GSA), the Society for Sedimentary Geology (SEPM) and the Geologists' Association (GA). Joint marketing agreements ensure that GSL Fellows may purchase these societies' publications at a discount. The Society's online bookshop (accessible from www.geolsoc.org.uk) offers secure book purchasing with your credit or debit card.

To find out about joining the Society and benefiting from substantial discounts on publications of GSL and other societies worldwide, consult www.geolsoc.org.uk, or contact the Fellowship Department at: The Geological Society, Burlington House, Piccadilly, London W1J 0BG: Tel. +44 (0)20 7434 9944; Fax +44 (0)20 7439 8975; Email: enquiries@geolsoc.org.uk.

For information about the Society's meetings, consult *Events* on www.geolsoc.org.uk. To find out more about the Society's Corporate Affiliates Scheme, write to enquiries@geolsoc.org.uk.

Published by The Geological Society from:
The Geological Society Publishing House
Unit 7, Brassmill Enterprise Centre
Brassmill Lane
Bath BA1 3JN,
UK

(*Orders*: Tel. +44 (0)1225 445046
Fax +44 (0)1225 442836)
Online bookshop: http://bookshop.geolsoc.org.uk

British Library Cataloguing in Publication Data
A catalogue record for this book is available from the British Library.

ISBN 1–86239–106–8

Typeset by Type Study, Scarborough, UK
Printed by Hobbs the Printers, UK.

Distributors
USA
AAPG Bookstore
PO Box 979
Tulsa
OK 74101–0979
USA
Orders: Tel. +1 918 584-2555
Fax +1 918 560-2652
E-mail *bookstore@aapg.org*

India
Affiliated East–West Press PVT Ltd
G-1/16 Ansari Road, Daryaganj,
New Delhi 110 002
India
Orders: Tel. +91 11 327-9113
Fax +91 11 326-0538
E-mail *affiliat@nda.vsnl.net.in*

Japan
Kanda Book Trading Co.
Cityhouse Tama 204
Tsurumaki 1-3-10
Tama-shi
Tokyo 206–0034
Japan
Orders: Tel. +81 (0)423 57-7650
Fax +81 (0)423 57-7651

Contents

Preface

Biodiversity change is now one of the most important topics of investigation for biogeographer and palaeobiogeographer alike. Demonstrably, great shifts in the numbers of plant and animal taxa are occurring on certain parts of the Earth's surface at the present day, and much effort is currently being expended to determine why this should be so. Climate change in particular is held by many to be the most likely cause of dramatic range shifts and local extinctions.

With their longer time perspective, palaeontologists are continually surveying the fossil record for signs of global biodiversity change in the past. Much attention has rightly been paid to studying the spectacular mass extinctions during the Phanerozoic, but what happens in between them? Many palaeobiologists now believe that there may have been two pronounced intervals when life radiated (i.e. diversified) spectacularly: the Ordovician Period, and the mid-Mesozoic–Cenozoic eras. These episodes mark the steepest sustained rises on the 'curve of Life through time'; the intervening Silurian–Jurassic interval is the much flatter plateau, punctuated by mass extinctions and their recovery intervals.

Both the scale of these spectacular diversity increases and their probable causes are currently topics of intense debate. To the geologist, there is an intriguing link here between dispersed continents (as opposed to the presence of supercontinents), changing climates (both intervals ended with widespread glaciations), and the proliferation of life. We now know that the accumulation of Life on Earth is extremely complex; it is not just a matter of packing more taxa into any one habitat, but also of packing more habitats within a province, more provinces within a region, and more regions within the biosphere. Were the Ordovican and mid-Mesozoic–Cenozoic two intervals of time when there was a fundamental reorganization of biodiversity on a hierarchy of biogeographical scales?

Further study of these two great evolutionary radiations by a mixed group of geologists, palaeontologists and biogeographers was the theme of the 2001 Lyell Meeting convened on behalf of the Joint Committee for Palaeontology representing the Palaeontological Association, Geological Society of London and British Micropalaeontological Society. Entitled 'Palaeobiogeography and Biodiversity Change', it was held at Burlington House on 21 February 2001 and received generous financial support from the Geological Society and Palaeontological Association. Some 75 scientists attended and contributed to a lively debate on the fundamental causes of major evolutionary radiations. With half of the programme devoted to the Ordovician radiation, and half to the Mesozoic–Cenozoic one, it was possible to compare and contrast these two great evolutionary events.

The results of this meeting are presented in this volume as a series of individual papers. These represent by no means the last word on this important topic, but serve instead as an introduction to some of the key issues involved. In particular it is hoped that, collectively, they demonstrate how long periods of time and plate tectonic movements can have a fundamental influence on the generation and maintenance of major biodiversity patterns.

We are indebted to many colleagues for their help in refereeing the papers in the volume, including: H. A. Armstrong, R. Askin, C. Babin, M. J. Benton, M. Caron, A. Clarke, E. N. K. Clarkson, S. Damborenea, W. T. Dean, A. Hallam, R. Hill, P. Janvier, S. Lidgard, C. Little, R. A. Livermore, A. I. Miller, T. McCormick, J. A. Rasmussen, M. Romano, A. W. A. Rushton, D. J. Siveter, M. P. Smith, S. P. Tunnicliff, J. Whittaker and P. B. Wignall.

Alistair Crame and Alan Owen

Palaeobiogeography and the Ordovician and Mesozoic–Cenozoic biotic radiations

A. W. OWEN[1] & J. A. CRAME[2]
[1]*Division of Earth Sciences, University of Glasgow, Gregory Building, Lilybank Gardens, Glasgow G12 8QQ, UK (e-mail: a.owen@earthsci.gla.ac.uk)*
[2]*British Antarctic Survey, High Cross, Madingley Road, Cambridge CB3 0ET, UK (e-mail: A.Crame@bas.ac.uk)*

Abstract: Curves of taxonomic diversity through geological time consistently show major evolutionary radiations during the Ordovician Period and from the mid-Mesozoic to the present day. Both intervals were characterized by marked biotic provincialism, reflecting episodes of major continental break-up and global tectonism, and their later histories featured steep global climatic gradients. The Ordovician radiation can be recognized at a wide spectrum of taxonomic levels from species to class, and the biogeographical patterns associated with the radiation of individual clades reflect a complex combination of plate distribution, tectonic activity, sedimentary environment, sea-level rise and, ultimately, glaciation. The true scale of the mid-Mesozoic–Cenozoic biotic radiation is currently a topic of intense debate but there is no doubt that it affected plants and animals in both the marine and terrestrial realms. The role of land bridges and ocean gateways in controlling the formation of biodiversity patterns has been a persistent theme in Mesozoic–Cenozoic biogeography, and a complex set of Neogene tectonic events probably aided the development of both latitudinal and longitudinal provinces during the Cenozoic. The present volume highlights some of the successes across a spectrum of approaches to unravelling the Ordovician and Mesozoic–Cenozoic radiations within the context of palaeobiogeography.

'... geologic history is the pacemaker of biologic diversity' (Schopf 1979, p.454)

Post-Cambrian curves of taxonomic diversity through geological time consistently show a major evolutionary radiation during the Ordovician Period, and a second, larger one from the mid-Mesozoic to the present day. Between these steep upward slopes the Silurian–Jurassic interval usually appears as a distinct plateau punctuated by mass extinctions and their recovery intervals (e.g. Sepkoski 1981, 1993, 1997; Niklas *et al.* 1985; Padian & Clemens 1985; Benton 1995, 2000, 2001; Conway Morris 1998). Although this general pattern seems to be robust, it is becoming increasingly clear that it may include biases introduced by both the quality of the fossil record and the sampling procedures employed. For example, Miller & Foote (1996) have shown how the use of more rigorous taxonomic and diversity metrics can alter the perception of the Ordovician radiation, and Alroy (2000; Alroy *et al.* 2001), using a technique that standardizes sampling to absolute time intervals, has cast doubt on the scale of the mid-Cretaceous–Recent one. Potential biases on a variety of spatial and temporal scales have been further highlighted by Smith (2000; Smith *et al.* 2001) and McCormick & Owen (2001).

Nevertheless, the general trends are still prominent across a wide range of faunal and floral groups (Benton 1999), and have been the focus of intense scientific debate. The Ordovician radiation reflects the rise of Sepkoski's (1981) Palaeozoic Evolutionary Fauna and the mid-Mesozoic to Recent one, at least in the marine realm, the full development of the Modern Evolutionary Fauna. Sheehan (2001*b*) has recently summarized the gross ecological changes associated with the rise of these Evolutionary Faunas. The mid-Mesozoic to Recent was also the time of spectacular radiation of major terrestrial groups such as the flowering plants (angiosperms), birds and mammals (Hallam 1994). Why have there been these two great pulses in the diversification of life? Are there any obvious similarities or differences between them that may provide clues as to what triggers biological diversification on a global scale? As both these intervals were characterized by marked biotic provincialism, reflecting episodes of major continental break-up and global tectonism, it has become particularly important to ask whether there may be a key link between palaeogeographical configuration and biotic radiations. This is especially so as the latter parts of both intervals featured steep global climatic gradients.

From: CRAME, J. A. & OWEN, A. W. (eds) 2002. *Palaeobiogeography and Biodiversity Change: the Ordovician and Mesozoic–Cenozoic Radiations*. Geological Society, London, Special Publications, **194,** 1–11.
0305-8719/02/$15.00

In comparing the Vendian–Ordovician and Mesozoic marine diversifications, Erwin *et al.* (1987) highlighted the much lower number of taxonomically high level (order and above) originations in the later event. They attributed this to differences in the occupation of 'adaptive space' with the early Phanerozoic radiation reflecting the greater opportunity for the appearance of the novel morphologies used to diagnose higher level taxa. In the present volume, the accent is very much on diversification at lower taxonomic levels, and the Late Cenozoic–Recent interval in particular is characterized by the proliferation of species-rich clades (Crame 2001).

It is also clear that to make headway in the study of taxonomic diversity patterns, on either temporal or spatial scales, we have to be consistent in what is being measured and there is some confusion in terminology within the literature. Much of the rigorous definition of diversity measurement has been in relation to terrestrial environments. There are two basic categories of measurement: inventory diversity (*sensu* Whittaker 1977), that records the numbers of taxa per unit area (and may be weighted to take account of proportional abundances), and differentiation diversity that provides a measure of difference (or similarity) between levels of inventory diversity. Alpha (or within-habitat) diversity is the most common form of inventory diversity and records the number of taxa per area of homogenous habitat and so reflects species packing within a community. Beta (or between-habitat) diversity is the category of differentiation diversity that measures the variation in taxonomic composition between areas of alpha diversity (Magurran 1988). Whittaker (1977) used the terms 'gamma diversity' to reflect the number of taxa in an island or distinctive landscape and 'epsilon diversity' for the inventory diversity of a large biogeographic region. Using that terminological scheme, the term 'delta diversity' is used for the variation between areas of gamma diversity within an area of epsilon diversity (Magurran 1988). However, in palaeontological analyses of marine faunas, many workers (e.g. Sepkoski 1988 and references therein), have adopted a simpler scheme whereby gamma diversity is viewed as a measure of differentiation diversity at a larger spatial scale than beta, measuring taxonomic differentiation between geographical regions and thus a reflection of provinciality or endemicity.

Miller (e.g. 1997*a*, *b*, 1998, 2000) has consistently emphasized that it is essential to dissect the global diversity curves in order for them to be understood. If the patterns can be comprehended, then the processes that drive them can be addressed. Even understanding the partitioning of diversity change through the areal scales of its measurement is a significant challenge. Thus for the Ordovician, for example, the alpha diversities of a major clade may remain constant, in contrast to the global diversity change (e.g. Westrop & Adrain 1998; Adrain *et al.* 2000), increasing alpha diversities may be set against decreasing beta trends within a palaeogeographical region (e.g. Miller & Mao 1998) and measured increases in alpha and beta diversities may be insufficient to account for the scale of global biodiversity increase (Sepkoski 1988). To what extent therefore do gamma and delta diversity levels (i.e. provinciality) hold the key to understanding global diversity trends through time?

Biogeography and biodiversity change

Connections between biodiversity change and provinciality are well demonstrated in the literature (see Jablonski *et al.* 1985 for review). Boucot (1975, see also 1983) suggested that provincialism may be an important factor underlying diversity change in the Silurian and Devonian. Valentine (1973) and Schopf (1979) established a strong link between changes in endemism and the dramatic rise in taxa produced by the mid-Mesozoic–Cenozoic radiation. Valentine *et al.* (1978) attributed high marine species diversity in the Cenozoic to the marked rise in provinciality and simulated Phanerozoic marine diversity in terms of changing provincial patterns. They suggested as much as a five-fold increase in provinciality since the Late Palaeozoic, although this figure has been disputed (Bambach 1990). More recently, Smith (1988, fig. 8; see also Brenchley & Harper 1998, fig. 8.13) combined the schematic representation of the major continental plates and global marine familial diversity curve of Valentine & Moores (1970, 1972) to highlight the correspondence between diversity peaks and plate dispersal. Such a compilation would now be equally appropriate using Sepkoski's (1997) familial or genus-level curves.

Miller (1997*c*) drew attention to work by Jablonski (1987) on Cretaceous molluscs and other studies that show that geographical ranges were strongly correlated with stratigraphical durations, but these ranges were established very early in the histories of individual species. Miller showed a general increase in the longevity of Ordovician genera as well as an overall range expansion and suggested that by analogy with the Cretaceous mollusc data, the

Ordovician genera may have become more species-rich. Geographical range is also important in terms of differential survival during extinction events. Geographically widespread clades have a much higher probability of survival, as shown for example by Westrop (1991) for the latest Cambrian, Owen & Robertson (1995) and Sheehan *et al.* (1996) for the end-Ordovician and Jablonski (1986) for the end-Cretaceous extinction event.

Ordovician biogeography and biodiversity

In the introductory paper to the volume arising from the 1982 International Ordovician Symposium, Jaanusson (1984) highlighted three reasons why he considered the Ordovician to be 'so special': (i) extensive development of epicontinental seas, (ii) marked changes in the composition of skeleton-bearing faunas consequent on the appearance and diversification of major taxonomic groups, and (iii) pronounced biogeographical differentiation. The second of these has become the focus of considerable international attention in recent years, partially formalized under IGCP Project 410, 'The Great Ordovician Biodiversification Event' (see Webby *et al.* 1999). It is pertinent to assess the influence of the first and especially the third of Jaanusson's 'special' attributes of the Ordovician to that diversification.

The Ordovician radiation can be recognized at a wide spectrum of taxonomic levels from species to class. The major component of the diversification took place in clades that typified the Palaeozoic Fauna (Sepkoski 1981, 1995) both in groups that had a Cambrian history such as rhynchonelliformean ('articulate') brachiopods (e.g. Bassett *et al.* 2002) and vertebrates (e.g. Smith *et al.* 2002), and in groups that first appeared during the Period (e.g. bryozoans and some major echinoderm clades; Sepkoski 1995). Some elements of the Modern Fauna such as bivalve molluscs (e.g. Cope 2002) also radiated at this time as did clades within groups that characterize the Cambrian Fauna, in particular groups of 'inarticulated' brachiopods (Sepkoski 1995) and a substantial set of trilobite families constituting the so-called 'Whiterock Fauna' (Adrain *et al.* 1998).

Sepkoski (1988) argued that changes in alpha and beta diversities accounted for only about half of the increase in global genus-level diversity during the Ordovician and that much of the remaining increase was probably not attributable to changes reflecting provinciality or endemicity. Instead, he attributed much of the remaining diversity increase to 'hidden sources of beta diversity' such as the appearance and/or expansion of communities that were low in overall extent but rich in species, such as organic reefs, hardground communities, bryozoan thickets and crinoid gardens. This view was endorsed by Zhuravlev (2001) in a thoughtful comparison of the Cambrian and Ordovician radiations. Such communities reflect changes at a fairly high level in the ecological architecture of marine life (see Droser *et al.* 1997, 2000) but the lack of correlation between significant biodiversity change and palaeobiogeography during the Ordovician is surprising. Sepkoski considered that there was no change in the number of provinces from the Cambrian into the early Ordovician and that the degree of provinciality decreased during late Ordovician. However, his assessment did not take account of major palaeogeographical changes such as the fragmentation of the Gondwanan margin (Dalziel 1997; van Staal *et al.* 1998; Cocks 2001; McNamara *et al.* 2001, Bassett *et al.* 2002). Coupled with this, the plate tectonic dynamism of the Ordovician may have been crucial in generating hitherto unsuspected provinciality.

Miller (1997*b*) compared genus-level diversity curves from six palaeocontinents using both raw and rarefied (see Miller & Foote 1996) data. The extreme ends of the spectrum of palaeocontinental variation in diversity patterns were those of Laurentia and South China which both lay at low latitudes. Moreover, in terms of the partitioning of overall diversity change, Miller & Mao (1998) showed that whilst mean alpha diversities showed an overall slight increase from the Tremadoc to the Ashgill in both regions, there was a fall in mean beta diversity and this was much more marked in South China than in Laurentia. In itself, the fall in beta diversity shown in both regions set against rising alpha diversity is surprising, and was provisionally explained as reflecting the increasing environmental and geographical range of many genera throughout the Ordovician (see also Miller 1997*c*), possibly along with an increase in the species richness of these genera. This would reduce the beta diversity through time as measured at genus level as individual genera became more widespread causing a decrease in the distinction between assemblages at this taxonomic level. Miller & Mao (1998, p. 306) also argued that the increase in beta diversity recognized by Sepkoski (1988), largely on the basis of Laurentian data, represented a net change from the Cambrian to the Ordovician and was not in direct contradiction of the decreasing trend that they recorded through the Ordovician.

In a clever reversal of Sepkoski's (1988)

question about the partitioning of diversity during the Ordovician biodiversification, Brenchley *et al.* (2001) recently addressed the issue of where the diversity was lost during the two phases of extinction at the end of that Period. They argued that the considerable reduction in alpha diversity had a significant impact on both beta diversity and provinciality. Given the speed of the extinctions, it was the major loss of endemic taxa that significantly reduced the number of provinces, rather than plate movements.

Emerging patterns

The biogeographical patterns associated with the radiation of individual clades during the Early Palaeozoic are only just starting to come to light and reflect a complex combination of plate distribution, tectonic activity, sedimentary environment, sea-level rise and, ultimately, climatic gradient. Smith *et al.* (2002) argue that the latest common ancestor of all 'ostracoderms' and jawed vertebrates was Laurentian but thereafter the early Palaeozoic diversification of the vertebrates reflected complex biogeographical patterns of dispersal, vicariance and tectonic convergence. In contrast, an 'out of Gondwana' radiation for many invertebrate groups is a recurring theme. However, the picture is complex as that continent extended from the south pole to north of the equator (Cocks 2001; Dalziel 1997) and encompassed both siliciclastic- and (at lower latitudes) carbonate-dominated shelf environments. Thus while Bassett *et al.* (2002) have shown that the mid-Cambrian precursors of the typical Palaeozoic Fauna rhynchonelliformean ('articulate') brachiopod communities first appeared in shallow-water carbonate settings in north and east Gondwana, the early Ordovician radiation of the bivalve molluscs took place in the shallow-water siliciclastic shelves of Gondwana (Babin 1995; Cope 2002) at all latitudes. Only late in the Ordovician did bivalves become established on the low latitude carbonate shelves of Laurentia and Baltica. Preliminary analysis suggests that the rise to dominance of the Whiterock Fauna of trilobite families (Adrain *et al.* 1998) took place a little earlier on the fairly high latitude siliciclastic margins of Gondwana (Owen & McCormick in press) than on low latitude Laurentia. A cladistic analysis by Turvey (2002) of one of the groups belonging to the Whiterock Fauna, the Reedocalymeninae (one of the archetypical Gondwanan groups), points to a series of biogeographical events affecting large areas of that plate during the early Ordovician.

The nature of the depositional environment clearly played a significant role in the diversification of clades during the Ordovician. Miller's (1997*b*) preliminary analysis of diversity curves from six palaeocontinents identified differences in the scale and timing of changes both in terms of total diversity and that of individual major clades representing all three Evolutionary Faunas. These differences were linked to the amount and timing of siliciclastic input, including the effects of uplift caused by orogeny and volcanism (see also Miller & Mao 1995, 1998), and therefore emphasize the importance of plate tectonic activity to the promotion of biodiversity change. Miller (1997*b*) suggested that the spread of siliciclastic sediment after the Taconic Orogeny triggered the flourishing of bivalves in Laurentia following their earlier radiation in siliciclastic sedimentary environments in Gondwana.

The Taconic Orogeny and the resultant increase in weathering of siliceous rocks in the late Ordovician may also have been responsible for the profound change in climate leading to the Hirnantian glaciation (Kump *et al.* 1999). This very short-lived glaciation during what was essentially a greenhouse world was the proximal cause of the complex series of events that produced the end-Ordovician mass extinctions (see reviews in Brenchley *et al.* 2001; Sheehan 2001*a*). The climatic decline (see Armstrong & Coe 1997; Kump *et al.* 1999) that led up to it may also have had some effect on biodiversity. Cope (2002) recognized an increase in bivalve endemicity as climatic gradients steepened during the late Ordovician. An analysis of conodont biofacies by Armstrong & Owen (2002) suggests that diversity fell in low latitude Laurentia coincident with the onset of ocean cooling in the early Ashgill, whereas on Avalonia the decline took place a little later when that microcontinent drifted northwards into tropical latitudes.

Plate tectonic activity clearly played a significant role in controlling biodiversity change during the Ordovician. On a local scale, but with possible wider consequence, Botting (2002) has provided a possible mechanism for diversification as a result of repeated volcanic activity whereby ash falls produce overturn of the stratified watermass. This overturn could cause blooms in the plankton and benthos and the generation of localized, temporary benthic 'islands' within the otherwise dysaerobic shelf which were the sites of high speciation rates. On a much larger scale, Bassett *et al.* (2002) link the early and mid-Ordovician spread and diversification of the rhynchonelliformean brachiopods typical of the Palaeozoic Evolutionary Fauna to a combination of global sea-level rise and fragmentation of the Gondwanan margin. Harper & MacNiocaill (2002) reach a similar conclusion and also demonstrate the importance of intra-oceanic sites in the development, divergence and survival of taxa.

Mesozoic–Cenozoic biogeography and biodiversity

The true scale of the mid-Mesozoic–Cenozoic biotic radiation is currently a topic of intense debate. Whereas a new compilation of generic rather than familial data suggests that it may not have been as marked as Sepkoski (1981, 1993) originally indicated (Alroy *et al.* 2001), a detailed taxonomic investigation of just one small part of the Late Cenozoic tropics suggests that it was considerably greater (Jackson & Johnson 2000, 2001). One thing that is beyond dispute about this radiation is that it affected plants and animals in both the marine and terrestrial realms. The spectacular radiation of the angiosperms, mammals and certain reptile groups (such as the snakes) on land is matched by that of the planktonic foraminifera, neogastropods, heteroconch bivalves, cheilostome bryozoans, decapod crustaceans and teleost fish in shallow seas (Hallam 1994). This co-radiation of such disparate taxa is often taken as strong evidence for a real evolutionary phenomenon (Signor 1990; Benton 1999). This was also the time when some of the largest (in terms of species richness) clades evolved. For example, some of the massive eudicot angiosperm clades have between 20 000 and 25 000 extant taxa, and both the neogastropod subfamily Coninae and the heteroconch bivalve family Veneridae have approximately 500 living species each (Crame 2001).

Some geographical constraints on the radiation

It now seems almost certain that the extant planktonic formainifera evolved from benthonic ancestors in Tethyan (i.e. tropical/subtropical) facies in the Early Jurassic (Toarcian) (Hart *et al.* 2002). A biostratigraphic analysis has indicated some expansion within the North Atlantic–European region in both the Bajoican–Bathonian and Aptian, but it was not until the latest Albian that a near-global distribution was achieved. There is a strong link here between phases of diversification as revealed by stratigraphic analysis, continental fragmentation, and the expansion of shelf seas. Interestingly enough, it would appear that latitudinal gradients in taxonomic diversity had been established by the Campanian–Maastrichtian (Huber 1988), and these have persisted through to the present day (Bé 1977).

The role of land bridges and ocean gateways in controlling the formation of biodiversity patterns has been a persistent theme in Mesozoic–Cenozoic biogeography (e.g. Hallam 1981). Have they, perhaps, created bottlenecks in the distribution of biotas that are large enough to be discernible in the fossil record? This forms the central theme in the papers by Aberhan (2002) and Cantrill & Poole (2002), and is discussed in a more general way by Crame & Rosen (2002).

In his study of Early Jurassic bivalve biodiversity, Aberhan (2002) has been particularly concerned with the role of the Hispanic Corridor, a putative narrow seaway across central Pangaea linking the eastern Pacific with western Tethys. Was there a preferential extinction of endemic South American taxa across the Pliensbachian–Toarcian boundary due to a sudden influx of cosmopolitan Tethyan forms, or was a renewed rise in late Toarcian–Aalenian species diversity in NW Europe attributable to an influx from Andean South America in the other direction? Using a comprehensive species-level database, Aberhan is able to demonstrate that, in fact, in both regions early Pliensbachian to Aalenian immigration rates through the corridor remained low. In comparison, the within-region origination of new species played a much more important role in the recovery of post-extinction faunas. The relative importance of immigration versus in situ radiation in the generation of large-scale biodiversity patterns throughout the fossil record is still a controversial issue (e.g. Vermeij 1993).

By the time of radiation of crown group angiosperms (flowering plants) in the Early Cretaceous, the major southern hemisphere continents had largely separated from one another and Antarctica was rapidly becoming isolated. It has long been thought that, by Aptian–Albian times, the only major terrestrial connection between western Gondwana (South America and Africa) and eastern Gondwana (India, Antarctica and Australasia) was the volcanic archipelago of the Antarctic Peninsula region. However, in a new analysis of southern, high latitude Cretaceous floras, Cantrill & Poole (2002) cast some doubt on this theory. Whereas the earliest known angiosperms from the Antarctic Peninsula are Early Albian in age, there are angiospermous pollen grains from Australia that are Early Aptian. Quite what the alternative immigration route may have been is uncertain, but Cantrill & Poole point to the proven existence of Turonian terrestrial sediments on the Kerguelen Plateau (Shipboard Scientific Party 2000) and Late Cretaceous dinosaur finds that link Madagascar, South America and India (Sampson *et al.* 1998). It is clear that the earliest Antarctic angiosperms occupied areas of disturbance and were understorey colonizers. Their major pulse of radiation in the Turonian was coincident with the acme of Cretaceous global warmth.

The relative importance of tectonics and climate

In a wide-ranging review, Crame & Rosen (2002) take as their starting point the major extant biodiversity patterns, and then work back through time. They address the reason for there being discrete tropical high-diversity foci in both the marine and terrestrial realms, with steep latitudinal diversity gradients leading away from them. Whatever the ultimate cause of tropical high diversity *per se*, it is clear that a critical sequence of Early Miocene (*c.* 20 Ma) tectonic events fractured a once homogeneous tropical biota into several distinct centres; the Indo-West Pacific (IWP) and the Atlantic–Caribbean–East Pacific (ACEP) are the most important of these in the marine realm. Together with the physical isolation of Antarctica, these same plate tectonic movements contributed significantly to the global reorganization of major ocean current systems and Cenozoic cooling. This in turn led to a marked increase in provincialism and biotic differentiation on a regional scale. Even so, something else may have been involved in the development of the steepest diversity gradients during the Late Cenozoic. Rates of origination may have been driven by an increase in Milankovitch cyclicity over the last 10–15 Ma and there may be some crucial links here between the geographical range of an organism and its propensity to allopatric speciation. However, the issue is a complex one and is yet to be fully resolved. The more frequent nature of climatic change in high latitude regions might mean that ecological isolates are less likely to persist or, alternatively, the more constant environment of the tropics could mean that ranges can be more easily split by slight temperature shifts (Chown & Gaston 2000; Dynesius & Jansson 2000).

The importance of databases in biodiversity studies

With the widespread availability of relational database and GIS software, it is now much easier to build a comprehensive biodiversity database on a desktop computer. However, in a methodological review of just how such a database should be constructed, Markwick & Lupia (2002) emphasize the need to take into account variations in scale, inconsistency in the data, and potential errors (inaccuracy). Scale is a particularly critical issue, for, whereas 'large-scale' to a biologist refers to a large area or duration, a 'large-scale map' is usually one of great detail but small area. As this can lead to confusion when using a GIS for biological or palaeobiological purposes, Markwick & Lupia advocate the adoption of two terms from landscape ecology. The first of these is 'grain', which is the minimum resolution or scale of an observation, and the second is 'extent', which is the total amount of space or time observed. Thus, a large-scale map is fine-grained but of limited extent. It is the heteorogeneities in scale (grain) in the fossil record that are potentially the major causes of problems in the construction of palaeontological databases.

Markwick (2002) goes on to show how a comprehensive database of Recent North American, European, South African and Australian non-avian tetrapods can be used to tackle some fundamental problems in the study of latitudinal diversity gradients. Pooling taxa from all these regions it can be shown that there is a strong, monotonic relationship with both absolute latitude and mean annual temperature (MAT); the highest diversities are clearly associated with the highest MATs and lowest latitudes. Nevertheless, it is clear that there are strong differences in the form of the latitudinal gradients between the ectotherms (reptiles and amphibians) and endotherms (mammals). Whereas the former show comparatively simple, steep gradients, reflecting their dependence on the immediate environment, the latter show a more complex pattern, based on an indirect dependence on the environment (through various feeding strategies). Markwick concludes that patterns of terrestrial species diversity are not simply a function of the amount of available energy, but how that energy is procured. As the proportion of ectotherms to endotherms shows a strong linear relationship to temperature, it can be used to retrodict MATs in the fossil record. An example is given for the Middle Eocene Messel fauna of Germany that shows close agreement with other palaeotemperature estimates.

Synopsis

The Ordovician Period witnessed the rise of Palaeozoic Evolutionary Fauna, comprising elements such as the rhynchonelliformean brachiopods, bryozoans, echinoderms, primitive vertebrates and many families of trilobites. The Mesozoic–Cenozoic radiation includes the rise of such groups as the flowering plants, birds, mammals, marine molluscs, teleost fish and decapod crustaceans. Full substantiation of these patterns remains a major task and will involve a massive coordinated effort in taxonomic palaeontology (Kerr 2001; Jackson & Johnson 2001).

If these two steep increases in taxonomic diversity are not artefacts of the fossil record, a challenge of equal dimensions is to establish what has been driving them. One of the simplest explanations to consider is that they represent two intervals in which there were fundamental increases in within-habitat species richness (i.e. alpha diversity). In a seminal study of Phanerozoic marine communities, Bambach (1977) suggested that there may well have been periodic expansion of available ecospace through the development of new or increased resource supplies (see also Bambach 1993). Rosenzweig (1995, p. 306), for example, has suggested that the Ordovician was the first time in Earth history that muddy bottoms were extensively colonized by marine organisms. Precisely why this may have been so is uncertain, but dissolved oxygen levels may have increased until they reached some sort of threshold and allowed widespread colonization. Similarly, Bambach (1977, 1993) linked the late Mesozoic–Cenozoic rise of both terrestrial and marine organisms to the blossoming of the angiosperms. Here was an abundant source of both new food and habitat space for a wide variety of organisms. In addition, Bambach (1977) pointed to both the late Ordovician and Cenozoic glaciations as a possible source of oceanic nutrient recycling. Changes in thermohaline circulation brought about by climate change could have led to substantial oceanic mixing, which in turn affected the food resources of shelf seas. In both instances though, the global diversity increase was well underway and in the case of the Ordovician may have already reached its plateau by the time such processes began.

In contrast to attempts to seek a limited number of global controls on biodiversity change, Miller (e.g. 2000) has argued that it is more realistic to try to understand the regional patterns and the processes likely to have driven them. None the less, some form of periodic increase in resource supply could lead to a rapid expansion in the numbers of taxa within some habitats and regions. Once an adaptive threshold has been breached it may be possible to pack more taxa into a locality or region within a comparatively short space of time. Patterns of alpha and gamma diversity could be expected to increase periodically rather than continuously. The concept of non-hierarchical, non-additive levels of ecological change introduced by Droser *et al.* (1997, see also 2000) promises to be a useful way of addressing step changes in biodiversity and/or the utilization of ecospace, the two not necessarily changing in tandem.

However, it is clear that something other than a sheer rise in numbers must be involved. There is abundant evidence that differentiation of biotas has occurred and on a variety of geographical scales. For example, in their review of the Cenozoic diversification event, Crame & Rosen (2002) indicate that a complex set of Neogene tectonic events probably aided the development of both latitudinal and longitudinal provinces. There are indications to suggest that between-habitat diversity (i.e. beta diversity) increases in tandem with alpha diversity but as well as this, between-community and between-province diversity must be increasing too. Similarly Harper & MacNiocaill (2002) argue that the break-up of the Gondwanan margin promoted a rise in between-province diversity amongst the rhynchonelliformean brachiopods and this break-up, together with sea-level rise that promoted migration over the continental shelves, also produced a rise in alpha diversities as established communities were augmented by immigrants. Subsequent to this, beta diversities rose as communities became increasingly developed in deeper water environments. This pattern of nearshore innovation and offshore expansion of communities at the expense of the existing community types during the Ordovician was documented by Jablonski *et al.* (1983; see also Bassett *et al.* 2002). However, Westrop & Adrain (1998) and Adrain *et al.* (2000) have also demonstrated that for the trilobites, alpha diversities remained remarkably constant across the shelf throughout the Period and so, whilst their diversity relative to the rapidly diversifying clades might have declined, they were not simply being displaced by them and 'pushed' into deeper water refuges. Jablonski & Bottjer (1991) also showed that the origins of post-Palaeozoic benthonic orders were largely in nearshore environments, but Jacobs & Lindberg (1998) have argued that this only applied prior to the Turonian after which offshore bottom waters became more widely oxic and amenable to originations. That change also marked the demise of dysoxic faunas, the expansion of deep burrowers into offshore environments and the offshore expansion of bryozoan diversity.

Coupled with any purely tectonic definition of provinces, it is clear that biotic differentiation was assisted by the steepened latitudinal temperature gradients that became established towards the end of each of these intervals. The studies by Cope (2002), Armstrong & Owen (2002), Harper & MacNiocall (2002), Crame & Rosen (2002), Cantrill & Poole (2002) and Markwick (2002) all emphasize that climate change has assisted range shifts into higher or lower latitudes. The differentiation of biotas along environmental gradients undoubtedly enhanced the process of global biodiversification during both the Ordovician and Mesozoic–Cenozoic. In the case of the latter, there are still areas of the Earth's surface, such as the very highest and lowest latitudes, about which very little is known, which must bias both the observed patterns and their interpretations. Tropical systems are particularly poorly known for many intervals in time, and one can only guess at how much 'hidden diversity' is locked up within them (Sepkoski 1988; Jackson & Johnson 2001).

Clearly, there is still a great deal to learn about the precise mechanisms underpinning the major biotic radiations and it is unlikely that they will be attributable to a single, simple cause. The palaeobiologist is uniquely placed to study the interaction of physical and biological forces that have shaped the major patterns of life on Earth and understanding the palaeobiogeographical context is crucial. It encompasses a multitude of controls on the distribution and ecology of organisms, from the location of continents, islands and oceans to the global climate. Importantly, especially from a geological perspective, it represents a series of very dynamic systems. The present volume reflects some of the wide spectrum of approaches to studying biodiversity change on a variety of spatial and temporal scales, from taxonomy to the analysis of large databases, and emphasizes their links to the components of these systems in the two most sustained episodes of overall biotic radiation in the Phanerozoic.

We thank the referees, E. Clarkson and P. Wignall for their helpful comments. A.W.O. acknowledges support from NERC Grant GR3/11834 and this paper is a contribution to IGCP 410 'The Great Ordovician Biodiversification Event'. J.A.C. is supported by BAS Core Project, 'Antarctic marine biodiversity: a historical perspective'.

References

ABERHAN, M. 2002. Opening of the Hispanic Corridor and Early Jurassic bivalve biodiversity. *In*: CRAME, J. A. & OWEN, A. W. (eds) *Palaeobiogeography and Biodiversity Change: the Ordovician and Mesozoic–Cenozoic Radiations*. Geological Society, London, Special Publications, **194**, 127–139.

ADRAIN, J. M., FORTEY, R. A. & WESTROP, S. R. 1998. Post-Cambrian trilobite diversity and evolutionary faunas. *Science*, **280**, 1922–1925.

ADRAIN, J. M., WESTROP, S. R., CHATTERTON, B. D. E. & RAMSKÖLD, L. 2000. Silurian trilobite alpha diversity and the end-Ordovician mass extinction. *Paleobiology*, **26**, 625–464.

ALROY, J. 2000. Successive approximations of diversity curves: ten more years in the library. *Geology*, **28**, 1023–1026.

ALROY, J., MARSHALL, C. R., BAMBACH, R. K., *ET AL*. 2001. Effects of sampling standardization on estimates of Phanerozoic marine diversification. *Proceedings of the National Academy of Science of the USA*, **98**, 6261–6266.

ARMSTRONG, H. A. & COE, A. L. 1997. Deep-sea sediments record the geophysiology of the late Ordovician glaciation. *Journal of the Geological Society, London*, **154**, 929–943.

ARMSTRONG, H. A. & OWEN, A. W. 2002. Euconodont diversity changes in a cooling and closing Iapetus Ocean. *In*: CRAME, J. A. & OWEN, A. W. (eds) *Palaeobiogeography and Biodiversity Change: the Ordovician and Mesozoic–Cenozoic Radiations*. Geological Society, London, Special Publications, **194**, 85–98.

BABIN, C. 1995. The initial Ordovician mollusc radiations on the western Gondwanan shelves. *In:* COOPER, J. D., DROSER, M. L. & FINNEY, S. C. (eds) *Ordovician Odyssey: Short Papers for the Seventh International Symposium on the Ordovician System*. SEPM Pacific Section, Fullerton California, 491–498.

BAMBACH, R. K. 1977. Species richness and marine benthis habitats through the Phanerozoic. *Paleobiology*, **3**, 152–167.

BAMBACH, R. K. 1990. Late Palaeozoic provinciality in the marine realm. *In*: MCKERROW, W. S. & SCOTESE, C. R. (eds) *Palaeozoic Palaeogeography and Biogoegraphy*. Geological Society, London, Memoir, **12**, 307–323.

BAMBACH, R. K. 1993. Seafood through time: changes in biomass, energetics and productivity in the marine ecosystem. *Paleobiology*, **19**, 372–397.

BASSETT, M. G., POPOV, L. E. & HOLMER, L. E. 2002. Brachiopods: Cambrian–Tremadoc precursors to Ordovician radiation events. *In*: CRAME, J. A. & OWEN, A. W. (eds) *Palaeobiogeography and Biodiversity Change: the Ordovician and Mesozoic–Cenozoic Radiations*. Geological Society, London, Special Publications, **194**, 13–23.

BÉ, A. W. H. 1977. An ecological, zoogeographic and taxonomic review of recent planktonic foraminifera. *In*: RAMSAY, A. T. S. (ed.) *Oceanic Micropalaeontology Volume 1*. Academic Press, London, 1–88.

BENTON, M. J. 1995. Diversification and extinction in the history of life. *Science*, **268**, 52–58.

BENTON, M. J. 1999. The history of life: large databases in palaeontology. *In*: HARPER, D. A. T. (ed.) *Numerical Palaeobiology*. Wiley, Chichester, 249–283.

BENTON, M. J. 2000. *Vertebrate Palaeontology*. Blackwell Science, Oxford.

BENTON, M. J. 2001. Biodiversity on land and in the sea. *Geological Journal*, **36**, 211–230.

BOTTING, J. P. 2002. The role of pyroclastic volcanism in Ordovician diversification. *In*: CRAME, J. A. & OWEN, A. W. (eds) *Palaeobiogeography and Biodiversity Change: the Ordovician and Mesozoic–Cenozoic Radiations*. Geological Society, London, Special Publications, **194**, 99–113.

BOUCOT, A. J. 1975. *Evolution and Extinction Rate Controls*. Elsevier, Amsterdam.

BOUCOT, A. J. 1983. Does evolution take place in an ecological vacuum? II. *Journal of Paleontology*, **57**, 1–30.

BRENCHLEY, P. J. & HARPER, D. A. T. 1998. *Palaeoecology: Ecosystems, Environments and Evolution*. Chapman & Hall, London.

BRENCHLEY, P. J., MARSHALL, J. D. & UNDERWOOD, C. J. 2001. Do all mass extinctions represent an ecological crisis? *Geological Journal*, **36**, 329–340.

CANTRILL, D. J. & POOLE, I. 2002. Cretaceous patterns of floristic change in the Antarctic Peninsula. *In*: CRAME, J. A. & OWEN, A. W. (eds) *Palaeobiogeography and Biodiversity Change: the Ordovician and Mesozoic–Cenozoic Radiations*. Geological Society, London, Special Publications, **194**, 141–152.

CHOWN, S. L. & GASTON, K. J. 2000. Areas, cradles and museums: the latitudinal gradient in species richness. *Trends in Ecology & Evolution*, **15**, 311–315.

COCKS, L. R. M. 2001. Ordovician and Silurian global geography. *Journal of the Geological Society, London*, **158**, 197–210.

CONWAY MORRIS, S. 1998. The evolution of diversity in ancient ecosystems: a review. *Philosophical Transactions of the Royal Society of London, B*, **353**, 327–345.

COPE, J. C. W. 2002. Diversification and biogeography of bivalves during the Ordovician Period. *In*: CRAME, J. A. & OWEN, A. W. (eds) *Palaeobiogeography and Biodiversity Change: the Ordovician and Mesozoic–Cenozoic Radiations*. Geological Society, London, Special Publications, **194**, 35–52.

CRAME, J. A. 2001. Taxonomic diversity gradients through geological time. *Diversity and Distributions*, **7**, 175–189.

CRAME, J. A. & ROSEN, B. R. 2002. Cenozoic palaeogeography and the rise of modern biodiversity patterns. *In*: CRAME, J. A. & OWEN, A. W. (eds) *Palaeobiogeography and Biodiversity Change: the Ordovician and Mesozoic–Cenozoic Radiations*. Geological Society, London, Special Publications, **194**, 153–168.

DALZIEL, I. W. D. 1997. Neoproterozoic-Paleozoic geography and tectonics: review, hypothesis, environmental speculation. *Geological Society of America Bulletin*, **109**, 16–42.

DROSER, M. L., BOTTJER, D. J. & SHEEHAN, P. M. 1997. Evaluating the ecological architecture of major events in the Phanerozoic history of marine invertebrate life. *Geology*, **25**, 167–170.

DROSER, M. L., BOTTJER, D. J., SHEEHAN, P. M. & MCGEE, J. R. JR. 2000. Decoupling of taxonomic and ecological severity of Phanerozoic marine mass extinctions. *Geology*, **28**, 675–678.

DYNESIUS, M. & JANSSON, R. 2000. Evolutionary consequences of changes in species' geographical distributions driven by Milankovitch climate oscillations. *Proceedings of the National Academy of Sciences USA*, **97**, 9115–9120.

ERWIN, D. H., VALENTINE, J. W. & SEPKOSKI, J. J. 1987. A comparative study of diversification events: the early Paleozoic versus the Mesozoic. *Evolution*, **41**, 1177–1186.

HALLAM, A. 1981. Relative importance of plate movements, eustasy and climate in controlling major biogeographical changes since the Early Mesozoic. *In:* NELSON, G. & ROSEN, D. E. (eds) *Vicariance Biogeography. A Critique*. Columbia University Press, New York, 303–330.

HALLAM, A. 1994. *An Outline of Phanerozoic Biogeography*. Oxford University Press, Oxford.

HARPER, D. A. T. & MAC NIOCAILL, C. 2002. Early Ordovician rhynchonelliformean brachiopod biodiversity: comparing some platforms, margins and intra-oceanic sites around the Iapetus Ocean. *In*: OWEN, A. W. & CRAME, J. A. (eds) *Palaeobiogeography and Biodiversity Change: the Ordovician and Mesozoic–Cenozoic Radiations*. Geological Society, London, Special Publications, **194**, 25–34.

HART, M. B., OXFORD, M. J. & HUDSON, W. 2002. The early evolution and palaeobiogeography of Mesozoic planktonic foraminifera. *In*: OWEN, A. W. & CRAME, J. A. (eds) *Palaeobiogeography and Biodiversity Change: the Ordovician and Mesozoic–Cenozoic Radiations*. Geological Society, London, Special Publications, **194**, 115–125.

HUBER, B. T. 1988. Upper Campanian – Paleocene foraminifera from the James Ross Island region, Antarctic Peninsula. *In*: FELDMANN, R. M. & WOODBURNE, M. O. (eds) *Geology and Paleontology of Seymour Island, Antarctic Peninsula*. Geological Society of America, Memoir, **169**, 163–252.

JAANUSSON, V. 1984. What is so special about the Ordovician? *In:* BRUTON, D. L. (ed.) *Aspects of the Ordovician System*. Palaeontological Contributions from the University of Oslo, **295**, Universitetsforlaget, Oslo, 1–3.

JABLONSKI, D. 1986. Background and mass extinctions: the alternation of macroevolutionary regimes. *Science*, **231**, 129–133.

JABLONSKI, D. 1987. Heritability at the species level: analysis of geographic ranges of Cretaceous mollusks. *Science*, **238**, 360–363.

JABLONSKI, D. & BOTTJER, D. J. 1991. Environmental patterns in the origins of higher taxa: the post-Paleozoic marine record. *Science*, **252**, 1831–1833.

JABLONSKI, D., SEPOSKI, J. J., BOTTJER, D. J. & SHEEHAN, P. M. 1983. Onshore–offshore patterns in the evolution of Phanerozoic shelf communities. *Science*, **222**, 1123–1125.

JABLONSKI, D., FLESSA, K. W. & VALENTINE, J. W. 1985. Biogeography and paleobiology. *Paleobiology*, **11**, 75–90.

JACKSON, J. B. C. & JOHNSON, K. G. 2000. Life in the last few million years. *In*: ERWIN, D. H. & WING, S. L. (eds) *Deep Time: Paleobiology's Perspective. Paleobiology*, Supplement, **26**(4), 221–235.

JACKSON, J. B. C. & JOHNSON, K. G. 2001. Measuring past biodiversity. *Science*, **293**, 2402–2404.

JACOBS, D. K. & LINDBERG, D. R. 1998. Oxygen and evolutionary patterns in the sea: onshore/offshore trends and recent recruitment of deep sea faunas. *Proceedings of the National Academy of Science of the USA*, **95**, 9396–9401.

KERR, R. A. 2001. Putting limits on the diversity of life. *Science*, **292**, 1481.

KUMP, L. R., ARTHUR, M. A., PATZKOWSKY, M. E., GIBBS, M. T., PINKUS, D. S. & SHEEHAN, P. M. 1999. A weathering hypothesis for glaciation at high atmospheric pCO_2 during the Late Ordovician. *Palaeogeography, Palaeoclimatology, Palaeoecology*, **152**, 173–187.

MCCORMICK, T. & OWEN, A. W. 2001. Assessing trilobite biodiversity change in the Ordovician of the British Isles. *Geological Journal*, **36**, 279–290.

MCNAMARA, A. K., MAC NIOCAILL, C., VAN DER PLUIJM, B. A. & VAN DER VOO, R. 2001. West African proximity of the Avalon terrane in the latest Precambrian. *Geological Society of America Bulletin*, **113**, 1161–1170.

MAGURRAN, A. E. 1988. *Ecological Diversity and its Measurement*. Croom Helm, London.

MARKWICK, P. J. 2002. Integrating the present and past records of climate, biodiversity and biogeography: implications for palaeoecology and palaeoclimatology. *In*: CRAME, J. A & OWEN, A. W. (eds) *Palaeobiogeography and Biodiversity Change: the Ordovician and Mesozoic–Cenozoic Radiations*. Geological Society, London, Special Publications, **194**, 179–199.

MARKWICK, P. J. & LUPIA, R. 2002. Palaeontological databases for palaeobiogeography, palaeoecology and biodiversity: a question of scale. *In*: CRAME, J. A. & OWEN, A. W. (eds) *Palaeobiogeography and Biodiversity Change: the Ordovician and Mesozoic–Cenozoic Radiations*. Geological Society, London, Special Publications, **194**, 169–178.

MILLAR, A. I. 1997*a*. Dissecting global diversity patterns. *Annual Review of Ecology and Systematics*, **28**, 85–104.

MILLAR, A. I. 1997*b*. Comparative diversification dynamics among palaeocontinents during the Ordovician radiation. *Geobios*, **20**, 397–406.

MILLAR, A. I. 1997*c*. A new look at age and area: the geographical and environmental expansion of genera during the Ordovician. *Paleobiology*, **23**, 410–419.

MILLAR, A. I. 1998. Biotic transitions in global marine diversity. *Science*, **281**, 1157–1160.

MILLAR, A. I. 2000. Conversations about Phanerozoic global diversity. *In*: ERWIN, D. H. & WING, S. L. (eds) *Deep Time: Paleobiology's Perspective. Paleobiology*, Supplement, **26**(4) 53–73.

MILLAR, A. I. & FOOTE, M. 1996. Calibrating the Ordovician radiation of marine life: implications for Phanerozoic diversity trends. *Paleobiology*, **22**, 304–309.

MILLAR, A. I. & MAO, SHUGUANG. 1995. Association of orogenic activity with the Ordovician radiation of marine life. *Geology*, **23**, 305–308.

MILLAR, A. I. & MAO, SHUGUANG. 1998. Scales of diversification and the Ordovician radiation. *In*: MCKINNEY, M. L. & DRAKE, J. A. (eds) *Biodiversity Dynamics: Turnover of Populations, Taxa and Communities*. Columbia University Press, New York, 288–310.

NIKLAS, K. J., TIFFNEY, B. H. & KNOLL, A. H. 1985. Patterns in vascular land plant diversification: an analysis at the species level. *In*: VALENTINE, J. W. (ed.) *Phanerozoic Diversity Patterns: Profiles in Macroevolution*. Princetown University Press, Princetown, 97–128.

OWEN, A. W. & MCCORMICK, T. (in press). Ordovician trilobite biodiversity change in the Anglo-Welsh sector of Avalonia – a comparison of regional and global patterns. *Special Papers in Palaeontology*.

OWEN, A. W. & ROBERTSON, D. B. R. 1995. Ecological changes during the end-Ordovician extinction. *Modern Geology*, **20**, 21–39.

PADIAN, K. & CLEMENS, W. A. 1985. Terrestrial vertebrate diversity: episodes and insights. *In*: VALENTINE, J. W. (ed.) *Phanerozoic Diversity Patterns. Profiles in Macroevolution*. Princetown University Press, Princetown, 41–96.

ROSENZWEIG, M. L. 1995. *Species Diversity in Space and Time*. Cambridge University Press, Cambridge.

SAMPSON, S. D., WITMER, L. M., FORSTER, C. A., KRAUSE, D. W., O'CONNOR, P. M., DODSON, P. & RAVOAVY, F. 1998. Predatory dinosaur remains from Madagascar: implications for the Cretaceous biogeography of Gondwana. *Science*, **280**, 1048–1051.

SCHOPF, T. J. M. 1979. The role of biogeographical provinces in regulating marine faunal diversity through geologic time. *In*: GRAY, J. & BOUCOT, A. J. (eds) *Historical Biogeography, Plate Tectonics, and the Changing Environment*. Oregon State University Press, Corvallis, 449–457.

SEPKOSKI, J. J. 1981. A factor analytical description of the Phanerozoic marine fossil record. *Paleobiology*, **7**, 36–53.

SEPKOSKI, J. J. 1988. Alpha, beta, or gamma: where does all the diversity go? *Paleobiology*, **14**, 221–234.

SEPKOSKI, J. J. 1993. Ten years in the library: new data confirm paleontological patterns. *Paleobiology*, **19**, 43–51.

SEPKOSKI, J. J. 1995. The Ordovician radiations: diversification and extinction shown by global genus-level taxonomic data *In*: COOPER, J. D., DROSER, M. L. & FINNEY, S. C. (eds) *Ordovician Odyssey: Short Papers for the Seventh*

International Symposium on the Ordovician System. SEPM Pacific Section, Fullerton, California, 393–396.

SEPKOSKI, J. J. 1997. Biodiversity: past, present, and future. *Journal of Paleontology*, **71**, 533–539.

SHEEHAN, P. M. 2001*a*. The late Ordovician mass extinction. *Annual Review of Earth and Planetary Sciences*, **29**, 331–364.

SHEEHAN, P. M. 2001*b*. History of marine biodiversity. *Geological Journal*, **36**, 231–249.

SHEEHAN, P. M., COOROUGH, P. J. & FASTOVSKY, D. E. 1996. Biotic selectivity during the K/T and Late Ordovician extinction events. *In:* RYDER, G., FASTOVSKY, D. E. & GARTNER, S. (eds) *The Cretaceous–Tertiary Event and other Catastrophies in Earth History*. Geological Society of America, Special Papers, **307**, 477–489.

SHIPBOARD SCIENTIFIC PARTY, 2000. Leg 183 summary: Kerguelen Plateau – Broken Ridge – A large igneous province. *In:* COFFIN, M. F., FREY, F. A., WALLACE, P. J. *et al.* (eds) *Proceedings Ocean Drilling Program, Initial Reports,* **183**, 1–101.

SIGNOR, P. W. 1990. The geologic history of diversity. *Annual Reviews of Ecology and Systematics*, **21**, 509–539.

SMITH, A. B. 2000. Large-scale heterogeneity of the fossil record: implications for Phanerozoic biodiversity studies. *Philosophical Transactions of the Royal Society of London, B,* **356**, 351–367.

SMITH, A. B., GALE, A. S. & MONKS, N. E. A. 2001. Sea-level change and rock-record bias in the Cretaceous: a problem for extinction and biodiversity studies. *Paleobiology*, **27**, 241–253.

SMITH, M. P., DONOGHUE, P. C. J. & SANSOM, I. J. 2002. The spatial and temporal diversification of Early Palaeozoic vertebrates. *In*: CRAME, J. A. & OWEN, A. W. (eds) *Palaeobiogeography and Biodiversity Change: the Ordovician and Mesozoic–Cenozoic Radiations*. Geological Society, London, Special Publications, **194**, 69–83.

SMITH, P. L. 1988. Palaeobiogeography and plate tectonics. *Geoscience Canada*, **15**, 261–279.

TURVEY, S. T. 2002. Phylogeny of the Reedocalymeninae (Trilobita): implications for Early Ordovician biogeography of Gondwana. *In*: CRAME, J. A. & OWEN, A. W. (eds) *Palaeobiogeography and Biodiversity Change: the Ordovician and Mesozoic–Cenozoic Radiations*. Geological Society, London, Special Publications, **194**, 53–68.

VALENTINE, J. W. 1973. *Evolutionary Paleoecology of the Marine Biosphere*. Prentice-Hall, Englewood Cliffs, NJ.

VALENTINE, J. W. & MOORES, E. M. 1970. Plate tectonic regulation of biodiversity and sea level: a model. *Nature*, **228**, 657–659.

VALENTINE, J. W. & MOORES, E. M. 1972. Global tectonics and the fossil record. *Journal of Geology*, **80**, 167–184.

VALENTINE, J. W., FOIN, T. C. & PEART, D. 1978. A provincial model of Phanerozoic marine diversity. *Paleobiology*, **4**, 55–66.

VAN STAAL, C. R., DEWEY, J. F., MAC NIOCAILL, C. & MCKERROW, W. S. 1998. The Cambrian–Silurian tectonic evolution of the northern Appalachians and British Caledonides: history of complex, west and southwest Pacific-type segment of Iapetus. *In*: BLUNDELL, D. J. & SCOTT, A. C. (eds) *Lyell: the Past is the Key to the Present*. Geological Society of London, Special Publications, **143**, 199–242.

VERMEIJ, G. J. 1993. The biological history of a seaway. *Science*, **260**, 1603–1604.

WEBBY, B. D., PARIS, F. & DROSER, M. L. 1999. IGCP Project No. 410: The great Ordovician biodiversification event: implications for global correlation and resources. *Acta Universitatis Carolinae – Geologica* **43**: xi–xiii.

WESTROP, S. 1991. Intercontinental variation in mass extinction patterns: influence of biogeographic structure. *Paleobiology*, **17**, 363–368.

WESTROP, S. & ADRAIN, J. 1998. Trilobite alpha diversity and the reorganization of Ordovician benthic marine communities. *Paleobiology*, **24**, 1–16.

WHITTAKER, R. H. 1977. Evolution of species diversity in land communities. *Evolutionary Biology*, **10**, 1–67.

ZHURAVLEV, A. Y. 2001. Biotic diversity and structure during the Neoproterozoic–Ordovician transition. *In*: ZHURAVLEV, A. Y. & RIDING, R. (eds) *The Ecology of the Cambrian Radiation*. Columbia University Press, New York, 173–199.

Brachiopods: Cambrian–Tremadoc precursors to Ordovician radiation events

MICHAEL G. BASSETT [1], LEONID E. POPOV[1] & LARS E. HOLMER[2]

[1]*Department of Geology, National Museum of Wales, Cathays Park, Cardiff CF10 3NP, UK (e-mail:* mike.bassett@nmgw.ac.uk*)*

[2]*Institute of Earth Sciences, Department of Historical Geology and Palaeontology, University of Uppsala Norbyvägen 22, S-75236, Uppsala, Sweden (e-mail:* Lars.Holmer@pal.uu.se*)*

Abstract: Brachiopod-dominated palaeocommunities incorporating a structure typical of faunal groups within the Palaeozoic Evolutionary Fauna were already present in North and East Gondwana and associated terranes as early as the mid-Cambrian, confined exclusively to shallow marine, inshore environments. The late Cambrian and Tremadoc record of these faunas is incomplete, because of pronounced global sea-level lowstand and subsequent break-up and destruction of the Cambrian Gondwanan margin. It is likely, however, that those groups later forming the core of the Palaeozoic Evolutionary Fauna evolved originally in shallow-water environments of low-latitude peri-Gondwana, and dispersed widely when favourable ecological conditions developed. Conspicuous sea-level rise through the early to mid-Arenig provided newly available habitats in the expanding epeiric seas, where the new faunas evolved and diversified by the mid-Ordovician, when rapid drift separated the early Palaeozoic continents. Relatively short-lived precursor and transitional brachiopod assemblages can be identified on most of the main palaeocontinents prior to the Ordovician radiation of the Palaeozoic Evolutionary Fauna.

The Ordovician evolutionary radiation of marine metazoans was the second significant biodiversification event in Phanerozoic biotic history, following the origin and explosive radiation of skeletonized faunas at the beginning of the Cambrian. Ordovician events resulted not only in a significant global increase of taxonomic diversity, but also in substantial changes of benthic community structure on marine shelves, where the dominant trilobite–lingulate brachiopod associations of the Cambrian Evolutionary Fauna were replaced mainly by more advanced and structured benthic assemblages of the Palaeozoic Evolutionary Fauna, dominated by filter-feeders and especially by rhynchonelliformean brachiopods, bryozoans and pelmatozoan echinoderms (Sepkoski 1981, 1995). There was substantial overlap between these two biotopes in the earlier Ordovician, but generally on some major Lower Palaeozoic plates, including Laurentia, Baltica and Siberia, fully developed communities of the Palaeozoic Evolutionary Fauna have little in common with transitional faunas in which direct descendants of local Cambrian lineages predominate, or where transformation of the assemblages occurred almost abruptly so that newly emerging benthic faunas had little evident linkage with their Cambrian and early Ordovician predecessors (Sokolov 1982; Sepkoski & Sheehan 1983; Popov 1993; Patzkowsky 1995).

Rhynchonelliformean brachiopod assemblages that we identify and name below as transitional to the Palaeozoic Evolutionary Fauna (e.g. *Clarkella* Fauna and various syntrophinidine and *Tritoechia–Protambonites* associations) require particular discussion and definition. They already formed trophic structures (dominance of suspension-feeders) and tiering (presence of at least two levels) characteristic of the Palaeozoic Fauna, but the taxonomic composition of brachiopod genera and families retains distinct links to the ancestral Cambrian faunas, whereas their relationship with succeeding brachiopod faunas is less evident. These transitional assemblages also lack ostracodes and bryozoans, which are usually important components of Ordovician biotopes.

The nature of faunal patterns and replacement was especially significant on the shallow shelves of Gondwana and its closely associated marginal terranes, where the development of distinctive trophic and taxonomic community structures was rooted deep in the Cambrian. This suggests that increased faunal exchange between palaeoplates at the beginning of the Ordovician could have been an important factor in triggering subsequent changes in community structure more widely across marine shelves,

From: CRAME, J. A. & OWEN, A. W. (eds) 2002. *Palaeobiogeography and Biodiversity Change: the Ordovician and Mesozoic–Cenozoic Radiations*. Geological Society, London, Special Publications, **194**, 13–23.
0305-8719/02/$15.00

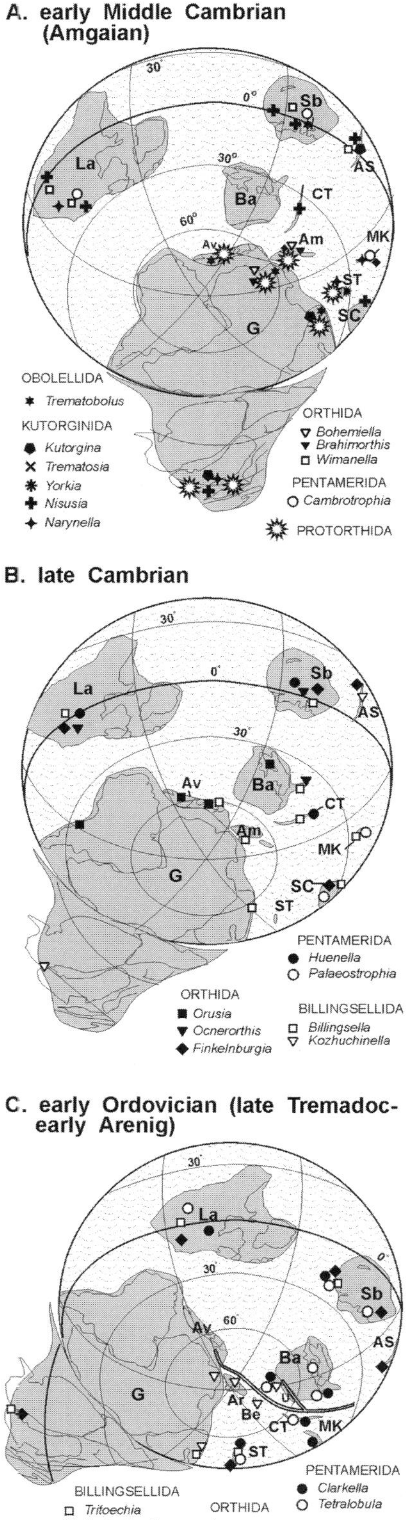

resulting in the significant increases in biodiversity that characterize the Ordovician Period.

This chapter traces major steps in the construction of Cambrian rhynchonelliformean (organocalcitic shelled) brachiopod assemblages and their environmental settings, and identifies their main biogeographical patterns (Fig. 1) together with the possible causes of change that resulted in their subsequent geographical and environmental expansion. We have elsewhere (Bassett *et al.* 1999*a*) documented patterns of evolutionary origins and extinctions in organophosphatic shelled 'inarticulated' linguliformean brachiopods through the Cambrian–Ordovician, so an additional aim here is to add comparative data and interpretations for the whole phylum. In identifying these distinctive pre-Palaeozoic Evolutionary Fauna precursor and transitional assemblages, we use the term 'Fauna' to embrace biotas containing several closely related brachiopod assemblages in a particular region, and the term 'Association' to include low-diversity biotas generally in a particular environmental setting (Fig. 2).

Cambrian brachiopod-dominated faunas

Several short-lived rhynchonelliformean groups, such as the Obolellata and Kutorginata, were distinctive components of the Cambrian Evolutionary Fauna, and were among the first

Fig. 1. Biogeographical distribution of characteristic rhynchonelliformean brachiopods through the Cambrian to early Ordovician interval. Palaeogeographical reconstructions modified after Torsvik (1998). During the mid-Cambrian, protorthides had a peri-Gondwanan distribution (**A**). In the late Cambrian (**B**) and early Ordovician (**C**), pentameride distribution was confined mainly to low and mid-latitudes, whereas North Gondwana was characterized by an abundance of billingselloideans and polytoechioideans. The reconstruction for the early Ordovician (**C**) shows the possible position of a newly initiated spreading zone (double line) resulting from the separation of the Uralian margin of Baltica and the separation of Avalonia. Am, Armorica; AS, Altay–Sayany island arc systems; Av, Avalonia; Ba, Baltica; Be, Bohemia (Perunica); CT, Chingiz–Tarbagatay island arc (now eastern Kazakhstan, presumably active margin of Baltica in the Cambrian–early Ordovician); G, Gondwana; La, Laurentia; MK, Malyi Karatau (Aisha-Bibi) terrane; S, Siberia; ST, South Tien Shan island arc systems; SC, South China; U, East Uralian (Mugodzhary) microcontinent.

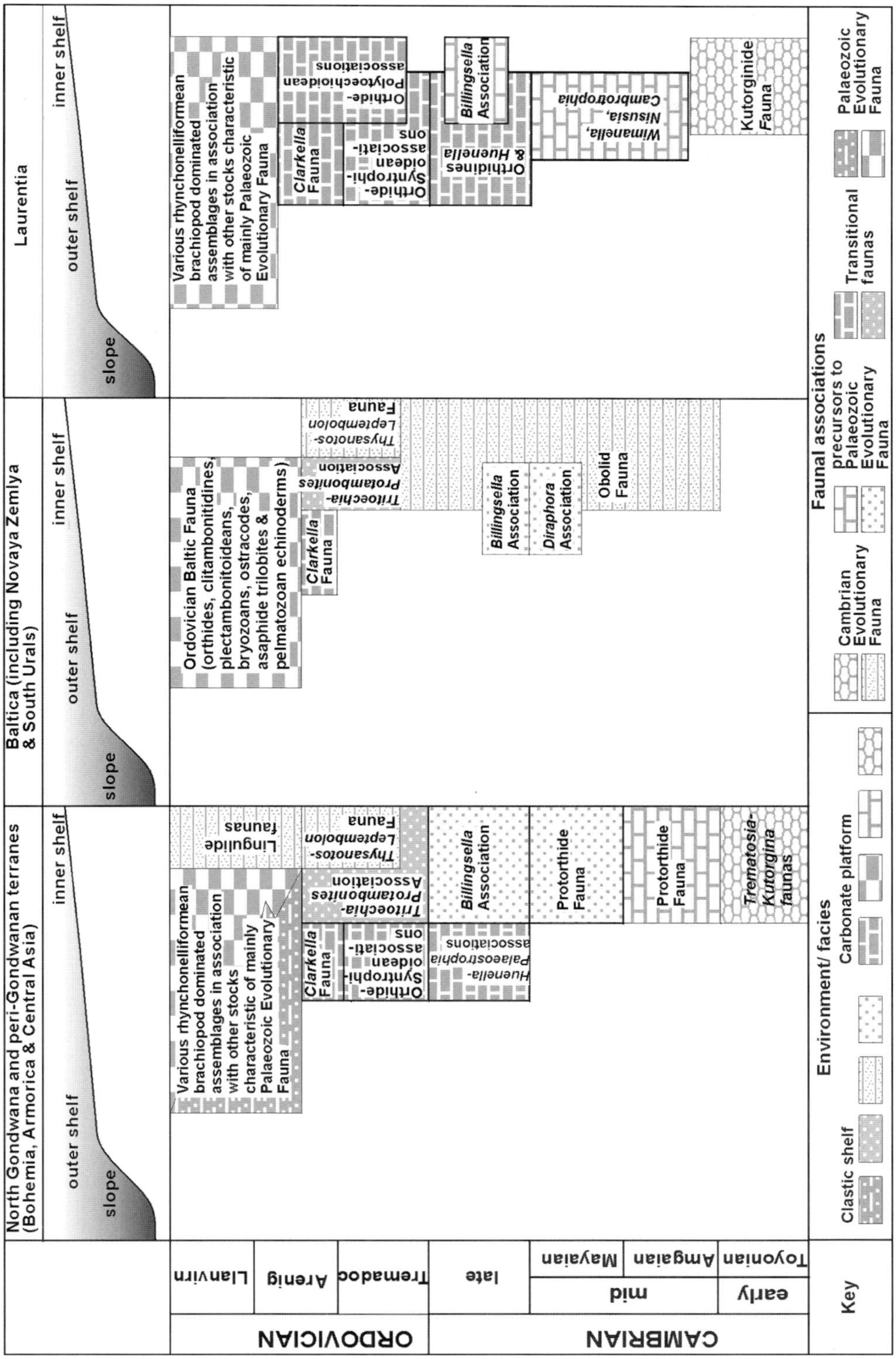

Fig. 2. Succession and environmental distribution of distinctive Cambrian–early Ordovician rhynchonelliformean brachiopod assemblages across different palaeoplates.

animals to form distinctive communities with essentially the same trophic and environmental characteristics as brachiopod-dominated communities of the Palaeozoic Evolutionary Fauna. The kutorginides *Kutorgina* and *Trematosia* and the obolellide *Trematobolus* formed the oldest known rhynchonelliformean brachiopod-dominated assemblages (Figs 1, 2). These assemblages were almost invariably of low taxonomic diversity, generally comprising or dominated by a single species, living in low-energy, shallow shelf environments affected by storm events, on low-latitude carbonate platforms usually in or adjacent to areas of hypersaline carbonate accumulation. Two illustrative examples are the marginal part of the Turukhansk–Irkutsk facies belt on the Siberian Plate, characterized by deposition of evaporites (Pelman 1992), and the Burj Formation on the eastern Dead Sea coast of Jordan together with the closely adjacent and coeval Nimra Formation of the southern Negev desert, Israel (Cooper 1976; and our unpublished studies), deposited on northern peri-Gondwana. These successions also contain the oldest known brachiopod coquinoid concentrations, usually in storm beds, which in some cases are of remarkable abundance and extent (Powell 1989). With the extinction of obolellides and kutorginides at the end of the early Cambrian, rhynchonelliformean brachiopod-dominated assemblages and shell bed accumulations declined during the mid-Cambrian in Siberia and Laurentia, but they remained a characteristic feature of shallow shelf environments across Gondwana and peri-Gondwanan regions, as discussed below.

By mid-Cambrian times, benthic palaeocommunities with a structure and dominant taxonomic composition typical of the Palaeozoic Evolutionary Fauna occurred widely across Gondwana (Middle East, Australia) and neighbouring areas, where they were confined exclusively to shallow, inshore marine environments (e.g. Cooper 1976; Roberts & Jell 1990). Two characteristic assemblages can be recognized.

The first is named here as the Protorthide Fauna (Fig. 2), comprising a medium-diversity association of archaic orthidines and protorthides and often including some obolellide, kutorginide and linguliformean taxa (Fig. 1A). We interpret protorthides as a stem group to pentamerides and clitambonitoideans (unpublished studies; see also Williams *et al.* 1996). This fauna embraced a wide variety of mid-Cambrian brachiopod associations containing characteristic protorthide components (e.g. *Protorthis* in New Brunswuck; *Jamesella* in Bohemia, Spain and North Africa; *Glyptoria* and *Psiloria* in Israel and Kyrgizstan; *Arctochedra* in Kyrgizstan and Australia, etc.). Nearly all known protorthide occurrences (Fig. 1A) except Kyrgizstan and Alaska were parts of Gondwana through the Cambrian (Dalziel 1997; Torsvik & Rehnström 2001). Lower to Middle Cambrian deposits of South Tien Shan in Kyrgizstan probably originated in an adjacent peri-Gondwanan island arc (Fig. 1A; Holmer *et al.* 2000). The occurrence of the protorthide *Arctochedra* in Alaska (Cooper 1936) does not seemingly indicate expansion of the group to Laurentia, but in our view was probably related to an exotic terrane of uncertain origin.

The second assemblage is defined here as the *Billingsella* Association (Fig. 2). It is a low-diversity association dominated by *Billingsella* itself or by related billingsellid genera (e.g. *Cymbithyris*), and shows patterns characteristic of opportunistic life strategies (Alexander 1977), e.g. limited areal distribution, high density clustering in thin, widespread isochronous horizons, overwhelming numerical dominance in the assemblage and remarkable abundance in atypical facies, mostly related to nearshore depositional environments. Billingsellids often formed inshore coquinoid accumulations. The most spectacular examples known to us are extensive shell beds in the Upper Cambrian Kyjandy Formation of north-central Kazakhstan (Nikitin 1956) and in the Middle to Upper Cambrian Derenjal Formation of east-central Iran (Fig. 3a). We interpret billingsellides as the ancestral stock of the Strophomenata including strophomenides, orthotetides and polytoechioideans (see also Williams *et al.* 1996). The *Diraphora* Association is a generally monospecific assemblage which immediately precedes the *Billingsella* Association in regions such as Novaya Zemlya (Fig. 2; Popov 1984); the ecological setting of both these associations appears to be very similar.

The Protorthide Fauna was confined exclusively to the mid-Cambrian of peri-Gondwana, whereas the *Billingsella* Association became almost cosmopolitan in the late Cambrian (Fig. 1). The rhynchonelliformean taxa in both assemblages can be traced phylogenetically into Ordovician descendants, and both the Protorthide Fauna and *Billingsella* Association probably represent nuclei that gave rise to stocks characteristic of the Palaeozoic Evolutionary Fauna.

In the Tremadoc–early Arenig, the *Billingsella* Association *sensu stricto* was replaced across the shallow clastic shelves of North Gondwana (Fig. 2; Iran, North Africa, Armorica) by low-diversity assemblages

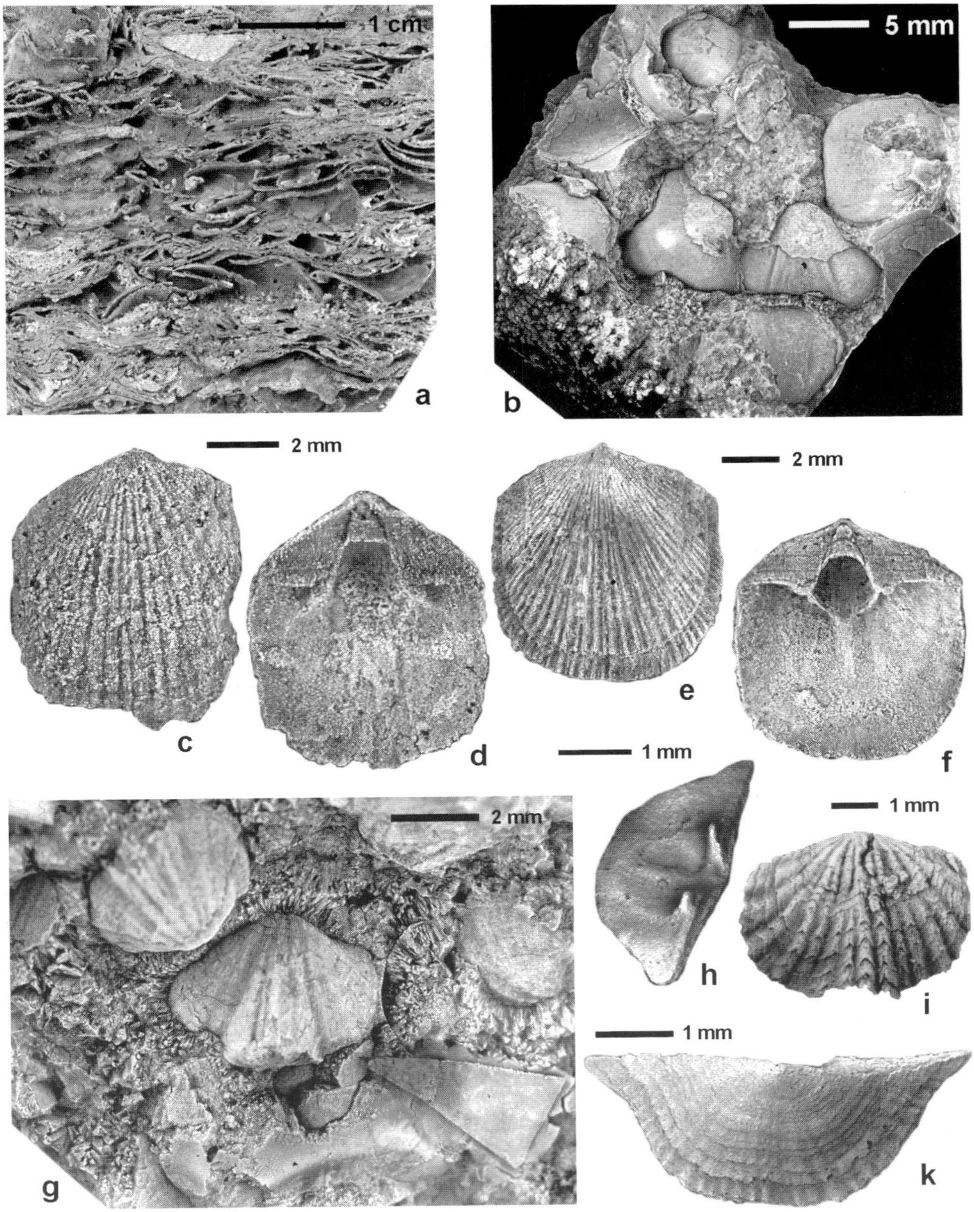

Fig. 3. Examples of late Cambrian–early Ordovician brachiopods that occur typically as coquinoid accumulations. (**a**) Storm bed formed by the accumulation of *Billingsella* sp., NMW 2001.45G.1, Upper Cambrian, Derenjal Formation, Shirgesht, Tabas Region, East Central Iran. (**b**) Shell bed comprising the syntrophioidean *Clarkella supina* Nikitin, NMW 98.67G.95, Lower Ordovician, Koagash Formation, Koagash River, South Urals, Kazakhstan. (**c, d**) *Billingsella* sp., NMW 2001.45G.2, interior and exterior of ventral valve, horizon and locality as in (**a**). (**e, f**) *Protambonites* sp. NMW 2001.45G.3, Lower Ordovician, Shirgesht Formation, Shirgesht, Tabas Region, East Central Iran. (**g**) Bedding surface with accumulation of *Huenella texana* (Walcott), USNM 52494, Upper Cambrian, Packsaddle Mountain, Llano County, Texas, USA. (**h, k**) Characteristic protorthide taxa from coquina accumulations in the Middle Cambrian, Amgaian, Arpatektyr Mountains, South Kyrgizstan: (**h, k**) NMW 98.69G.22, dorsal valve interior and NMW 98.69G.24 ventral valve of *Arctohedra pyramidalis* Aksarina; (**i**) CNIGR Museum 22/12761, dorsal valve exterior of *Glyptoria gulchensis* Popov and Tikhonov. NMW, National Museum of Wales, Cardiff; USNM, United States National Museum, Washington DC; CNIGR, Central Geological Research and Exploration Museum, St Petersburg.

dominated by the polytoechioideans *Protambonites* (Fig. 3e–f) or *Tritoechia* (e.g. Serre de los Cabos area of NW Spain, Shirgesht Formation of central Iran; Villas *et al.* 1995; Bassett *et al.* 1999*b*). These assemblages also spread to the Uralian margin of Baltica near the Tremadoc–Arenig boundary (Popov *et al.* 2001). Polytoechioideans were probably derived directly from the *Billingsella* Association (Popov *et al.* 2001) and demonstrate similar environmental requirements (Figs 1C, 2). In Iran and the South Urals they co-occur with a *Thysanotos–Leptembolon* Association of linguliformean brachiopods (Fig. 2; Popov & Holmer 1994, 1995; Bassett *et al.* 1999*b*).

Another distinctive group of late Cambrian–Tremadoc rhynchonelliformean assemblages was formed by syntrophioideans. These early pentamerides were most typical of low-latitude carbonate platforms on microcontinents and island arcs associated with Gondwana (e.g. South and North China, Central Asia, etc.) and Laurentia (Figs 1B, 2). The core of these assemblages was generally formed by a single taxon, usually *Huenella* (Fig. 3g) or *Palaeostrophia*. Associated brachiopods were mostly orthides (e.g. *Apheoorthis*, *Eoorthis*, etc.) and less abundant *Billingsella*. The co-occurrence of these faunas with pelmatozoan echinoderms such as eocrinoids is fairly typical, suggesting a relatively complex trophic structure involving at least two tiering levels. By the end of the Tremadoc (lower *Paroistodus proteus* Biozone), the syntrophioidean-dominated faunas were transformed to medium-diversity associations typified by the co-occurrence of the pentamerides *Clarkella*, *Diaphelasma*, *Glyptotrophia*, *Tetralobula*, etc. Here we identify this as the *Clarkella* Fauna (Figs 1C, 2, 3b). Brachiopods from the Lower Ordovician Olenty Formation of north-central Kazakhstan described by Nikitin (1956) include a good example of this newly emerging fauna. The *Clarkella* Fauna is unknown in West Gondwana, which by then had drifted to a high latitude (Fig. 1C), but it is characteristic of low latitude Laurentia and of the numerous terranes now incorporated into the complex tectonic collage of Central Asia and Kazakhstan, and is also recorded in South China and the Uralian margin of Baltica, located at that time in temperate latitudes. In all these regions, assemblages of the *Clarkella* Fauna occur in shallow-water carbonate depositional environments; they also incorporate descendants of local late Cambrian lineages (e.g. *Finkelnburgia*, *Apheoorthis*, *Eoorthis,* etc.) and polytoechioideans, but show little in common with succeeding Ordovician brachiopod associations.

Laurentia, Baltica and Siberia: general patterns of faunal replacement in the early Ordovician

The general characters of faunal replacement prior to the early Ordovician radiation are well documented and analysed only for Laurentia (Sepkoski & Sheehan 1983; Droser & Sheehan 1997), where the transition from benthic community types of the Cambrian Evolutionary Fauna to those characteristic of the Palaeozoic Evolutionary Fauna occurred in the Ibexian–early Whiterockian. Notable diversification of echinoderms (e.g. Sprinkle 1995) in the Ibexian represents the earliest indication of the Ordovician radiation in Laurentia (Guensberg & Sprinkle 1992), whereas the diversity of rhynchonelliformean brachiopods remained relatively low, represented mostly by families transitional from the Cambrian (Patzkowsky 1995). The abundance of polytoechiides and syntrophiidines diagnostic of the *Clarkella* Fauna is also characteristic. Typical Ordovician rhynchonelliformeans (e.g. camerelloideans and plectambonitoideans) together with bryozoans became increasingly abundant during the Whiterockian (Wilson *et al.* 1992), which was the interval when benthic assemblages with a structure and composition characteristic of the Palaeozoic Evolutionary Fauna became fully formed and widespread across Laurentia (Patzkowsky 1995; Sepkoski & Sheehan 1983; Droser & Sheehan 1997).

Available data from Baltica demonstrate that a benthic structure typical of the Palaeozoic Evolutionary Fauna arose there as early as the Billingenian (mid-Arenig, *Prioniodus elegans* Biozone), including the earliest representatives of plectambonitoideans, camerelloideans, endopunctate orthides and clitambonitidines among brachiopods, plus bryozoans, ostracodes, pelmatozoan echinoderms and asaphide trilobites (Fig. 2; Popov 1993; Pushkin & Popov 1999). This newly emergent fauna had no obvious origins in the low-diversity obolid-dominated assemblages of shallow shelf environments or in the predominantly dysoxic outer shelf olenid trilobite faunas characteristic of Baltica in the late Cambrian–Tremadoc. The sharp nature of faunal replacement in Balto-scandia suggests regional extinction and subsequent fairly rapid immigration of new taxa, concomitant with significant environmental changes including the onset of continuous carbonate sedimentation and the development of numerous hardground surfaces (Dronov *et al.* 1996).

In Siberia, early Ordovician (Tremadoc to

early Arenig) benthic assemblages retained major characters of the Cambrian Evolutionary Fauna, with abundant trilobites and with low-diversity orthide brachiopod assemblages represented by families transitional from the late Cambrian, accompanied by syntrophioideans (Sokolov 1982). The transitional interval through the late Arenig–Llanvirn was characterized by the presence of an essentially endemic low-diversity fauna of trilobites and rhynchonelliformeans, the problematical molluscs *Angarella* and *Moyeronia*, gastropods, locally abundant ostracodes, and the rare tabulate coral *Cryptolichenaria* (Kanygin *et al.* 1984). Other distinctive components of the Palaeozoic Evolutionary Fauna, such as rhynchonelliformean brachiopods of the orders Strophomenida, Triplesiida and Rhynchonellida, bryozoans and crinoids were introduced into Siberia only during the late Llanvirn–Llandeilo, where their sudden appearance suggests a major immigration from Laurentia (Rozman 1977).

Brachiopod radiation and palaeogeography

Within the newly emergent Ordovician faunas, rhynchonelliformean brachiopods were among the most important components, which in significant part defined the structure of benthic communities. Characters of faunal turnover and replacement in the early Ordovician of Laurentia, Baltica and Siberia – the three major early Palaeozoic plates outside Gondwana – suggest the importance of large-scale immigration of new brachiopod taxa, such as strophomenides, endopunctate orthides, clitambonitidines, triplesiides, and possibly camerelloideans, often of uncertain geographical and evolutionary origin, which comprise a significant component of the new faunas. Present-day rhynchonelliformeans and many bryozoans have a very low dispersal potential because of the very short free-swimming duration of their lecithotrophic larvae (Hyman 1959). Recent phylogenetic studies suggest that lecithotrophic larvae evolved early in ancestors of craniiformean and rhynchonelliformean brachiopods (Williams *et al.* 1996), and data from Palaeozoic rhynchonelliformean brachiopod biogeography also point to a relatively low ability for migration across oceans (Boucot 1975; Williams 1973, 1976). Their rapid expansion across the shelves of all major palaeoplates in the early to mid-Ordovician may therefore reflect a unique combination of environmental, geographical and tectonic factors that allowed them to cross oceanic divides separating major continents and then to settle, replace or integrate successfully into previously aboriginal benthic assemblages.

Following the early Cambrian radiation, when the trilobite and lingulate brachiopod-dominated Cambrian Evolutionary Fauna was established and diversified, the mid-Cambrian to Tremadoc interval was a time of relative evolutionary and environmental conservatism. By comparison with the succeeding Ordovician, this interval was characterized by pronounced global sea-level lowstand with insignificant sea-level fluctuations, as indicated recently by the analysis of almost continuous shallow marine deposition in the East Baltic (Artyushkov *et al.* 2000), accompanied by minimal volcanic activity, while the expansion of Iapetus led to increasing separation of Baltica from Laurentia. By contrast, the Arenig–Llanvirn was a period of remarkable tectonic activity, which included the break-up of the Gondwanan margin, resulting in particular in the origin of Avalonia as a separate microplate (Torsvik 1998) and separation of the East Uralian microcontinent from Baltica (Şengör & Natal'in 1996). Onset of arc volcanism in the Tremadoc of Eastern Avalonia (e.g. Kokelaar *et al.* 1984) and the late Cambrian to early Tremadoc rapid subsidence of the Welsh basin (Prigmore *et al.* 1997) possibly suggest initial development of rifting of Avalonia from Gondwana at about that time. Prigmore *et al.* (1997) supposed that the wider separation of Avalonia from Gondwana possibly coincided with a second phase of subsidence in the Arenig–early Llanvirn, although only by the late Llanvirn had evident biogeographical differences developed between Avalonian and Gondwanan shallow-shelf benthic faunas (Cocks 1999). Available data for the Uralian margin of Baltica are less focused, but they suggest initiation of late Cambrian to early Ordovician rifting of the Uralian margin of Baltica (Zonenshain *et al.* 1984), and possibly the detachment of island arc systems east of Baltica related to backarc spreading (Şengör and Natal'in 1996).

The possible position of Baltica in relation to North Gondwana (Fig. 1C) suggests that this chain of events may have resulted in the formation of a new spreading zone comparable with that of the present-day North Atlantic oceanic ridge, followed by a tectonically induced sea-level rise and a change of direction in the rotation of Baltica, leading to initiation of the closure of Iapetus. Volcanism commenced in Avalonia by at least the late Tremadoc (Kokelaar *et al.* 1984), and the presence of an active volcanic arc in the proximity of the Caledonian margin of Baltica at about that time became evident (Sturt & Roberts 1991). Evidence of increased early Ordovician island arc volcanism in the orogenic belts of Central and East Asia (Nikitin *et al.* 1991; Apollonov 2000) also suggests that the relatively narrow oceanic divide between the eastern margin of Baltica and equatorial Gondwana was occupied by chains of volcanic island arcs and microcontinents acting possibly as faunal 'bridges' between these two continents.

There is little doubt that the Ordovician Period mainly documents an interval of substantial sea-level rise, with a maximum in the Caradoc when most palaeoplates were flooded (Barnes *et al.* 1996). Superimposed second-order eustatic cycles corresponding approximately to the Tremadoc, Arenig and Llanvirn also can be recognized (Fortey 1984). The first episode of sea-level rise, corresponding to the transgressive phase of the Black Mountain eustatic event (mid-Tremadoc) of Miller (1984), is recorded on nearly all early Palaeozoic continents. This and the second transgressive phase in the Arenig reduced the influx of siliciclastic sediments into the large epeiric basins and provided good conditions for carbonate deposition, both in low (Laurentia, Siberia, North China) and temperate (Baltica, South China) latitudes. Areas of shallow marine carbonate deposition with numerous hardgrounds provided environments in which the earliest rhynchonelliformean brachiopod-dominated associations evolved.

Little is known of the characteristics of faunal replacement during the late Cambrian to early Ordovician of Gondwana and its associated terranes. In part this is a result of the marked period of sea-level lowstand through the late Cambrian to early Tremadoc, and the subsequent destruction of the Cambrian Gondwanan margin through the Palaeozoic, when areas of Cambrian shallow marine deposition were separated and later obscured within tectonic collages of Western Europe, the Middle East, and Central and Southeast Asia. By the beginning of the Ordovician, the North African and Middle Eastern sectors of Gondwana drifted into high latitudes and brachiopod assemblages linked to areas of carbonate deposition mainly disappeared (Fig. 1C). However, the late Cambrian *Billingsella* Association and succeeding early Ordovician polytoechioidean associations remained characteristic of some regions, for example in Iran and Armorica (Fig. 2). It is likely that in the late Cambrian–Tremadoc, benthic assemblages transitional to the Palaeozoic Evolutionary Fauna persisted mainly in faunas of peri-Gondwanan terranes, such as the South and North China plates, or on island arcs and microplates incorporated into the orogenic belts of Central Asia, where the *Clarkella* Fauna became widespread in the late Tremadoc–early Arenig (Nikitin 1956; Holmer *et al.* 2001) and also colonized the Uralian margin of Baltica (Bondarev 1968) before that plate began its rapid northwesterly drift away from Gondwana. The early Arenig brachiopod-dominated fauna of the Billingenian Regional Stage (mid-Arenig) in the East Baltic contains all the main features of the Palaeozoic Evolutionary Fauna, including such characteristic groups as bryozoans and ostracodes (Popov 1993; Pushkin & Popov 1999). Assemblages of similar character appeared in Laurentia somewhat later, in the mid-Whiterockian (Wilson *et al.* 1992). Similar faunas in Siberia, which was possibly the most isolated continent of that time, appeared only in the late Llanvirn–Llandeilo. The abrupt nature of the faunal replacement in Baltica and Siberia, together with the abundance of new groups of high taxonomic rank and of taxa unrooted in indigenous lineages, suggests that the newly appearing faunas were mature assemblages that had evolved and immigrated from elsewhere. By contrast, faunal replacement in Laurentia was passive and non-competitive in nature (Westrop & Adrain 1998), and newly appearing taxa of rhynchonelliformean brachiopods and bryozoans were incorporated into the transformed benthic assemblages together with locally evolved faunas of trilobites and echinoderms.

Synopsis

The pre-Arenig nuclei of benthic communities dominated by suspension-feeders with a structure and composition characteristic of the Palaeozoic Evolutionary Fauna most probably evolved on the Cambrian shallow carbonate shelves of equatorial Gondwana and peri-Gondwanan terranes. Dispersion of these faunas outside Gondwana is evident from the late Cambrian–Tremadoc, when syntrophioidean, *Billingsella* and polytoechiid associations reached Laurentia, Siberia and the Uralian margin of Baltica, but wider significant transformation of benthic shelf assemblages was delayed until the Arenig–Llanvirn and then occurred diachronously across the major palaeocontinents. Substantial sea-level rise from the Tremadoc to Caradoc, coupled with extensive carbonate deposition and the development of associated hardgrounds in the shallow epeiric seas of low and temperate latitudes, established increased ecospace for colonization by the newly emerging benthic assemblages of the Palaeozoic

Evolutionary Fauna. These faunas included new stocks of rhynchonelliformean brachiopods, and possibly bryozoans (Hu & Spjeldnæs 1991), which were preadapted to these environments. Thus the well known onshore–offshore pattern in ecological expansion of rhynchonelliformean brachiopod assemblages in the Ordovician (Sepkoski 1991) reflects the fact that areas of shallow carbonate deposition were the primary environments in which these brachiopods flourished in the Cambrian, and which they then occupied first in their geographical expansion in the early Ordovician.

As noted by Sepkoski & Sheehan (1983), 'there was no immediately obvious physical trigger for such a great burst of evolutionary activity' to drive the Ordovician biodiversification. However, significant intensification of plate tectonic activity at the beginning of the Ordovician, including rifting of the Gondwanan margin and the Uralian margin of Baltica, resulted in the origin of extensive new spreading zones and substantial increase of island arc volcanism and eustatic sea-level rise, which provided a physical background for the remarkable transformations of marine biodiversity.

Reconstructions of Baltica and Laurentia during the Tremadoc–early Arenig place them in the relative proximity of Gondwana (Dalziel 1997; Torsvik 1998; Torsvik & Rehnström 2001). Benthic faunal groups with low dispersal potential thus had enhanced opportunities to cross the intervening oceans; at the same time, newly emergent volcanic arcs formed 'island hopping bridges' between the major continents, to further develop the pathways for such migrations. Conspicuous sea-level rise at the beginning of the Ordovician led to the development of vast new shelves and platforms in the expanding epeiric seas, which the new faunas invaded, then evolved and diversified by mid-Ordovician times when rapid drift began the separation of the early Palaeozoic continents.

M.G.B. and L.E.P. acknowledge support from the Royal Society of London and the National Museum of Wales for work on early Palaeozoic brachiopods. L.E.H. completed part of his contribution in Cardiff under a European Science Exchange Programme funded by the Royal Society and Royal Swedish Academy, awarded jointly with M.G.B. L.E.H. and L.E.P. acknowledge the receipt of grants from the Swedish Natural Science Research Council (NFR). This paper is a contribution to IGCP Project 410. We are grateful to A. W. A. Rushton for his most helpful comments that have improved the manuscript, and to A. W. Owen for his scientific and editorial comments.

References

ALEXANDER, R. R. 1977. Growth, morphology and ecology of Paleozoic and Mezosoic opportunistic species of brachiopods from Idaho-Utah. *Journal of Paleontology*, **51**, 1133–1149.

APOLLONOV, M. 2000. Geodinmicheskaya evolyutsiya Kazakhstana v rannem paleozoe (s pozitsyi klassicheskoi tektoniki plit) [Geodynamic evolution of Kazakhstan in the early Palaeozoic (from classic plate tectonic positions)]. *In*: BESPAEV, H. A. (ed.) *Geodinamika i mineregeniya Kazakhstana*, **1**, VUC Publishing House, Almaty, 46–63 (in Russian).

ARTYUSHKOV, E.V., LINDSTRÖM, M. & POPOV, L. E. 2000. Relative sea-level changes in Baltoscandia in the Cambrian and early Ordovician: the predominance of tectonic factors and the absence of large scale eustatic fluctuations. *Tectonophyisics,* **320**, 375–407.

BARNES, C. R., FORTEY, R. A. & WILLIAMS, S. H. 1996. The pattern of global bio-events during the Ordovician Period. *In*: WALLISER, O. T. (ed.) *Global Events and Event Stratigraphy in the Phanerozoic.* Springer, Berlin-Heidelberg, 139–172.

BASSETT, M. G., POPOV, L. E. & HOLMER, L. E. 1999*a*. Organophosphatic brachiopods: patterns of biodiversification and extinction in the early Palaeozoic. *Geobios*, **32**, 145–163.

BASSETT, M. G., DASTANPOUR, M. & POPOV, L. E. 1999*b*. New data on Ordovician faunas and stratigraphy of the Kerman and Tabas regions, east-central Iran. *Acta Universitatis Carolinae, Geologica*, **43**, 483–486.

BONDAREV, V. I. 1968. Stratigrafiya i kharakternye brakhiopody ordovikskikh otlozhenii yuga Novoi Zemli, ostrova Vaigach i severnogo Pai-Khoia [Stratigraphy and characteristic brachiopods of the Ordovician deposits of southern Novaya Zemlya, Isle of Vaigach and northern Pai-Khoi]. *Trudy Nauchno- Issledovatel'skogo Instituta Geologii Arktiki (NIIGA)*, **157**, 3–144 (in Russian).

BOUCOT, A. 1975. *Evolution and Extinction Rate Controls*. Developments in Paleontology and Stratigraphy, Elsevier, Amsterdam.

COCKS, L. R. M. 1999. The Early Palaeozoic geography of Europe. *Journal of the Geological Society, London*, **157**, 1–10.

COOPER, G. A. 1936. New Cambrian brachiopods from Alaska. *Journal of Paleontology*, **10**, 210–214.

COOPER, G. A. 1976. Lower Cambrian brachiopods from the Rift Valley (Israel and Jordan). *Journal of Paleontology*, **50**, 269–289.

DALZIEL, I. W. D. 1997. Neoproterozoic-Paleozoic geography and tectonics: Review, hypothesis, environmental speculation. *Geological Society of America Bulletin*, **109**, 16–42.

DRONOV, A. Ju., SAVITSKY, Ju. V., FEDOROV, P. V. & TSYGANOVA, E. A. 1996. Detailed lithostratigraphy of the Ordovician lower Volkhovian limestone along the eastern part of the Baltic-Ladoga Glint, northwestern Russia. *GFF*, **118**, 19–24.

DROSER, M. L. & SHEEHAN, P. M. 1997. Palaeoecology of the Ordovician radiation: resolution of large-scale patterns with individual clade histories, palaeogeography and environments. *Geobios*, **20**, 221–229.

FORTEY, R. A. 1984. Global earlier Ordovician transgressions and regressions and their biological implications. *In*: BRUTON, D. L. (ed.) *Aspects of the Ordovician System*. University of Oslo Palaeontological Contributions, **295**, 37–50.

GUENSBERG, T. E. & SPRINKLE, J. 1992. Rise of echinoderms in the Paleozoic fauna: Significance of paleoenvironmental controls. *Geology*, **20**, 407–410.

HOLMER, L. E., POPOV, L. E. & BASSETT, M. G. 2000. Early Ordovician organophosphatic brachiopods with Baltoscandian affinities from the Alay Range, southern Kyrgyzstan. *GFF*, **122**, 367–375.

HOLMER, L. E., POPOV, L. E., KONEVA, S. P. & BASSETT, M. G. 2001. Cambrian – early Ordovician brachiopods from Malyi Karatau, the western Balkhash Region, and northern Tien Shan, Central Asia. *Special Papers in Palaeontology*, **65**, 1–180.

HU, ZHAO-XUN & SPJELDNÆS, N. 1991. Early Ordovician bryozoans from China. *Bulletin de la Société des Sciences Naturelle Ouest France*, Hors Serie **1**, 179–189.

HYMAN, L. H. 1959. *The Invertebrates 5. Smaller Coelomate Groups*. McGraw Hill, New York.

KANYGIN, A. V., MOSKALENKO, T. A., OBUT, A. M. & SENNIKOV, N. V. 1984. Ordovik [Ordovician]. *In*: Fanerozoi Sibiri. 1. Vend, paleozoi. *Trudy Instituta geologii i geifiziki Sibirskogo otdeleniya Akademii Nauk SSSR*, **595**, 60–88 (in Russian).

KOKELAAR, B. P., HOWELLS, M. F., BEVINS, R. E., ROACH, R. A. & DUNKLEY, P. N. 1984. The Ordovician marginal basin of Wales. *In*: KOKELAAR, B. P. & HOWELLS, M. F. (eds) *Marginal Basin Geology: Volcanic and Associated Sedimentary and Tectonic Processes in Modern and Ancient Marginal Basins*. Geological Society, London, Special Publications, **16**, 245–269.

MILLER, J. F. 1984. Cambrian and earliest Ordovician conodont evolution, biofacies and provincialism. *In*: CLARK, D. L. (ed.) *Conodont Biofacies and Provincialism*. Geological Society of America, Special Papers, **196**, 43–68.

NIKITIN, I. F. 1956. *Brakhiopody kembriya i nizhnego ordovika severo-vostoka Tsentral'nogo Kazakhstana* [Cambrian and Lower Ordovician brachiopods of north-eastern Central Kazakhstan]. Nauka, Alma-Ata (in Russian).

NIKITIN, I. F., FRID, N. M. & ZVONTSOV, V. S. 1991. Paleogeography and main features of volcanicity in the Ordovician of Kazakhstan and North Tien-Shan. *Geological Survey of Canada,* Paper **90–9**, 259–270.

PATZKOWSKY, M. E. 1995. Ecologic aspects of the Ordovician radiation of articulate brachiopods. *In*: COOPER, J. D., DROSER, M. L. & FINNEY, S. C. (eds) *Ordovician Odyssey: Short Papers for the Seventh International Symposium on the Ordovician System*. Pacific Section, Society for Sedimentary Geology (SEPM), Fullerton, California, 413–414.

PELMAN, YU. L. 1992. Biostratigraficheskoye raschleneniye kembriya Severnoi Azii po brakhiopodam. 1. *Sibirskaya* Platforma [Biostratigraphical subdivision of the Cambrian of north Asia based on brachiopods. 1. Siberian Platform.] *In*: REPINA, L. N. & ROZANOV, A. YU. (eds) *Drevneishie brakhiopody territorii Severnoi Evrazii*. Rossiiskaya Akademiya Nauk, Paleontologicheskii Institut, Sibirskoye otdeleniye, Obyedinennyi institut geologii, geofiziki i mineralogii, Novosibirsk, 10–21 (in Russian).

POPOV, L. E. 1984. Kembriiskie zamkovye brakhiopody iz yuzhnogo berega guby Gribovoi, Yiuzhnyi ostrov [Cambrian articulate brachiopods from the southern coast of the Gribovaja Harbor, Southern Island]. *In*: BONDAREV, V. I. (ed.) *Novaya Zemlia na Rannikh Etapakh Geologicheskogo Razvitiya*. Sevmorgeologiya, Leningrad, 27–34 (in Russian).

POPOV, L. E. 1993. Dinamika raznoobraziya bentosnoi fauny na granitsre kembriya i ordovika na territorii Vostochno-Evropeiskoi platformy i ee obramleniya [Dynamics of faunal biodiversity across the Cambrian-Ordovician boundary on the territory of the East European Plate and adjacent areas]. *Stratigrafiya i korrelyatsyia*, **1**, 33–38 (in Russian).

POPOV, L. E. & HOLMER, L. E. 1994. Cambrian–Ordovician lingulate brachiopods from Scandinavia, Kazakhstan, and South Ural Mountains. *Fossils and Strata*, **35**, 1–156.

POPOV, L. E. & HOLMER, L. E. 1995. Distribution of brachiopods across the Cambrian–Ordovician boundary on the East European Plate and adjacent areas. *In*: COOPER, J. D., DROSER, M. L. & FINNEY, S. C. (eds) *Ordovician Odyssey: Short Papers for the Seventh International Symposium on the Ordovician System*. Pacific Section, Society for Sedimentary Geology (SEPM), Fullerton, California, 117–120.

POPOV, L. E., VINN, O. & NIKITINA, O. I. 2001. Brachiopods of the redefined family Tritoechiidae from the Ordovician of Kazakhstan and South Urals. *Geobios*, **32**, 131–155.

POWELL, J. H. 1989. Stratigraphy and sedimentation of Phanerozoic rocks in Central and South Jordan. Part A, Ram and Khreim Groups. *1: 50000 Mapping Series Geological Bulletin*, **11**, 1–72. Amman, Jordan.

PRIGMORE, J. K., BUTLER, A. J. & WOODCOCK, N. H. 1997. Rifting during separation of Eastern Avalonia from Gondwana Evidence from subsidence analysis. *Geology*, **25**, 203–206.

PUSHKIN, V. I. & POPOV, L. E. 1999. Early Ordovician bryozoans from north-western Russia. *Palaeontology*, **42**, 171–189.

ROBERTS, J. & JELL, P. A. 1990. Early Middle Cambrian (Ordian) brachiopods of the Coonigan Formation, western New South Wales. *Alcheringa*, **14**, 257–309.

ROZMAN, Kh. S. 1977. Biostratigrafiya i zoogeo-grafiya verkhnego ordovika Severnoi Azii i Severnoi Ameriki (po brakhiopodam) [Biostratigraphy and zoogeography of the Upper Ordovician of North Asia and North America (based on brachiopods)]. *Trudy Geologicheskogo instituta Akademii Nauk SSSR*, **305**, 1–171 (in Russian).

ŞENGÖR, A. M. C. & NATAL'IN, B. A. 1996. Paleotectonics of Asia: fragments of a synthesis. *In*: YIN, A. & HARRISON, M. (eds) *The Tectonic Evolution of Asia*. Cambridge University Press, Cambridge, 486–640.

SEPKOSKI, J. J. Jr. 1981. A factor analysis description of the Phanerozoic marine fossil record. *Paleobiology*, **7**, 36–53.

SEPKOSKI, J. J. Jr. 1991. A model of onshore-offshore change in faunal diversity. *Paleobiology*, **17**, 157–176.

SEPKOSKI, J. J. Jr. 1995. The Ordovician radiations: diversification and extinction shown by global genus level taxonomic data. *In*: COOPER, J. D., DROSER, M. L. & FINNEY, S. C. (eds) *Ordovician Odyssey: Short Papers for the Seventh International Symposium on the Ordovician System*. Pacific Section, Society for Sedimentary Geology (SEPM), Fullerton, California, 493–496.

SEPKOSKI, J. J. Jr & SHEEHAN, P. M. 1983. Diversification, faunal change and community replacement during the Ordovician radiations. *In*: TEVESZ, M. J. S. & MCCALL, L. L. (eds) *Biotic Interpretations in Recent and Fossil Communities*. Plenum Press, New York, 673–717.

SOKOLOV, B. S. (ed.) 1982. Ordovik Sibirskoi platformy (opornyi razrez na reke Kulyumbe) [Ordovician of Siberian Platform (key section on the Kulyumbe River)]. *Trudy Instituta geologii i geifiziki Sibirskogo otdeleniya Akademii Nauk SSSR*, **506**, 60–88 (in Russian).

SPRINKLE, J. 1995. Do eocrinoids belong to the Cambrian or to the Paleozoic evolutionary fauna? *In*: COOPER, J. D., DROSER, M. L. & FINNEY, S. C. (eds) *Ordovician Odyssey: Short Papers for the Seventh International Symposium on the Ordovician System*. Pacific Section, Society for Sedimentary Geology (SEPM), Fullerton, California, 397–400.

STURT, B. A. & ROBERTS, D. 1991. Tectonostratigraphic relationships and obduction histories of Scandinavian ophiolitic terranes. *In*: PETERS, T. *et al.* (eds) *Ophiolite Genesis and Evolution of the Oceanic Lithosphere*. Ministry of Petroleum and Minerals, Sultanate of Oman; Kluver, Amsterdam, 745–769.

TORSVIK, T. H. 1998. Palaeozoic palaeogeography: A North Atlantic viewpoint. *GFF*, **120**, 109–118.

TORSVIK, T. H. & REHNSTRÖM, E. F. 2001. Cambrian palaeomagnetic data from Baltica: implications for true polar wander and Cambrian palaeogeography. *Journal of the Geological Society, London*, **158**, 321–329.

VILLAS, E., ARBIZU, M., BERNARDEZ, E., MENDEZ-BEDIA, I. & ARAMBURU, C. 1995. *Protambonites primigenus* (Brachiopoda, Clitambonitidina) y el limite Cambrico-Ordovicico en la Serre de los Cabos (Zona Asturoccidental-leonesa, NO de Espana). *Española de Paleontologia*, **10**(2), 140–150.

WESTROP, S. R. & ADRAIN, J. M. 1998. Trilobite alpha diversity and the reorganization of Ordovician benthic marine communities. *Paleobiology*, **24**, 1–16.

WILLIAMS, A. 1973. Distribution of brachiopod assemblages in relation to Ordovician paleogeography. *In*: HUGHES, N. F. (ed.) *Organisms and Continents Through Time*. Special Papers in Palaeontology, **12,** 241–269 (also published as *Systematics Association Publication*, **9**).

WILLIAMS, A. 1976. Plate tectonics and biofacies evolution as factors in Ordovician correlation. *In*: BASSETT, M. G. (ed.) *The Ordovician System: Proceedings of a Palaeontological Association Symposium, Birmingham, September 1974*. University of Wales Press and National Museum of Wales, Cardiff, 29–66.

WILLIAMS, A., CARLSON, S. J., BRUNTON, C. H. C., HOLMER, L. E. & POPOV, L. E. 1996. A supraordinal classification of the Brachiopoda. *Philosophical Transactions of the Royal Society, Biological Sciences*, **35**, 1171–1193.

WILSON, G. B., PALMER, T. J., GUENSBERG, T. E., FINTON, C. D. & KAUFMAN, L. E. 1992. The development of an early Ordovician hardground community in response to rapid sea-floor calcite precipitation. *Lethaia*, **25**, 19–34.

ZONENSHAIN, L. P., KORINEVSKY, V. G., KUZMIN, M. I., PECHERSKY, D. M., KHAIN, V. V. & MATVEENKOV, V. V. 1984. Plate tectonic model of the South Urals development. *Tectonophysics*, **109**, 95–135.

Early Ordovician rhynchonelliformean brachiopod biodiversity: comparing some platforms, margins and intra-oceanic sites around the Iapetus Ocean

DAVID A.T. HARPER[1] & CONALL MAC NIOCAILL[2]
[1]*Geological Museum, Øster Voldgade 5–7, DK-1350 Copenhagen, Denmark*
(e-mail: dharper@savik.geomus.ku.dk*)*
[2]*Department of Earth Sciences, University of Oxford, Parks Road, Oxford, UK*
(e-mail: Conall.MacNiocaill@earth.ox.ac.uk*)*

Abstract: During the Arenig–Llanvirn interval a series of radiations across all the major clades, established the Brachiopoda as a major component of the Palaeozoic benthos. Radiations on Baltica and Laurentia during the Arenig formed the basis for two distinct biogeographical provinces with contrasting articulated brachiopod faunas. These platform provinces were supplemented by the marginal Celtic and Toquima-Head provinces; the latter included both marginal and intra-ocean island sites. These marginal and intra-oceanic sites may have served as both 'cradles and museums' alternately providing sources for radiations on the platforms and refugia for otherwise relict taxa. Such terranes also partitioned oceanic circulation patterns within the Iapetus Ocean and provided mosaics of rapidly changing, nearshore unstable environments. In contrast to later Ordovician brachiopod faunas, many early Ordovician genera are reported from only one or two sites in the Iapetus region; narrow geographical ranges are characteristic of many taxa. The strong biogeographic differentiation at the generic level is less marked at the familial and higher levels suggesting a series of late Cambrian and early Ordovician migrations prior to the more regional development of the Arenig–Llanvirn brachiopod biofacies across the Iapetus terranes. However, the origination of many brachiopod taxa apparently occurred outside the Iapetus region suggesting that the initial stages of the Ordovician radiations here were first spiked by a series of immigrations.

The Ordovician brachiopod faunas of the Greater Iapetus Region include arguably the best-documented assemblages of the period. The faunas are biostratigraphically and taxonomically well constrained while their palaeoenvironmental and palaeogeographical settings have been significantly refined during the last 30 years. Recent reviews have emphasized the resolution of current palaeogeographical models (e.g. Cocks 2000, 2001) and the quality of faunal data available for the early Ordovician (e.g. Harper *et al.* 1996). Harper & Sandy (2001) have placed Ordovician provinciality within the context of the changing biogeographical patterns of the entire Palaeozoic. During the early Ordovician, brachiopod provinciality was marked in comparison with subsequent Silurian faunas (Boucot 1990); comparative provinciality was, however, re-established during the mid-Devonian (Boucot 1993, 1999). This enhanced biogeographical differentation was associated with high diversities which together provided opportunities to test palaeogeographical models for the mid-Devonian (McKerrow *et al.* 2000), similar to those provided by early Ordovician faunas. But although data for Lower Ordovician faunas are characteristically dispersed and uneven, current information is sufficient to frame and test various models for the early Ordovician radiation within the Iapetus region.

Ordovician radiation

The Ordovician radiation is now firmly recognized as one of the most significant events in metazoan history (Droser & Sheehan 1997). The detailed anatomy of the radiation is currently an area of intense investigation. New Bambachian megaguilds (Droser *et al.* 1997), associated with deep burrowing and complex tiering, are features of the radiation together with the establishment of new community types as well as the appearance of stromatoporoid reef structures (Bottjer *et al.* 2001). In addition, the capacity of each Bambachian megaguild was expanded to their Palaeozoic levels during a relatively short period of time.

Initial detailed research, mainly in the Great Basin of Laurentia (see Droser & Sheehan 1997 for summary), has indicated the importance of the faunal turnover at the base of the Whiterock

From: CRAME, J. A. & OWEN, A. W. (eds) 2002. *Palaeobiogeography and Biodiversity Change: the Ordovician and Mesozoic–Cenozoic Radiations*. Geological Society, London, Special Publications, **194**, 25–34.
0305-8719/02/$15.00

(equivalent to a level within the middle Arenig and close to the base of the Volkhov). Here shell beds are reported to have escalated in thickness with brachiopods rather than trilobites becoming the dominant components of such accumulations (Droser & Sheehan 1997) together with a shift from echinoderm to bryozoan-dominated hardground palaeocommunities; the diversification of echinoderms associated with soft substrates occurred later. Although trilobites were the major component of the Cambrian Evolutionary Fauna, a further group of families, the Whiterock Fauna, provided a mid-Ordovician expansion of the trilobite clade mainly at low latitudes (Adrain *et al.* 1998). Two diversification events within the Brachiopoda were signalled by Droser & Sheehan (1997, fig. 6): the continued diversification of 'early' brachiopods during the mid-Ibex (late Tremadoc) and a second diversification during the Llanvirn. Within the Great Basin the development of new orthid community types characterizes the radiation (Bottjer *et al.* 2001).

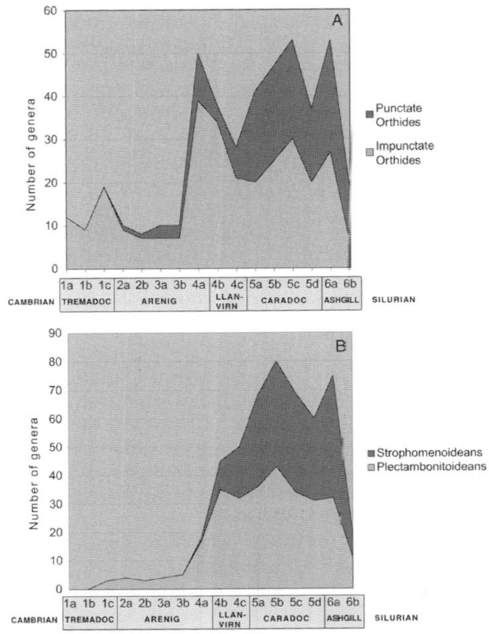

Fig. 1. Global biodiversity curves for the two main components of the early Ordovician brachiopod fauna: (**A**) Orthida (the two suborders Orthidina and Dalmanellidina are indicated); (**B**) Strophomenida (the two superfamilies Strophomenoidea and Plectambonitoidea are indicated). Data are from the revised *Treatise, Part H* (Williams *et al.* 2000).

Brachiopod diversifications

A number of step-wise radiations across the phylum Brachiopoda during the early to mid-Ordovician helped set the agenda for much of level-bottom life on the Palaeozoic seafloor (Harper *et al.* 2001). By the late Ordovician, with the exception of some of the more bizarre upper Palaeozoic taxa such as the lyttonioids and richthofenioids, the main ecogroups had evolved characterized by a variety of morphological adaptations (Harper & Wright 1996).

The organophosphatic linguliformeans had already diversified during the mid- and late Cambrian and represented an integral part of the Cambrian Evolutionary Fauna. During the Tremadoc–early Arenig, morphological and taxonomic diversity was restored following a marked decline in the latest Cambrian; during the early Ordovician the low-diversity lingulide communities retained a dominant position in nearshore environments on clastic shelves. The shallow-water epibenthic linguloideans were mainly replaced by infaunal linguloideans, discinoideans and bivalves during the Llanvirn. Radiation of the micromorphic acrotretides and siphonotretides, during the early and mid-Ordovician, provided the most significant additions to lingulate biodiversity at this time; moreover members of the group show a clear shift to more basinal environments (Bassett *et al.* 1999*a*). Concomitantly the craniformean brachiopods developed rapidly, with the morphological differentiation of the craniides and the trimerellides; both taxa significantly expanded their geographic ranges during the mid-Ordovician (Popov *et al.* 1999).

Rhynchonelliformean brachiopods represent one of the most important components of the developing Palaeozoic Evolutionary Fauna; distinctive community structures and a characteristic onshore–offshore biofacies distribution were established by the mid-Ordovician during late Arenig–early Llanvirn radiations. Among the deltidiodont rhynchonelliformeans, the orthides and strophomenides (Fig. 1) diversified into deeper water environments during the mid-Ordovician, the latter occupying niches within a soft-substrate biota; the pentamerides evolved rapidly at low latitudes globally, often forming the core of communities associated with carbonate build-ups. The origin of the cyrtomatodont articulation in the rhynchonellides during the Llanvirn formed the basis for another series of step-wise, but more restricted, radiations during the mid- to late Ordovician (Bassett *et al.* 1999*b*); ecological competition and displacement both within the cyrtomatodonts and between the deltidiodonts was marked. Among the spire-bearing brachiopods only the atrypides diversifed during the Caradoc in the tropics, whereas the athyridides and the spiriferides were not numerically significant components of the Ordovician radiations. The pentamerides and spire-bearers were to form the basis of subsequent radiations during the early Silurian, commonly associated with carbonate environments (Rong & Harper 2000; Harper & Rong 2001).

During the early Ordovician the contrasting escalation of the pedunculate orthides and the recumbent strophomenides was the most marked amongst the rhynchonelliformeans. Global datasets suggest initial (late Arenig–early Llanvirn) radiations were dominated by the orthides whereas the strophomenides first diversified during the early Llanvirn but most markedly during the Caradoc (Harper *et al.* 1999*a*).

The 'Greater Iapetus Region'

The Iapetus Ocean was bordered by a number of platform provinces and punctuated by a variety of marginal and oceanic terranes. This sector of the Earth's crust has provided a critical field area for the understanding of the dynamics and evolution of Early Palaeozoic brachiopod faunas. Terrane models for most parts of the region are relatively far advanced (Harper 1998) and the extent of most continents and their margins are now well defined. The margins of, for example, Avalonia (Cocks *et al.* 1997) and Baltica (Cocks & Fortey 1998) have been described in detail.

Perspectives on the development of the Iapetus Ocean system have changed markedly during the past three decades. The innovative orthogonal model for oceanic opening and closure (Wilson 1966), which has formed the basis for the Wilson Cycle, indicated the separation of North American and European brachiopod faunas during the Early Palaeozoic. Statistical analyses of a range of brachiopod faunas mainly associated with the Iapetus region (Williams 1969, 1973) suggested a number of provinces existed within the Greater Iapetus region. Nevertheless, as closure of the Iapetus Ocean progressed sequentially less mobile organisms were able to cross a narrowing seaway (McKerrow & Cocks 1976); for example, many brachiopod genera were present on both sides of the ocean by the Caradoc whereas similar brachiopod species were present in both Europe and North America during the Ashgill (McKerrow & Cocks 1976). During the early to mid-1980s a series of more complex models was developed for the ocean emphasizing, for example, the role of decreasing endemicity with depth (Cocks & Fortey 1982) and the participation of a more complex system of continents, including Laurentia, Avalonia and Baltica, within the cycle (Cocks & Fortey 1982; Fortey & Cocks 1986). Each continent, and to a lesser extent its margins, contained a distinctive early Ordovician brachiopod fauna defining a province and suggesting oceanic separation from coeval continental areas; distinctions become less clear during and following the early Caradoc *gracilis* transgression.

Neuman, in a series of papers (Neuman 1972, 1984; Neuman & Harper 1992), added greater realism to these palaeogeographical reconstructions with the addition of islands and archipelagos into otherwise barren oceanic tracts, based on data initially from the Appalachians. The importance of marginal and oceanic sites was confirmed and developed, on the basis of data from Scandinavia, by Bruton & Harper (1981, 1985). More focused statistical analyses of the brachiopod faunas from around and within the Iapetus Ocean (Neuman & Harper 1992; Harper *et al.* 1996) confirmed the status of the platform provinces but also flagged groups of marginal and ocean biotas associated with islands composed of both basement and volcanic material. For example, the Toquima-Table Head faunas developed circumferal and seaward of Laurentia at low latitudes whereas the Celtic faunas formed a high-latitude belt marginal to Gondwana (Harper *et al.* 1996). These models have been tested by seriation, using simulated annealing (Ryan *et al.* 1999). This algorithm clearly recognized clusters of associated faunas but was less successful in clarifying the relationships between the groupings. Possibly many of the early Ordovician high-latitude faunas were more isolated from coeval faunas elsewhere having fewer widespread taxa and more endemics than those faunas from low latitudes. That situation contrasts with the pattern for the latest Ordovician, when a clearer latitudinal gradient across more cosmopolitan faunas is apparent (Ryan *et al.* 1999).

Palaeogeographical templates for the early Ordovician of the Iapetus region have increasingly emphasized the role of marginal and oceanic tracts now entrained as a variety of terranes within the Appalachian–Caledonian mountain belt. The biological importance of these sites has been specified in a number of

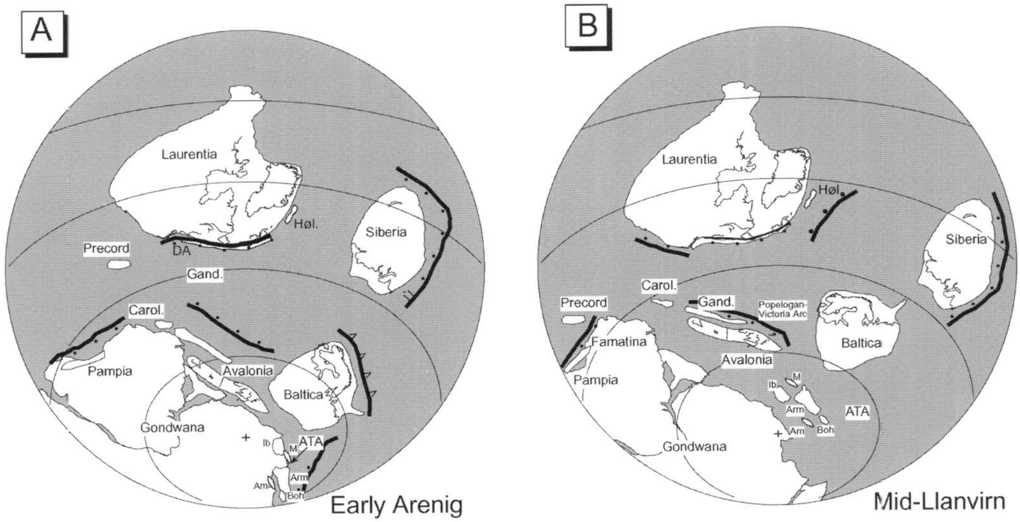

Fig. 2. Palaeogeographical reconstructions for (**A**) early Arenig and (**B**) the mid-Llanvrin. The early Arenig to mid-Llanvirn interval highlights the development of a series of peri-Gondwanan terranes, many forming the basis for the Celtic Province. Laurentia has been positioned according to the palaeomagnetic compilation of Mac Niocaill & Smethurst (1994); Baltica and Gondwana after Torsvik *et al.* (1996); Siberia after Smethurst *et al.* (1998); Avalonia has been palaeolatitudinally positioned after Mac Niocaill (2000). The locations of intra-Iapetan arcs are based on Mac Niocaill *et al.* (1997) and van Staal *et al.* (1998). Abbreviations used are as follows: Høl, Hølonda; DA, Dashwoods block; Precord, Precordillera; Gand, Gander terrane; Carol, Carolina terrane; ATA, Armorican Terrane Assemblage, the individual elements being as follows; Ib, Iberia; M, Meguma; Arm, Armorica; Boh, Bohemia [= Perunica]; Am, Alpine Massifs.

contributions (Neuman 1972, 1976; Bruton & Harper 1981, 1985; Harper *et al.* 1996); roles as centres for speciation, refugia during regression and stepping stones along migrational routes have been proposed. A number of islands clearly changed their faunal affinities during cross-latitudinal movement (Owen *et al.* 1992).

The addition of realistic ocean-circulation patterns (Christiansen & Stouge 1999) has enhanced understanding of the mechanics of the Iapetus oceanic system. Predicted warm- and cool-water gyres may have also provided a partition between the low-latitude Laurentian and Toquima-Table Head faunas and those at high latitudes associated with Avalonia, Gondwana and the Celtic group of faunas. The presence of island chains within the Iapetus Ocean may have helped develop the interface between the two current systems.

Platforms, margins and intra-oceanic sites

The early Ordovician Iapetus region contains a spectrum of continental and ocean terranes in a variety of tectonic settings (Fig. 2). The larger platform areas of Laurentia, Baltica and Gondwana are supplemented by marginal sites such as the Oslo region and Western Newfoundland and microcontinents such as Avalonia and Perunica. Although there are abundant brachiopod data from many of these sites there are few regional databases. Preliminary databases, at the generic level, have been established here as an initial test for a number of hypotheses regarding the early Ordovician brachiopod radiation. The East Baltic (Harper & Hints 2001), Morocco (Havlíček 1971), Welsh Basin (Lockley 1983) and the Prague Basin (Havlíček 1976, 1998) have been used as proxies for the Baltic, Gondwanan, Avalonian and Bohemian biogeographic units. Nevertheless such areas, for example Baltica, contain a range of facies, and initial studies (e.g. Jaanusson 1973, 1976; Harper 1986) have emphasized that the various facies belts reacted in different ways to changing latitude and waves of immigrants.

Critical to the early Ordovician radiations is the development of marginal and intra-oceanic terranes and their roles in the development, dispersal and survival of taxa, the partition of oceanic circulation patterns within the Iapetus Ocean and the provision of mosaics of rapidly changing, nearshore unstable environments.

Harper *et al.* (1996) identified nearly 20 early Ordovician intra-oceanic and marginal sites within the Iapetus region. A range of multivariate analyses confirmed the coherence of the Toquima-Table Head (low latitude) and Celtic (high latitude) groups of faunas. A number of authors have emphasized the biological importance of these sites (Neuman 1972, 1984; Bruton & Harper 1981, 1985; Fortey 1984; Harper 1992; Harper *et al.* 1996). These marginal and intra-oceanic sites may have served as both 'cradles' and 'museums' alternately providing sources for radiations on the platforms and refugia for otherwise relict taxa.

The early Ordovician rhynchonelliformean brachiopod fauna is dominated by billingsellides (including the clitambonitidines) and orthides. Analysis of the distribution of these taxa across the Celtic and Toquima-Table Head belts, targeting the marginal and intra-oceanic sites, suggests that 32% ($N = 45$) of the brachiopods at these sites were endemic to either a single locality or to a group of adjacent localities (Fig. 3). Moreover 30% of the taxa represent the first occurrence of a genus and 25% the last. Only 13% of the occurrences represent within-range spikes. The available data tend to confirm the role of these sites as the locus for endemics and possible roles as cradles and refugia within the Iapetus Ocean system.

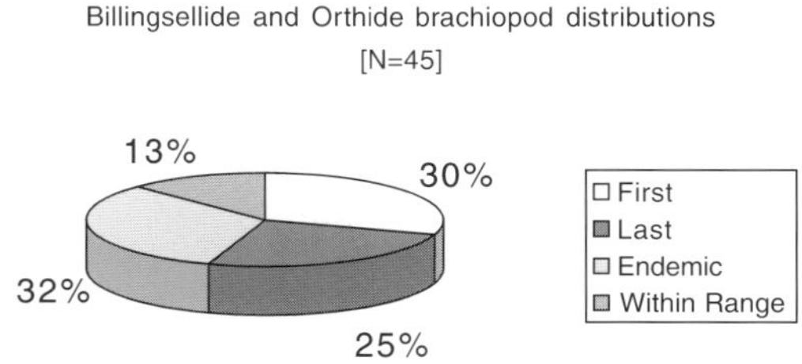

Fig. 3. Proportions of endemics, first, last and mid-range occurrences of billingsellide and orthide genera across the marginal Celtic and Table Head faunas. Data are modified from Harper *et al.* (1996).

Early to mid-Ordovician development of the Iapetus brachiopod fauna

The initial break-up of the northern margins of Gondwana and the arcs around Laurentia created a diversity of terranes (Fig. 2). This disparate array of terranes formed an ideal environment for the development of gamma diversity (inter-province) across the Iapetus Ocean. The isolation of these terranes was enhanced by the late Arenig regression, when there was a concentration of island biotas (Fortey 1984). Many islands and archipelagos may have been emergent, with narrow shelf areas with relatively small populations. The majority of brachiopod faunas inhabited shallow-water, nearshore environments, probably no deeper than Benthic Assemblage zones 3–4 (Cocks 1996). By comparison with the non-articulates, dispersion of the rhynchonelliformeans was relatively limited; the latter probably possessed relatively short, nektobenthic larval stages restricting the rapid spread of many articulated groups (Harper & Sandy 2001).

Subsequent transgressions promoted migration and the subsequent colonization of the shelf areas of the platform provinces (Figs 4 and 5). These mechanisms provided for the escalation of alpha diversity (within-community) as established regional palaeocommunities were supplemented by immigrants. These diversifications were most marked in shallow-water palaeocommunities, for example in the inner, nearshore facies of Avalonia and the Baltic province (Harper & Hints 2001). This diversity spike is most obvious along the margins of Baltica, for example in the Oslo Region (Harper 1986). The faunas were dominated by larger pedunculate orthidine brachiopods. In addition to sea-level changes the approach of terranes, particularly from the peri-Gondwanan region, may have provided additional taxa to participate in these radiations (Holmer *et al.* 2000). This process of integration has also been suggested for the development of the Celtic terranes (Bruton & Harper 1985).

The most marked radiations, however, occurred during the Caradoc (Figs 4 and 5). This diversity spike is clear in global databases but is also obvious in the regional datasets for Avalonia, Baltica, Gondwana and Perunica. These later diversifications may have been associated with beta diversity changes as a variety of new palaeocommunities inhabited deep-water environments with a greater dominance of the recumbent strophomenides (Harper *et al.* 1999*b*; Rong *et al.* 1999).

Comparison of early and late Ordovician brachiopod biofacies

In contrast to later Ordovician brachiopod faunas, a large number of early Ordovician genera are reported from only one or two sites in the Iapetus region; narrow geographical ranges are characteristic of many early Ordovician taxa (Fig. 6). The distribution patterns of these taxa show a positive skewness, suggesting

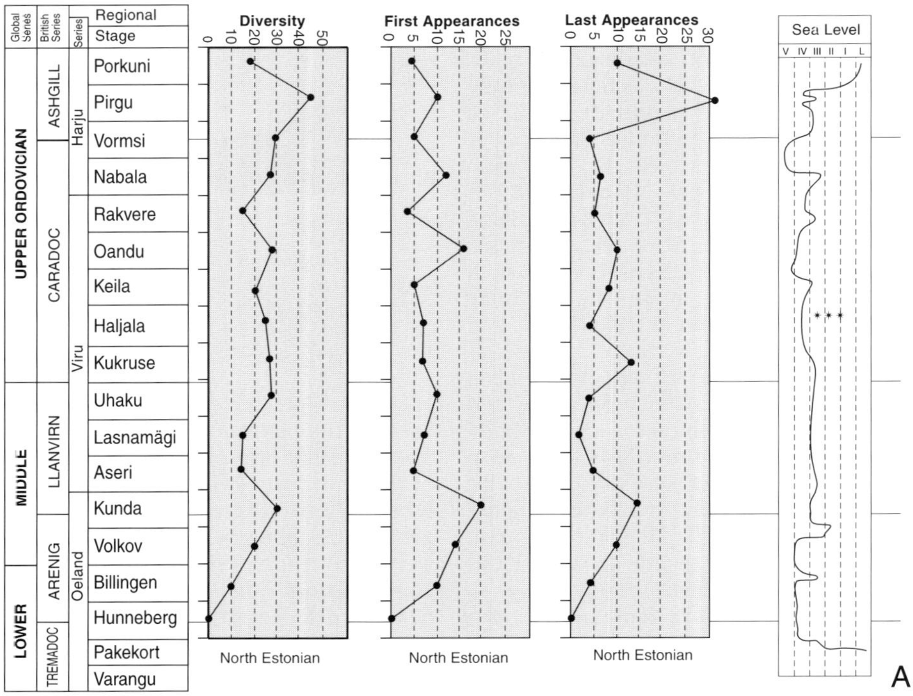

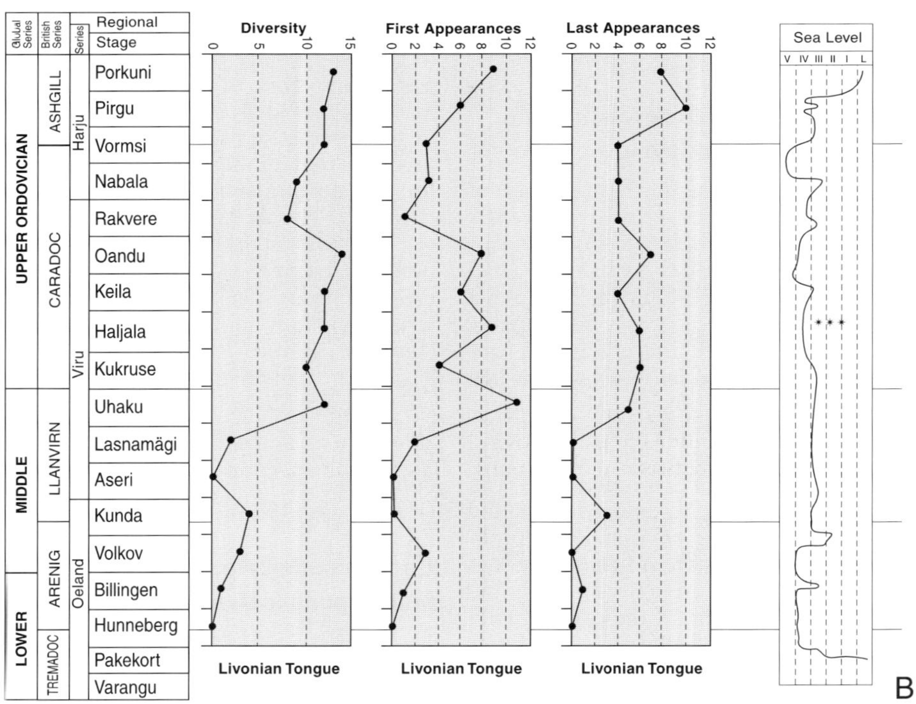

Fig. 4. Diversity profiles (unstandardized numbers of genera) through (**A**) the shallow-water North Estonian confacies belt and (**B**) the deeper-water Livonian Tongue (equivalent to the Central Baltoscandian confacies belt of Jaanusson 1976). Sea-level stands are calibrated in terms of Benthic Assemblage zones. Data are modified from Harper & Hints (2001).

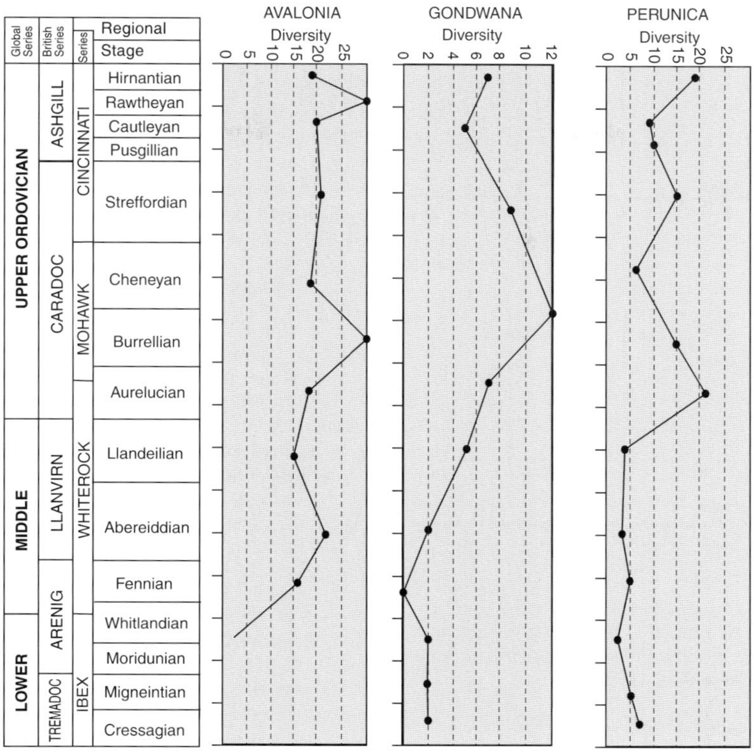

Fig. 5. Diversity profiles (unstandardized numbers of genera) from Avalonia (based on sources in Lockley (1983) and some additional data), part of the Gondwanan margin (based on Havlíček 1971) and Perunica (based on sources in Havlíček 1998).

that the majority of taxa are in fact reported from only one or two sites. This pattern contrasts with that for the late Ordovician where a less skewed distribution indicates fewer endemic taxa and more widespread genera. Seriation of these sites has emphasized these contrasting spatial patterns (Ryan *et al.* 1999). Early Ordovician faunas were successfully arranged into cohesive sets representing the main provincial groupings; the links between these groups were less clear. Faunal gradients are much clearer during the late Ordovician when a series of overlapping provinces can be recognized from the high-latitude Bani province, through the Kosov province to the low-latitude Edgewood province (Rong & Harper 1988; Owen *et al.* 1991; Ryan *et al.* 1999).

Conclusions

This study emphasizes the role of marginal and oceanic sites in the Ordovician radiation. Intra- and marginal Iapetus Ocean terranes acted as both cradles and museums for brachiopod taxa during the early Ordovician (late Arenig to early Llanvirn) phases of the diversification. Terrane fragmentation, particularly at high latitudes, was essential for the generation of an initial gamma diversity. Diversification, however, was more pronounced when individual platform communities first expanded in size with the development of alpha diversity, during the Llanvirn. These diversifications may have helped propel the offshore movement of communities during the Caradoc as beta diversity intensified.

During the early Ordovician many brachiopod families were relatively widespread: biogeographical signals provided by generic distributions. This suggests that the origins of many Ordovician brachiopod families may be tracked back into the Cambrian outside the Iapetus region. The circum-Iapetus platforms and margins provided opportunities but not the origins for a mid-Arenig radiation. The early Ordovician dispersion of the continents and associated outboard terranes provided the

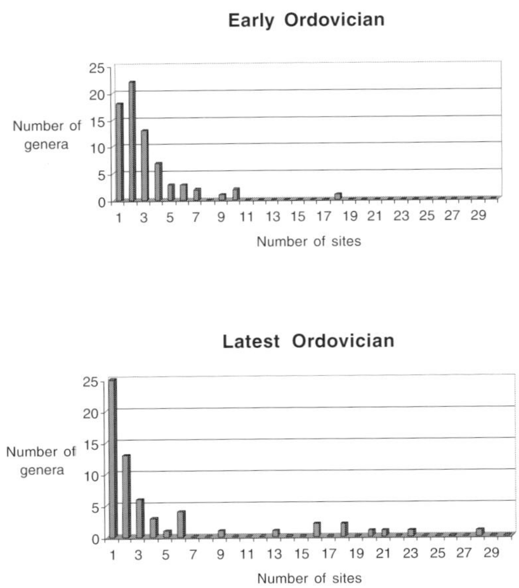

Fig. 6. Comparison of some of the distributional properties of early and late Ordovician brachiopod faunas. Data are modified from sources in Harper *et al.* (1996) and Owen *et al.* (1991).

conditions for geographic isolation and speciation. Migration of taxa onto the platforms during the early Llanvirn generated the first increases in within-community diversity that may have driven the development of new, particularly deeper-water, community types especially during the Caradoc.

D.A.T.H. thanks the Danish Natural Science Council (SNF) for financial support. This paper is a contribution to IGCP 410 'The Great Ordovician Biodiversification Event'. Helpful comments from T. McCormick, A. Owen and an anonymous reviewer improved the manuscript.

References

ADRAIN, J. M., FORTEY, R. A. & WESTROP, S. R. 1998. Post-Cambrian trilobite diversity and evolutionary faunas. *Science*, **280**, 1922–1925.

BASSETT, M. G., POPOV, L. E. & HOLMER, L. E. 1999*a*. Organophosphatic brachiopods: Patterns of biodiversification and extinction in the Early Palaeozoic. *Geobios*, **32**, 145–163.

BASSETT, M. G., POPOV, L. E. & SOKIRIN, E. V. 1999*b*. Patterns of diversification in Ordovician cyrtomatodont rhynchonellate brachiopods. *Acta Universitatis Carolinae, Geologica*, **43**, 329–332.

BOTTJER, D. J., DROSER, M. L., SHEEHAN, P. M. & MCGHEE JR., G. R. 2001. The ecological architecture of major events in the Phanerozoic history of marine invertebrate life. *In*: ALLMON, W. D. & BOTTJER, D. J. (eds) *Evolutionary Paleoecology*. Columbia University Press, New York, 35–61.

BOUCOT, A. J. 1990. Silurian biogeography. *In*: MCKERROW, W. S. & SCOTESE, C. R. (eds) *Palaeozoic Palaeogeography and Biogeography*. Geological Society, London, Memoir, **12**, 191–196.

BOUCOT, A. J. 1993. Paleozoic paleogeography and biogeography. *Revista Española de Paleontologia, Nº Extraordinario*, 15–20.

BOUCOT, A. J. 1999. Southern African Phanerozoic marine invertebrates: biogeography, palæoecology, climatology and comments on adjacent regions. *Journal of African Earth Sciences*, **28**, 129–143.

BRUTON, D. L. & HARPER, D. A. T. 1981. Brachiopods and trilobites of the early Ordovician serpentine Otta Conglomerate, south central Norway. *Norsk Geologisk Tidsskrift*, **61**, 151–181.

BRUTON, D. L. & HARPER, D. A. T. 1985. Early Ordovician (Arenig–Llanvirn) faunas from oceanic islands in the Appalachian-Caledonian orogen. *In*: GEE, D. G. & STURT, B. A. (eds) *The Caledonian Orogen – Scandinavia and Related Areas*. John Wiley & Sons, Chichester, 359–368.

CHRISTIANSEN, J. L. & STOUGE, S. 1999. Using palaeo-oceanical modelling in reconstructing Early Ordovician palaeogeography. *Acta Universitatis Carolinae, Geologica*, **43**, 515–518.

COCKS, L. R. M. 1996. Lower Palaeozoic brachiopod communities. *In*: COPPER, P. & JIN JSUO (eds) *Brachiopods*. A. A. Balkema, Rotterdam, 69–71.

COCKS, L. R. M. 2000. The Early Palaeozoic geography of Europe. *Journal of the Geological Society, London*, **157**, 1–10.

COCKS, L. R. M. 2001. Ordovician and Silurian global geography. *Journal of the Geological Society, London*, **158**, 197–210.

COCKS, L. R. M. & FORTEY, R. A. 1982. Faunal evidence for oceanic separation in the Palaeozoic of Britain. *Journal of the Geological Society, London*, **139**, 465–478.

COCKS, L. R. M. & FORTEY, R. A. 1998. The Lower Palaeozoic margins of Baltica. *GFF*, **120**, 173–179.

COCKS, L. R. M., MCKERROW, W. S. & VAN STAAL, C. R. 1997. The margins of Avalonia. *Geological Magazine*, **134**, 627–636.

DROSER, M. L. & SHEEHAN, P. M. 1997. Palaeoecology of the Ordovician radiation: resolution of large-scale patterns within individual clade histories, palaeogeography and environments. *Geobios*, **20**, 221–229.

DROSER, M. L., BOTTJER, D. J. & SHEEHAN, P. M. 1997. Evaluating the ecological architecture of major events in the Phanerozoic history of marine invertebrate life. *Geology*, **25**, 167–170.

FORTEY, R. A. 1984. Global earlier Ordovician transgressions and regressions and their biological implications. *In*: BRUTON, D. L. (ed.) *Aspects of the Ordovician System*. Palaeontological

Contributions from the University of Oslo, **295**, 37–50.

FORTEY, R. A. & COCKS, L. R. M. 1986. Marginal faunal belts and their stratigraphical implications. *Journal of the Geological Society, London*, **143**, 151–160.

HARPER, D. A. T. 1986. Distributional trends within the Ordovician brachiopod faunas of the Oslo Region, south Norway. *In*: RACHEBOEUF, P. R. & EMIG, C. C. (eds) *Les Brachiopodes fossils et actuels*. Biostratigraphie du Paleozoique, **4**, 465–475.

HARPER, D. A. T. 1992. Ordovician provincial signals from Appalachian-Caledonian terranes. *Terra Nova*, **4**, 204–209.

HARPER, D. A. T. 1998. Interpreting orogenic belts: principles and examples. *In*: DOYLE, P. & BENNETT, M. R. (eds) *Unlocking the Stratigraphical Record: Advances in Modern Stratigraphy*. John Wiley & Sons, Chichester, 491–524.

HARPER, D. A. T. & HINTS, L. 2001. Distribution and diversity of Ordovician articulated brachiopods in the East Baltic. *In*: BRUNTON, C. H. C., COCKS, L. R. M. & LONG, S. (eds) *Brachiopods Past and Present*. Taylor & Francis, New York & London, 315–326.

HARPER, D. A. T. & RONG JIA-YU 2001. Palaeozoic brachiopod extinctions, survival and recovery: patterns within the rhynchonelliformeans. *Geological Journal*, **36**, 317–328.

HARPER, D. A. T. & SANDY, M. R. 2001. Paleozoic brachiopod biogeography. *In*: CARLSON, S. J. & SANDY, M. R. (eds) *Brachiopods Ancient and Modern. A Tribute to G. Arthur Cooper*. Paleontological Society Papers, **7**, 207–222

HARPER, D. A. T. & WRIGHT, A. D. 1996. Brachiopods. *In*: HARPER, D. A. T. & OWEN, A. W. (eds) *Fossils of the Upper Ordovician*. Palaeontological Association, London, 63–94.

HARPER, D. A. T., MAC NIOCAILL, C. & WILLIAMS, S. H. 1996. The palaeogeography of early Ordovician Iapetus terranes: an integration of faunal and palaeomagnetic constraints. *Palaeogeography, Palaeoclimatology, Palaeoecology*, **121**, 297–312.

HARPER, D. A. T., RONG JIA-YU & SHEEHAN, P. M. 1999*a*. Ordovician diversity patterns in early rhynchonelliform (protorthide, orthide and strophomenide) brachiopods. *Acta Universitatis Carolinae, Geologica*, **43**, 325–327.

HARPER, D. A. T., RONG JIA-YU AND ZHAN REN-BIN 1999*b*. Late Ordovician development of deep-water brachiopod faunas. *Acta Universitatis Carolinae, Geologica*, **43**, 351–353.

HARPER, D. A. T., BASSETT, M. G., COCKS, L. R. M., COPPER, P., *ET AL*. 2001. Ordovician brachiopod diversity. *Paleobios*, **21** (2nd Supplement), 5–6.

HAVLÍČEK, V. 1971. Brachiopodes de L'Ordovicien du Maroc. *Notes et Memoires du Service Geologique*, **230**, 1–135.

HAVLÍČEK, V. 1976. Evolution of Ordovician brachiopod communities in the Mediterranean province. *In*: BASSETT, M.G. (ed.) *The Ordovician System*. University of Wales Press and the Palaeontological Association, 349–358.

HAVLÍČEK, V. 1998. Ordovician. *In*: CHULPÁČ, I., HAVLÍČEK, V. KŘÍŽ, J., KUKAL, Z. & ŠTORCH, P. (eds) *Palaeozoic of the Barrandian (Cambrian–Devonian)*. Czech Geological Survey, Prague, 41–79.

HOLMER, L. E., POPOV, L. E. & BASSETT, M. G. 2000. Early Ordovician organophosphatic brachiopods with Baltoscandian affinities from the Altay Range, southern Krygyzstan. *GFF*, **122**, 367–375.

JAANUSSON, V. 1973. Ordovician articulate brachiopods. *In*: HALLAM, A. (ed.) *Atlas of Palaeobiogeography*. Elsevier, Amsterdam, 19–26.

JAANUSSON, V. 1976. Faunal dynamics in the Middle Ordovician (Viruan) of Baltoscandia. *In*: BASSETT, M. G. (ed.) *The Ordovician System*. University of Wales Press and the Palaeontological Association, 301–326.

LOCKLEY, M. G. 1983. A review of the brachiopod dominated palaeocommunities from the type Ordovician. *Palaeontology*, **26**, 111–145.

MCKERROW, W. S. & COCKS, L. R. M. 1976. Progressive faunal migration across the Iapetus Ocean. *Nature*, **263**, 304–306.

MCKERROW, W. S., MAC NIOCAILL, C., AHLBERG, P. E., CLAYTON, G., CLEAL, C. J. & EAGAR, R. M. C. 2000. The Late Palaeozoic relations between Gondwana and Laurussia. *In*: FRANKE, W., HAAK, V., ONCKEN, O. & TANNER, D. (eds) *Orogenic Processes: Quantification and Modelling in the Variscan Belt*. Geological Society, London, Special Publications, **179**, 9–20.

MAC NIOCAILL, C. 2000. A new Silurian palaeolatitude for eastern Avalonia and evidence for crustal rotations in the Avalonian margin of southwestern Ireland. *Geophysical Journal International*, **141**, 661–671.

MAC NIOCAILL, C. & SMETHURST, M. A. 1994. Palaeozoic palaeogeography of Laurentia and its margins: a reassessment of palaeomagnetic data. *Geophysical Journal International*, **116**, 715–725.

MAC NIOCAILL, C., VAN DER PLUIJM, B. A. & VAN DER VOO, R. 1997. Ordovician paleogeography and the evolution of the Iapetus Ocean. *Geology*, **25**, 159–162.

NEUMAN, R. B. 1972. Brachiopods of early Ordovician volcanic islands. *International Geological Congress, Montreal, Section*, **7**, 297–302.

NEUMAN, R. B. 1976. Early Ordovician (Late Arenig) brachiopods from Virgin Arm, New World Island, Newfoundland. *Bulletin of the Geological Survey of Canada*, **261**, 11–61.

NEUMAN, R. B. 1984. Geology and paleobiology of islands in the Ordovician Iapetus Ocean. *Geological* Society *of America, Bulletin*, **95**, 1188–1201.

NEUMAN, R. B. & HARPER, D. A. T. 1992. Paleogeographic significance of Arenig–Llanvirn Toquima-Table Head and Celtic assemblages. *In*: WEBBY, B. D. & LAURIE, J. R. (eds) *Global Perspectives in Ordovician Geology*. Balkema, Rotterdam, 241–254.

OWEN, A. W., HARPER, D. A. T. & RONG JIA-YU 1991. Hirnantian trilobites and brachiopods in space and time. *In*: BARNES, C. & WILLIAMS, S.H. (eds.) *Advances in Ordovician Geology*. Geological Survey of Canada, Paper **90–9**, 179–190.

OWEN, A. W., HARPER, D. A. T. & ROMANO, M. 1992. The Ordovician biogeography of the Grangegeeth terrane and the position of the Iapetus suture in eastern Ireland. *Journal of the Geological Society, London*, **149**, 3–6.

POPOV, L.E., BASSETT, M.G., HOLMER, L.E. & GORJANSKY, V. YU. 1999. Ordovician patterns of diversification in craniiformean brachiopods. *Acta Universitatis Carolinae, Geologica*, **43**, 321–324.

RONG JIA-YU & HARPER, D. A. T. 1988. A global synthesis of the latest Ordovician Hirnantian brachiopod fauna. *Transactions of the Royal Society of Edinburgh: Earth Sciences*, **79**, 383–402.

RONG JIA-YU & HARPER, D. A. T. 2000. Brachiopod survival and recovery from the latest Ordovician mass extinctions. *Geological Journal*, **34**, 321–348.

RONG JIA-YU, ZHAN REN-BIN & HARPER, D. A. T. 1999. Late Ordovician (Caradoc–Ashgill) brachiopod faunas with *Foliomena* based on data from China. *Palaios*, **14**, 412–431.

RYAN, P. D., RYAN, M. & HARPER, D. A. T. 1999. Seriation: a new approach. *In*: HARPER, D.A.T. (ed.) *Numerical Palaeobiology*. John Wiley & Sons, Chichester, 433–449.

SMETHURST, M. A., KHRAMOV, A. & TORSVIK, T. H. 1998. Palaeomagnetic constraints on the Neoproterozoic–Palaeozoic drift history of the Siberian Platform: Rodinia to Pangea. *Earth Science Reviews*, **43**, 1–24.

TORSVIK, T. H., SMETHURST, M. A., MEERT, J. G., *ET AL.* 1996. Continental break-up and collision in the Neoproterozoic and Palaeozoic: A tale of Baltica and Laurentia. *Earth Science Reviews*, **40**, 229–258.

VAN STAAL, C. R., DEWEY, J. F., MAC NIOCAILL, C. & MCKERROW, W. S. 1998. The tectonic evolution of the Northern Appalachians and British Caledonides: history of a complex, west and southwest Pacific-type segment of Iapetus. *In*: BLUNDELL, D.J. & SCOTT, A.C. (eds) *Lyell: the Past is the Key to the Present*. Geological Society, London, Special Publications, **143**, 199–242.

WILLIAMS, A. 1969. Ordovician faunal provinces with reference to brachiopod distribution. *In*: WOOD, A. (ed.) *The Pre-Cambrian and Lower Palaeozoic Rocks of Wales*. University of Wales Press, Cardiff, 117–154.

WILLIAMS, A. 1973. Distribution of brachiopod assemblages in relation to Ordovician palaeogeography. *In*: HUGHES, N. F. (ed.) *Organisms and Continents through Time*. Special Papers in Palaeontology, **12**, 241–269.

WILLIAMS, A., BRUNTON, C. H. C. & CARLSON, S. J. (eds) 2000. *Treatise on Invertebrate Paleontology, Part H, Brachiopoda revised. Volume 2: Linguliformea, Craniiformea, and Rhynchonelliformea (part)*. Geological Society of America and University of Kansas Press, Boulder, Colorado and Lawrence, Kansas.

WILSON, J. T. 1966. Did the Atlantic close then reopen? *Nature*, **211**, 676–681.

Diversification and biogeography of bivalves during the Ordovician Period

JOHN C. W. COPE

Department of Earth Sciences, Cardiff University, PO Box 914, Cardiff CF10 3YE, UK
(e-mail: copejcw@cardiff.ac.uk*)*

Abstract: Bivalves have a wide distribution in the Lower and to a lesser extent Middle Cambrian rocks, but they have not yet been certainly identified in the Upper Cambrian. Recent discoveries have significantly increased our knowledge of Lower Ordovician bivalve faunas and their explosive radiation from the Early Ordovician apparently coincides with the evolution of the feeding gill. Early Ordovician faunas were confined to the siliciclastic facies of Gondwanan shelf seas; most genera were clearly latitudinally constrained, but others apparently migrated over wide latitudes. By the Mid-Ordovician, bivalves had begun to escape the confines of Gondwana and marked latitudinal differences in the composition of the faunas became apparent, with pteriomorphians showing clear preference for low latitudes, whilst heteroconchs clearly preferred median to high latitudes; surprisingly, nuculoids were both most diverse in terms of species and most abundant as a percentage of individuals within the bivalve populations at low latitudes. It was in the Late Ordovician that bivalves colonized the low-latitude carbonate platforms of Laurentia and Baltica, leading to a second major diversification particularly within the pteriomorphian bivalves, which developed semi-infaunal and epifaunal habits; they became the dominant low-latitude bivalve group. The latest-Ordovician eustatic regression which exposed the low-latitude carbonate platforms resulted in a major reduction in the epifaunal and semi-infaunal bivalves involving extinction of many genera, including the only Ordovician boring bivalves.

Thirty years ago Pojeta (1971) produced his classic work on Ordovician bivalves that illustrated something of the early diversity of this long-neglected fossil group. At the time of this publication, some revisionary work had been done on Late Ordovician faunas from North America (e.g. Pojeta 1962, 1966), but save the works on French faunas of the Montagne Noire (Thoral 1935) and Armorica (Babin 1966) little work had been done in the first 70 years of the twentieth century on Early Ordovician bivalve faunas. Phylogenetic links were necessarily sketchy (Pojeta 1971) and Cambrian bivalves were of doubtful interpretation. Pojeta's (1971) reservations about *Lamellodonta*, the earliest bivalve accepted by the *Treatise* (Cox *et al.* 1969–1971), were confirmed when the genus was shown to be based upon a distorted brachiopod (Havliček & Kříž 1978). Pojeta later (1973, 1975) went on to demonstrate the bivalve affinities of the widely dispersed Lower Cambrian *Fordilla*, whilst Jell (1980) described another Early Cambrian genus *Pojetaia* from Australia, subsequently found in Bornholm, China and Morocco and found to extend into the Mid-Cambrian. Other Mid-Cambrian forms have been described from New Zealand (McKinnon 1982), Bornholm (Berg-Madsen 1987; Hinz-Schallreuter 1995, 2000) and Morocco (Geyer & Streng 1998). The lack of bivalves from the Upper Cambrian remains an embarrassing hiatus in the study of early bivalve evolution. Berg-Madsen (1987) recorded a *Tuarangia* from a Polish erratic that contained conodonts she regarded as Upper Cambrian, but Hinz-Schallreuter (2000) pointed out that tricuspidate westergaardodinids also occur in the Middle Cambrian. Pojeta (1980) figured a rather featureless shell from the Upper Cambrian of Maryland, but this single specimen has no uniquely bivalve characteristics. Pojeta (2000) has recently reviewed the Cambrian history of the Bivalvia.

Lower Ordovician bivalves are much more diverse, much larger in size and more common as fossils. Cope (1995, 1997*b*) correlated these factors with the evolution of the feeding gill in the latest Cambrian or earliest Ordovician. Bivalves remain, however, very rare fossils in Lower Ordovician rocks. Thus from the Tremadoc, bivalves are known from four areas: Salta Province, Argentina (Harrington 1938); the Montagne Noire (Thoral 1935; Babin 1982*a*); central Australia (Pojeta & Gilbert-Tomlinson 1977; originally described as probably of Arenig age but later shown to be late Tremadoc by Shergold *et al.* 1991) and the Moroccan Anti-Atlas (pers. obs. 2001).

From: Crame, J. A. & Owen, A. W. (eds) 2002. *Palaeobiogeography and Biodiversity Change: the Ordovician and Mesozoic–Cenozoic Radiations*. Geological Society, London, Special Publications, **194**, 25–52.
0305-8719/02/$15.00

Altogether some nine or ten species are known so far from Tremadoc rocks. From the lower part of the Arenig Series, bivalves are known from Salta Province, Argentina (Harrington 1938); the Montagne Noire (Babin 1982*a*); Afghanistan (Desparmet *et al.* 1971); originally reported as Tremadoc but more probably of Arenig age; Babin & Gutiérrez-Marco 1991); the Moroccan Anti-Atlas (Babin & Destombes 1992); Ramsey Island, South Wales (Hicks 1873; Carter 1971); and most recently from Llangynog, South Wales (Cope 1996*b*), the latter locality yielding no fewer than 20 species belonging to 18 genera, representing well over half the species known hitherto from rocks of this age. After the Arenig, bivalves become commoner and Babin & Gutiérrez-Marco (1991) were able to report on Middle Ordovician bivalves collected from no fewer than 87 localities from Spain alone.

All the Early Ordovician bivalves known are from Gondwana and it now seems clear that they were thus geographically restricted (Avalonia being so close to the Gondwanan margins in Arenig times that, for the purposes of palaeobiogeography, it can still be considered as part of Gondwana). It is assumed that, although they were able to migrate freely around the Gondwanan margins, the width of the oceanic areas around Gondwana precluded bivalve larvae from crossing to other continental shelves before the Mid-Ordovician.

The major radiations of the Early Ordovician resulted in the appearance of most bivalve groups by the end of that time. Cope (2000) proposed a revised classification of the Bivalvia recognizing that the division into the subclasses based on gill grade, Protobranchia Pelseneer, 1889 and Autolamellibranchiata Grobben, 1894, long used by zoologists, could now be applied to the fossil record. The classification of the Bivalvia used in this paper is essentially that of Cope (2000) shown in Figure 1. The only difference between that figure and Cope (2000, fig. 2) is that the Nucinelloidea are not shown, as they do not appear until well after the Ordovician. Continental reconstructions for the Ordovician are shown in Figure 2.

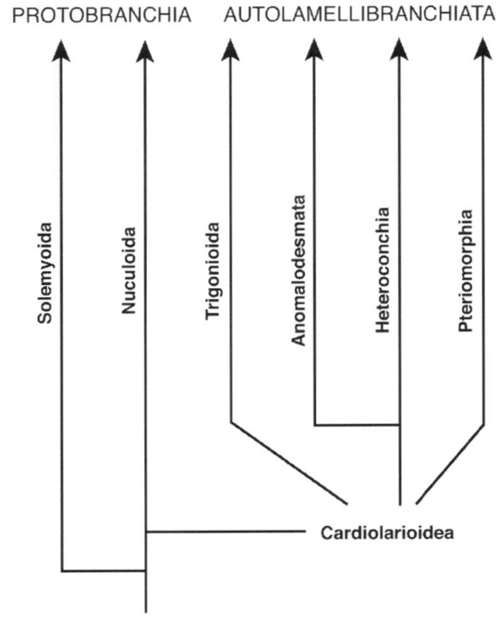

Fig. 1. Classification of the Bivalvia adopted herein, showing the phylogenetic links between the major bivalve groups. Modified after Cope (2000).

Protobranchia

Nuculoida

Amongst the protobranchs the Nuculoida are well represented in the Early Ordovician and range from low-latitude areas like Australia, to high-latitude areas such as the Montagne Noire and the Moroccan Anti-Atlas. The simplest forms, with undifferentiated taxodont dentition, belong to the family Praenuculidae McAlester, 1969 (in Cox *et al.* 1969–1971). The earliest praenuculids include forms like *Paulinea* and *Pensarnia* from the Early Arenig of South Wales (Cope 1996*b*); the latter also occurs in the Late Arenig of the Welsh Borderland (Cope 1999). Some other praenuculids also seem to be of restricted geographical range, such as *Fidera* Pojeta & Gilbert-Tomlinson, 1977, from the Middle to Upper Ordovician of Tasmania. However, *Praenucula* Pfab, 1934 itself occurs widely in the Middle Ordovician of Europe from Bohemia (e.g. *P. expansa* Pfab, 1934) westwards to Portugal (e.g. *P. ribeiro* (Sharpe, 1853)); and has also been recorded from the Lower, Middle and Upper Ordovician of the Moroccan Anti-Atlas (Babin & Destombes 1992). *Praenucula* and the praenuculid *Palaeoconcha* were recorded from the upper Middle or lower Upper Ordovician of Argentina by Sánchez (1990). Dzik (1994) figured a minute juvenile nuculoid from the Lower Llanvirn of Baltica that appears to be a *Praenucula*, as indicated by Babin (2000); this is notable in another context, as it is one of the earliest bivalves to have reached Baltica. *Praenucula* also occurs in the Upper Ordovician of Laurentia and *P. filistriata* (Ulrich, 1894) from Ohio was figured under the generic name

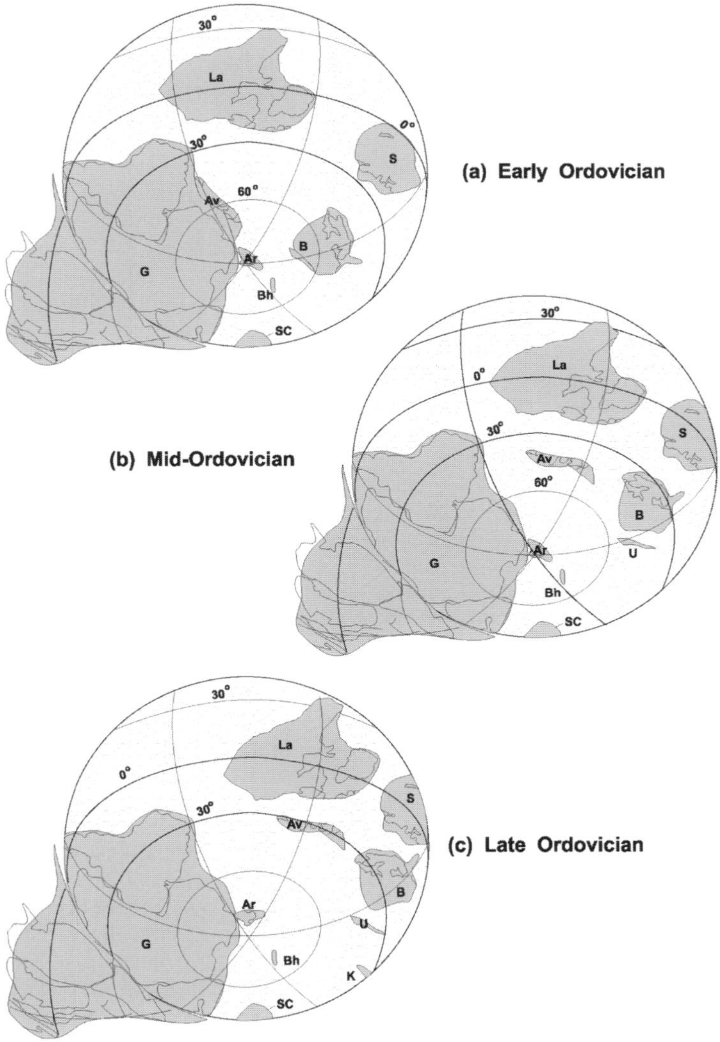

Fig. 2. Continental reconstructions for the Ordovician. (**a**) Early Ordovician. Note near-polar position of the Montagne Noire and the lower-latitude position of Avalonia which is still on the Gondwanan margin. Contrast with the lower latitudes of Argentina and Australia. (**b**) Mid-Ordovician. Note lower-latitude position of Avalonia, whilst Armorica stays at polar latitudes. (**c**) Late Ordovician. Note positions of Baltica and Laurentia, both areas of carbonate platforms.
Key to abbreviations: Ar, Armorica; Av, Avalonia; B, Baltica; Bh, Bohemia; G, Gondwana; K, Kazakhstania; La, Laurentia; S, Siberia; SC, South China; U, East Uralia. Reconstructions principally based on those of Torsvik (1998) with modifications by L. E. Popov.

Deceptrix by Pojeta (1978 pl. 2, figs 1–2), but Cope (1997*b*) showed that the latter genus was a cardiolariid and not a praenuculid (see below). Some of the species figured previously as *Deceptrix*, in addition to some of those figured by Pojeta (1971, 1978), including those figured by Tunnicliff (1982) and Cope (1996*a*), are now recognized as praenuculids and not cardiolariids. For those species Cope (1997*b*) proposed the praenuculid genus *Homilodonta* with *Arca subtruncata* Portlock, 1843 as type species; a full description and designation of a lectotype of this species was made by Tunnicliff (1982, pp. 60–61, pl. 9, figs 1–7, 9–11)

Similodonta Soot-Ryen, 1964, is a characteristic praenuculid genus of the Late Ordovician of Laurentia and Baltica. It is characterized by an abrupt 90° angle in the middle of the hinge-plate, although Soot-Ryen (1964), in her original diagnosis, claimed that the angle was near 80°. The genus is well exemplified by the type species *S. similis* (Ulrich, 1892) from the Richmond Group of Minnesota. Other Laurentian species include *S. recurva* (Ulrich, 1892) and *S. costata* (Branson, 1909) together with the Ashgill species *S. magna* (Lamont, 1946) and *S. collina* (Reed, 1946) refigured by Cope (1996*a*) both from Scotland, and a probable (but unidentified) species figured from Northern Ireland by Tunnicliff (1982). An as yet un-named new genus, clearly related to *Similodonta*, was figured by Pojeta (1978, pl. 2, figs 9–12). Late Ordovician species of *Similodonta* are also known from Baltica, and include *S. spjeldnaesi* (Soot-Ryen & Soot-Ryen, 1960) from the Upper Caradoc of the Oslo region; the same species was also recorded from the Ashgill of the Oslo region by Toni (1975). It could thus be considered that *Similodonta* was a genus characteristic of low-latitude areas (presumably with warm water). The origins of *Similodonta* are, however, probably Avalonian. Cope (1999) has figured the earliest known species, *S. ceryx*, from the Aurelucian Stage of the Lower Caradoc of mid-Wales and suggested that it may have evolved from the early Llanvirn praenuculid genus *Arcodonta* Cope, 1999, putting the origin of the genus in Avalonia. *Similodonta* also occurs in the Ashgill Series of North Wales (collections of the British Geological Survey). Another possible Gondwanan *Similodonta* was figured by Steinmann & Hoek (1912) from the Upper(?) Ordovician of Bolivia according to Babin (1993), but the text seems to make it clear that these specimens are distorted lingulid brachiopods. From the Precordillera of Argentina Sánchez (1999*a*) has figured what appear very clearly to be genera closely related to *Similodonta*. These include *Villicumia* and *Trigonoconcha*, the latter with a hinge-plate angle close to 65°; these clearly represent part of a local Caradoc bivalve diversification that was described by Sánchez (1999*b*); both these new genera could have been derived from *Similodonta*.

The Llanvirn *Tironucula* Morris & Fortey, 1976, from the Laurentian margin on Svalbard, was possibly the first bivalve to reach Laurentia. Its juvenile dentition is similar to that of many praenuculids, but it develops blade-like teeth ontogenetically. Babin (1982*a*) proposed the family Tironuculidae to accommodate this genus and the clearly related *Ekaterodonta*, originally described from the Lower Arenig of the Montagne Noire (Babin 1982*a*). This latter genus is now also known from the Lower Llanvirn of Spain (Babin & Gutiérrez-Marco 1991) and from the Upper Llanvirn of Bolivia (Babin & Branisa 1987) and, unlike *Tironucula*, was a high-latitude genus. Another tironuculid, *Natasia*, has been described from the Lower and Middle Arenig of NW Argentina by Sánchez (1996) and was subsequently assigned to a separate subfamily within the Tironuculidae (Sánchez 1997*a*).

The apparently wide geographical ranges of some other nuculoid forms may in some cases simply be due to inappropriate generic assignations. Thus the genus *Ctenodonta*, that has been widely reported from the low-latitude Upper Ordovician rocks, is based on the species *C. nasuta* (Hall, 1847) from the Upper Ordovician of the United States. However, many other species have been assigned to this genus in the past and, as remarked on by Pojeta (1971, p. 15), all that this generic name effectively denotes is that a nuculoid is being described. There do not appear to be any definitive records of this genus from the Early Ordovician although poorly preserved material from the Tremadoc of Argentina (*Palaeoneilo iruyensis* Harrington, 1938) was tentatively assigned to *Ctenodonta* by Pojeta & Gilbert-Tomlinson (1977), and Babin & Destombes (1992) figured two forms they compared to *Ctenodonta* from the Middle Arenig of Morocco. Definitive records of *Ctenodonta* are from the Middle Ordovician of low-latitude Gondwana (Australia; Pojeta & Gilbert-Tomlinson 1977) and the Upper Ordovician of Laurentia (Pojeta 1971) and Siberia (Krasilova 1976, 1979). Other Ordovician genera related to *Ctenodonta* include *Tancrediopsis*, the earliest known of which is from the Llanvirn of Mid-Wales (Cope 1999); the somewhat younger type species (*T. contracta*) was originally described from the Alumette Islands, Ontario, Canada, by Salter (1859). *Alococoncha* Pojeta & Gilbert-Tomlinson, 1977 is only known from the Middle Ordovician of the Amadeus Basin, Australia. *Siberoctenia* Krasilova, 1976, from the Upper Ordovician of Siberia, has a large number of small teeth.

Pojeta (1988, pp. 210–211) decided that the monotypic superfamily Ctenodontoidea and the family Ctenodontidae were poorly founded and of uncertain diagnosis and suggested that, for the time being, the few genera included in the family could be incorporated into the Nuculanoid family Malletiidae. Nuculanoids are nuculoids that have a posteriorly elongate shell and the family Malletiidae includes all the Ordovician nuculanoids. Excluding forms related to *Ctenodonta*, the earliest malletiids appear to be from the Gondwanan Middle Ordovician and include the genus *Cadomia* de Tromelin, 1877 (de Tromelin & Lebesconte 1877), first reported from Armorica, but more recently reported from the Middle Ordovician of Argentina (Sánchez 1986). The genus *Myoplusia* Neumayr, 1884, shows multiple muscle scars, and other malletiids include the endemic Australian genera *Johnmartinia*, *Lophoconcha*, *Sthenodonta* and *Zeehania* (Pojeta & Gilbert-Tomlinson 1977). The genera *Nuculites* and *Palaeoneilo* are very widely reported from the Upper Ordovician (and younger rocks) and appear cosmopolitan, but, as commented upon by Pojeta & Gilbert-Tomlinson (1977, p. 16), both genera are in need of revision and probably include several distinct generic level taxa.

Solemyoida

The other protobranch group to appear is the Solemyoida, known first from the early Arenig of South Wales (Cope 1996*b*), although as pointed out by Cope (1997*b*) the Montagne Noire specimen figured as a *Cymatonota*? by Babin (1982*a*, pl. 11, fig. 17) could possibly belong here too. The genus *Ovatoconcha* Cope, 1996*b* possesses the strongly anisomyarian anteriorly elongate shell and radial ornamentation typical of later solemyoids; it also apparently had a thick periostracal fringe that projected beyond the calcified margin of the valves, as in Recent solemyoids (Cope 2000). Pojeta (1988) discussed the origin of the solemyoids and derived them from ctenodontid nuculoids; the earliest form he identified with the solemyoids was the Baltic species *Dystactella aedilis* (Eichwald, 1856) from the early Late Ordovician of Estonia. Cope (2000) showed how the presence of ligamental nymphs in *Ctenodonta* and the Solemyoida, claimed as a synapomorphy to the two groups by Waller (1990, 1998), was independently derived, as Mid-Ordovician ctenodontids apparently lack the nymphs possessed by Late Ordovician ctenodontids. The nymphs of *Ctenodonta* were probably necessitated in the Late Ordovician as the umbones migrated rearwards, thus shortening the ligament and making stronger insertion essential (Cope 2000). This therefore confirmed Cope's earlier (1997*b*) view that Pojeta's claimed intermediate forms between solemyoids and ctenodontids must be persistent intermediate stocks, as fully evolved solemyoids were present at least as early as the Early Arenig. No pre-Arenig bivalves are known that could indicate the origins of *Ovatoconcha*.

Autolamellibranchia

Cardiolarioidea

Cope (1995) reported that the palaeotaxodont genus *Cardiolaria* had a dentition that was so similar in some respects to that of the heteroconch *Glyptarca* that they were likely to be phylogenetically related. He suggested that the asymmetrical hinge in *Cardiolaria* could have been evolved concomitantly with the evolution of the filibranch gill and that the hinge was designed to permit far wider valve opening than that permitted by the nuculoid hinge. Cope later (1997*b*) proposed the family Cardiolariidae for *Cardiolaria* and related genera, but still retained them within the Palaeotaxodonta. Subsequently as further cardiolariid taxa have been recognized, the Cardiolarioidea was proposed as a superfamily (Cope 2000) and it was established that it was then possible to classify Ordovician bivalve groups on the basis of their possessing either a protobranch or a more advanced gill type, thus reflecting the principal divisions of many zoological classifications of the bivalves. It was clear at this point that the subclass Palaeotaxodonta Korobkov, 1954 was a paraphyletic combination of protobranch nuculoids and autolamellibranch cardiolarioids, and should be abandoned.

The earliest cardiolarioids known hitherto are from the Arenig, but if the ideas of Cope (1995, 1997*b*) on the evolution of the filibranch gill are correct, they should predate other autolamellibranchs. *Cardiolaria* itself is best known from the Middle Ordovician, where it is widespread, particularly across European Gondwana, from Portugal to Bohemia and into Morocco. It is known earlier, from the Upper Arenig of Armorica (Babin 1966) and also occurs in the Arenig of Guadarranque, Extramadura, Spain (*fide* Babin 1993). As earlier Ordovician faunas are still known from so few localities, its lack of earlier occurrence is not seen as a major problem. *Cardiolaria* is also known from the early Caradoc of the Central Andean Basin of NW Argentina and Bolivia (Sánchez 1999*b*).

One cardiolarioid has an extraordinarily wide latitudinal range. Cope (2000) proposed the family Eritropidae to include the distinctive cardiolarioid *Eritropis*. *Eritropis* has a pronounced posterior carina and was first described from the Middle Ordovician (Llanvirn Series) of the Amadeus Basin, Australia, by Pojeta & Gilbert-Tomlinson (1977). The same genus occurs in the Llanvirn Series of mid-Wales (Cope 1999) showing that it was able to migrate for large distances around the Gondwanan margins to Avalonia, apparently quite independently of water temperature. *Eritropis* may also have reached Bolivia as *Arca gracilis* Hoek, 1912 (in Steinmann & Hoek 1912, pp. 248–249, pl. 8, fig. 17) appears remarkably similar in external morphology to *Eritropis*, but regrettably the dentition is unknown. The only other interpretation of this latter fossil is that it is a glyptarcoid bivalve related to either *Glyptarca* or *Hemiprionodonta*. Despite the wide latitudinal tolerance of *Eritropis*, the other eritropid genus described by Pojeta & Gilbert-Tomlinson (1977), *Inaequidens*, appears restricted to its type area.

Trigonioida

The Trigonioida are a distinctive autolamellibranch group that can be distinguished from the early Ordovician; they may be separated from the heteroconchs by their type of dentition and a short parivincular ligament with strong nymphae and deep ligament grooves (Johnston 1996; Johnston & Zhang 1998); this type of ligament is unlike that of the actinodonts, a group of heteroconchs to which the trigonioids have often been related (e.g. Pojeta 1978; Babin 1993; Waller 1998). Its earliest representative appears to be *Noradonta* from the Early Arenig of the Montagne Noire (Babin 1982*a*); this genus was first described from the Nora Formation (earliest Mid-Ordovician) of the Georgina Basin, Australia, by Pojeta & Gilbert-Tomlinson (1977) and this is again an example of a genus that was able to migrate from the very high latitudes of the Montagne Noire to the low latitude of Australia. The dentition of *Noradonta* has affinities with that of the cardiolarioids, suggesting derivation from them (Cope 2000). *Noradonta* is a genus which forms a suitable ancestor for *Tromelinodonta* (Babin 1982*b*) from the Late Arenig of Armorica. In turn, *Tromelinodonta*, by loss of posterior teeth, could produce the typical form of *Lyrodesma* that has a radiating subumbonal sheaf of crenulate teeth. The earliest *Lyrodesma* recorded hitherto is from the *murchisoni* Zone of the Llanvirn Series of mid-Wales (Cope 1999). Slightly younger ones are known from Spain (Gutiérrez-Marco & Babin 1999) and there are many species known from the Late Ordovician; these include forms from high-latitude Gondwana such as the Armorican *L. secure* de Tromelin & Lebesconte, 1876, or the *Lyrodesma* described from Bolivia by Sánchez & Suárez-Soruco (1996). Undescribed Avalonian forms occur in both the Caradoc and Ashgill of Wales and NW England (collections of the British Geological Survey). Low-latitude Laurentian forms include such species as *L. majus* (Ulrich, 1879) or the Baltican lyrodesmatid figured by Neuman *et al.* (1997). Unlike contemporary bivalves which favoured silty muds, lyrodesmatids seem characteristic of the unstable habitats represented by high-energy sands (Cope 1999) and it is noteworthy that this is similar to the habitat of Mesozoic trigoniids (Stanley 1977). Post-Ordovician lyrodesmatids were first reported from the Llandovery Series of the Laurentian Silurian by Harrison & Harrison (1975), but have subsequently been found to occur as late as the Middle Devonian (Pojeta & Zhang 1986). The ventrally flaring crenulate teeth and dorsal musculature with pedal retractors at the hinge-plate extremities provide good evidence of a phylogenetic link between *Lyrodesma* and the trigonioids. Upper Palaeozoic trigonioids lack the dental crenulation but have the same musculature as both the lyrodesmatids and Mesozoic trigonioids (Newell & Boyd 1975) and the juvenile dentition of *Lyrodesma* was found to be very similar to the schizodid dentition, suggesting paedomorphic retention of this feature in the Upper Palaeozoic schizodids (Harrison & Harrison 1975). Within the Late Ordovician the lyrodesmatids gave rise to the genus *Pseudarca* de Tromelin & Lebesconte, 1875. For over a century this genus was of doubtful affinities and was included in the Nuculanoidea by Cox *et al.* (1969), although Babin (1966) had suggested it was probably a lyrodesmatid. Later, Babin (1987) described the rediscovered type material, and, noting its peculiar dentition with a diverging sheath of short teeth, concluded that it was a lyrodesmatid. Tunnicliff (1987) described material from the Caradoc rocks of North Wales that included a species of *Pseudarca*, previously unknown in Britain, and independently came to the conclusion that its affinities were lyrodesmatid. *Brachilyrodesma* Pojeta & Gilbert-Tomlinson, 1977 is a Mid- to Late Ordovician lyrodesmatid from the Toko Group of the Georgina Basin, Australia, unknown from elsewhere.

Heteroconchia

The Heteroconchia Hertwig, 1895 is a combination of the subclasses Palaeoheterodonta Newell, 1965 and Heterodonta Neumayr, 1884, but as used here excludes the Trigonioida which were included in the Palaeoheterodonta by Cox *et al.* (1969).

The actinodontid heteroconchs include forms with a sheaf of ventrally radiating teeth. The first such form to be described was *Actinodonta cuneata* Phillips, 1848, from the Middle Llandovery (Aeronian) of South Wales. Pojeta (1971) wrongly recorded this species as from the Ordovician and a number of other authors have reported the genus from the Ordovician. For example Barrois (1891) figured several bivalves from the Grès Armoricain that he attributed to *Actinodonta*, and Babin (1966) also attributed some Armorican forms to the genus. The lectotype of *A. cuneata* was designated by Stubblefield (1938) and a latex cast of this specimen was figured by Pojeta (1978). The teeth branch out subumbonally and several of the posterior teeth are elongated. It has now become clear that *Actinodonta* is confined to the Silurian and that Ordovician forms ascribed to that genus will need new names. Within the Lower Ordovician, the genus *Carminodonta* Cope, 1996*b* approaches *Actinodonta* the closest in dentition; the former is only known from the Lower Arenig of the Llangynog Inlier, South Wales. Other actinodontids with this type of dentition include the Mid-Ordovician genus *Copidens* Pojeta & Gilbert-Tomlinson, 1977 from the Georgina Basin of Australia; this species was also recorded from South China by Guo (1988) as a new genus and species, *Zadimerodia fastigata*, but the two forms are clearly conspecific (Fang & Cope, unpublished work). *Ananterodonta* Babin & Gutiérrez-Marco, 1985 from the Lower Llanvirn of Spain, is a further actinodont of this type; this is still known only from a single specimen. The earliest *Cycloconcha* was recorded from the Middle Ordovician of Argentina by Sánchez (1986) but is subsequently known from Laurentia. Late Ordovician species of *Cycloconcha* show the typical actinodont subumbonal multiplicity of teeth; such a form is the *C. mediocardinalis* Miller, 1874, figured by Pojeta & Runnegar (1985, fig. 13A).

Heteroconchs with reduced dentition are more widely dispersed. The family Redoniidae Babin, 1966 comprises heteroconchs with much reduced subumbonal teeth and lacking anterior dentition. *Redonia* itself has a wide distribution, being recorded from the Lower Arenig of the Montagne Noire (Babin 1982*a*) and Morocco (Babin & Destombes 1992). The genus was also recorded from the lowest Middle Arenig of Argentina (Sánchez & Babin 1994; Sánchez 1997*b*). It occurs in the Upper Arenig of the Welsh Borderland (Cope 1999) and Turkey (Dean & Monod 1970) and in the Llanvirn of Morocco (Babin & Destombes 1992) and Bohemia (Barrande 1881). Other redoniids include the Avalonian genus *Moridunia* which is abundant in the Lower Arenig of South Wales (Cope 1996*b*) and probably occurs too in the poorly preserved Ramsey Island fauna (Hicks 1873), whence it was recorded as an *Actinodonta* (Carter 1971). In Spain another local genus is found in the Llanvirn, *Dulcineia* Babin & Gutiérrez Marco, 1991, that is characterized by crenulate teeth. Redoniids appear to be characteristic of high to median latitudes as no member of the family has yet been found in any of the low latitude areas.

Another group with very much reduced dentition, frequently related in the past to the heteroconchs, is the family Modiomorphidae. Pojeta (1971, p. 20) remarked upon the vague concept of this taxon and the fact that it probably included more than one family level group. Recent work has confirmed this view (Fang & Morris 1997), and it now seems that there are two separate bivalve groups involved, neither of which should be assigned to the heteroconchs. These are discussed more fully below.

The genus *Babinka* Barrande, 1881 is one of the most perplexing of Ordovician bivalves as far as its affinities are concerned. Its multiple muscle scars suggested that it was closely related to monoplacophorans to McAlester (1965, 1966) who decided that it was an early lucinoid. Starobogatov (1971) disagreed with McAlester's orientation of *Babinka* and thus its suggested affinities, but Pojeta (1978) agreed with the orientation and systematic placing. Other workers, however (e.g. Babin 1982*a*; Cope 1997*b*), have preferred to regard it as a heteroconch. More recently, Taylor & Glover (2000) have described the anterior respiratory pouch in the lucinoids; this is apparently absent in *Babinka* and Taylor & Glover (2000, p. 221) concluded that McAlester (1965) had exaggerated the lucinoid characters of the genus. The stratigraphical and biogeographical range of *Babinka* is also interesting. The earliest examples are from the Tremadoc of the Montagne Noire (Babin 1982*a*), but the genus is known from the Lower Arenig of Morocco (Babin & Destombes 1982), from the Upper Arenig of South Wales (Cope 1997*b*), the Lower Llanvirn of the Welsh Borderland (Cope 1999) and Sweden (Soot-Ryen 1969) and the Llanvirn of Bohemia (Barrande 1881). The genus is thus widely dispersed over the high to mid-latitudes of Gondwana and Avalonia and was one of the first bivalves to reach Baltica.

Babinka is far from the only Ordovician

heteroconch genus to display a multiplicity of muscle scars. Such features are common in other genera of heteroconchs, including *Cycloconcha* Miller, 1874, *Celtoconcha* Cope, 1996*b* and *Coxiconcha* Babin, 1966. Pojeta & Runnegar (1985) suggested that the multiple accessory muscle scars in both *Babinka* and *Coxiconcha* indicated derivation from *Cycloconcha*. This suggestion seems untenable in view of the fact that the former two genera have a well-established Early Ordovician history, whereas the latter does not appear until the Mid-Ordovician. *Coxiconcha* is widely dispersed, being known earliest from the Early Arenig of the Montagne Noire (Babin 1982*a*), but in the Mid-Ordovician is known in France, Spain and Portugal (Babin 1977), Morocco (Babin & Destombes 1992), and Bohemia (Kříž 1995), reaching Bolivia by the Late Llanvirn (Babin & Branisa 1987).

Another group assigned to the heteroconchs is the Superfamily Glyptarcoidea Cope, 1996*b*. Cope (1996*b*) pointed out that these differ from other heteroconchs in that their teeth radiate dorsally, rather than ventrally as in other heteroconchs, and in their cladistic analysis of bivalves Carter *et al.* (2000) concluded that *Glyptarca* was a pteriomorph. The type species of *Glyptarca*, *G. primaeva* Hicks, 1873, described from the Lower Arenig of Ramsey Island, South Wales, although abundant in the type locality, is poorly preserved and little of the dentition is known apart from a small sheaf of anterior teeth and a blade-like posterior tooth; no subumbonal dentition is visible. As a result of this view of its dentition it was thought to have a subumbonal lacuna in the dentition. The genus was thus viewed as an ideal ancestor for the cyrtodontids, which have curved dorsally flaring anterior teeth, a subumbonal dental lacuna and one or more straight posterior teeth (e.g. Morris 1978). Cope (1996*b*) figured a new species, *G. serrata*, from the Llangynog Inlier, some 60 km east of Ramsey Island, but at the same horizon, which showed perfectly preserved dentition, demonstrating that the genus had in reality a complex subumbonal overlap of the anterior and posterior teeth and thus there was no direct link to the cyrtodontids. Cope (1995, 1997*b*) showed that this dentition bears strong similarities to that of cardiolarioids and these might well have been ancestral to the glyptarcoids. The glyptarcoids include the genera *Glyptarca* and *Celtoconcha*, both known from the Lower Arenig of Wales (Cope 1996*b*), the former also occurring in the Llanvirn of Mid-Wales (Cope 1999) where it occurs together with another glyptarcoid, *Camnantia* Cope, 1999. The only other glyptarcoid known hitherto was widely reported as *Glyptarca*, but following Cope's (1996*b*) redefinition of the genus, is now known to belong to a separate genus, *Hemiprionodonta* Cope, 1996*b* (type species ?*Dolabra lusitanica* Sharpe, 1853). This genus is known widely from the Middle Ordovician of Spain (Babin & Gutiérrez-Marco 1991) and Armorica (Babin 1966) and has also been reported from the Middle Ordovician of Saudi Arabia (Fortey & Morris 1982). *Hemiprionodonta* is known from both the Late Llanvirn (Llandeilian) and the Caradoc of Morocco (Babin & Destombes 1992).

Pteriomorphia

Glyptarca has dentition closely related to that of the genus *Catamarcaia* Sánchez & Babin, 1993 from the upper part of the Middle Arenig of Argentina; this latter genus is the earliest form that has a duplivincular ligament. Sánchez & Babin (1993) believed that it was an early pteriomorph lacking the central edentulous area on the hinge-plate and Cope (1997*a*) decided that it was the earliest arcoid *sensu stricto* that he then included in the Neotaxodonta. More recently Cope (2000) accepted that the Neotaxodonta could be included within the Pteriomorpha, as they shared many features including the multiple ligamental insertions, but he retained the Order Arcoida in a restricted sense and recognized the Cyrtodontoida as a separate order. As *Catamarcaia* had a unique mix of characters, the monotypic family Catamarcaidae was proposed for it (Cope 2000). The deficient early fossil record of the Order Arcoida (*sensu* Cope 1997*a*) needs examination; after *Catamarcaia* in the Middle Arenig, no arcoids are known from the remainder of the Ordovician. The next record is that of *Alytodonta* Cope, 1997*b* from the Lower Llandovery of Girvan, Scotland, on the Laurentian margin, and then the Wenlock genera *Trecanolia* and *Uskardita* (Ratter & Cope 1998) to add to the previously known *Freja* Liljedahl, 1984. Judging from the occurrence of these later genera, it appears possible that arcoids may have inhabited some of the nearest inshore environments and this may explain their absence thus far from the Mid- and Late Ordovician records, which must be put down to collection failure.

In their cladistic review of early bivalve phylogeny Carter *et al.* (2000) preferred to derive the glyptarcoids from *Catamarcaia* and assigned them to the Pteriomorphia. Clearly *Glyptarca* lies close to the point of divergence of the heteroconchs and pteriomorphians and thus they show affinities to both groups and it could be that their placement is correct within the latter clade; there are many early arcoids and cyrtodonts that apparently do not have multiple ligamental insertions, including the earliest cyrtodontids known from the Tremadoc of Australia (*Cyrtodonta* and *Pharcidoconcha*; Pojeta & Gilbert-Tomlinson 1977), and from the Lower Arenig of Wales (*Cyrtodonta* and *Cyrtodontula*; Cope 1996*b*), together with the strongly ribbed genus *Falcatodonta* (Cope 1996*b*).

Cyrtodonta itself ranges from the Tremadoc Series of Australia through to the Silurian. In the Early Ordovician, cyrtodontids seem to have been little constrained by latitude; they appear first in the Tremadoc of Australia (Pojeta & Gilbert-Tomlinson 1977) and they also occur in the Lower Arenig of Avalonia (Cope 1996*b*). They have been recorded with a query from the higher latitudes of the Mid-Ordovician of both Spain (Babin & Gutiérrez-Marco 1991) and Morocco (Babin & Destombes 1992). In the last two cases no dentition is known and the generic assignment was made on the basis of shell shape. *Cyrtodontula* was also recorded from the Middle Ordovician of Argentina by Sánchez (1990). However, it is on the Late Ordovician low-latitude carbonate platforms that cyrtodonts became most diverse (Cope & Babin 1999) and by then clearly showing a preference for low latitudes. In Laurentia *Cyrtodonta* occurs together with such genera as *Cyrtodontula*, *Ortonella* and *Vanuxemia*. The earliest Laurentian occurrence of cyrtodontids appears to be the shallow water St Peter's Sandstone fauna of Minnesota first described by Sardeson (1896) from the late Chazyan (= Early Caradoc). This fauna includes two species each of *Cyrtodonta* and *Vanuxemia* (Sardeson 1939*a*, *b*). The latter genus is also found on the Late Ordovician carbonate platform of Baltica, whence Isberg (1934) also described the genera *Thorslundia* and *Warburgia*. Soot-Ryen & Soot-Ryen (1960) also record *Cyrtodontula* from the Oslo area. *Cyrtodonta* and *Vanuxemia* were recorded from Kazakhstania by Khalfin (1958) whilst *Cyrtodonta*, *Cyrtodontula*, *Plethocardia*, *Ortonella* and *Vanuxemia* were recorded from Siberia by Krasilova (1970, 1979)

Accompanying the cyrtodontids are a variety of genera of ambonychiids; these are pterioid bivalves that have their umbones at the anterior end of the shell and commonly have a byssal sinus. They are strongly anisomyarian or monomyarian in the adult stage and the posterior adductor is often displaced towards the centre of the valves. Such clearly epifaunal forms are, unsurprisingly, common in the low-latitude carbonate platforms. However, the earliest ambonychiid recorded hitherto is from Avalonia, from the Middle Arenig of South Wales (Cope 1996*b*), where it occurs in a sandstone; this is an undetermined species of *Cleionychia*, which is also the earliest ambonychiid genus to occur in the Upper Ordovician of North America (Pojeta 1966). Middle Ordovician ambonychiids are known from Australia whence Pojeta & Gilbert-Tomlinson (1977) described species of *Glyptonychia*, *Leconychia* and *Pteronychia*. Pojeta (1966) illustrated something of the large variety of North American ambonychiids and figured Late Ordovician species of *Allonychia*, *Ambonychiopsis*, *Anomalodonta*, *Ambonychia*, *Cleionychia*, *Eridonychia*, *Maryonychia*, *Opisthoptera* and *Psilonychia*; to these may be added *Claudeonychia* Pojeta, 1997. Some of these genera are also known from the Baltic Upper Ordovician carbonate platforms (Isberg 1934) where the fauna also includes genera such as *Anomalocoelia*, *Paramytilarca* and *Praeanomalodonta*. The ambonychiid faunas from around the continents of Kazakhstania and Siberia are not as varied as those of Laurentia and Baltica but include species of *Ambonychia* and *Cleionychia* (Khalfin 1958; Krasilova 1979).

Pterineids are also particularly characteristic of the Upper Ordovician, but the earliest is a *Palaeopteria* from the Lower Arenig of South Wales (Cope 1996*b*) only known from a single right valve. The next oldest genus is Gondwanan, in this case from the Amadeus Basin of Australia, the genus *Denticelox* Pojeta & Gilbert-Tomlinson, 1977, which, unlike later pterineids is biconvex. *Carotidens* Foerste, 1910 and *Palaeopteria* occur in the Upper Ordovician of the United States and Canada; both these genera have asymmetrical valves, with a flatter right than left valve.

The last group of pteriomorphian bivalves to evolve in the Ordovician was the limids. Tunnicliff (1987) described the genus *Myodakryotus* from the Caradoc of North Wales. This genus has characteristics in common with both cyrtodontids and later limids, suggesting that the group was derived from the cyrtodontids. Pojeta & Runnegar (1985, fig. 17) figured a limid species from the Late Ordovician of Canada and the United States, which they assigned with a query to the genus *Prolobella* Ulrich, 1894, regarded by Cox *et al.* (1969) as a palaeoheterodont. Ordovician limids lack the internal ligament of more recent forms (Pojeta & Runnegar 1985).

A further group of the pteriomorphians is the

superfamily Modiolopsoidea Fischer, 1887, formerly often classified with the heteroconchs, but shown by Carter & Seed (1998) to include some taxa that display multiply inserted non-parivincular ligaments; they may thus be related to the pteriomorphs. In the *Treatise* (Cox *et al.* 1969) the modiolopsoids are included in the order Modiomorphoida and the superfamily Modiomorphoidea. Fang & Morris (1997) revised the taxonomy of this problematic superfamily and showed that it included two distinct groups of bivalves: the modiomorphoids, typified by the Devonian genus *Modiomorpha*, had well-developed ligamental nymphs that allowed them to be distinguished from the Ordovician modiolopsoids, a group with which they had frequently been confused, but which lacked the ligamental nymphs. Fang & Morris (1997) also showed that the family Permophoridae Van de Poel, 1959 was a subjective junior synonym of the family Modiomorphidae Miller, 1877 and recommended that order Modiomorphoida be abandoned and that the superfamily Modiomorphoidea Miller, 1877, as used in the *Treatise* (Cox *et al.* 1969), should be replaced by the Modiolopsoidea Fischer, 1887, in which they recognized the families Modiolopsidae Fischer, 1887, Colpomyidae Pojeta & Gilbert-Tomlinson, 1977 and the Modiolodontidae Fang & Morris, 1997. Carter & Seed (1998) showed that some modiolopsoids had multiply inserted non-parivincular ligaments and suggested that their affinities thus lay with the pteriomorphs (see above). The modiomorphoids, on the other hand, possessed a periostracum with calcified spicules and Carter & Aller (1975) suggested that this feature, widely found in the anomalodesmatans, indicated that their affinities lay with that group.

Modiolopsids are recorded as far back as the Early Tremadoc of Argentina, whence Harrington (1938) figured a form he described as *Cosmogoniophorina tenuicostata* sp. nov. Harrington figured three syntypes, two of which (Harrington 1938, pl. 3, figs 1, 5) show an almost carinate shell and one (Harrington 1938, fig. 1) also shows clear radial ornamentation on the post-umbilical shoulder. There appears to be little quarrel that these are modiolopsids, but I find myself in agreement with Pojeta (1971, p. 17) that Harrington's other syntype (Harrington 1938, pl. 3, fig. 4) appears to show taxodont teeth, and a less marked post-umbonal shoulder and apparently lacks radial ornamentation, suggesting this specimen is a nuculoid. To clarify matters I designate the specimen figured by Harrington (1938, pl. 4, fig. 1) as lectotype of the species *Goniophorina (Cosmogoniophorina) tenuicostata* (Harrington), 1938. The same species has been recently recorded from the early Middle Arenig of Argentina by Sánchez (1997*b*) who noted that as Harrington did not describe the dentition, the generic and subgeneric assignments are uncertain.

The next oldest species are two species of *Modiolopsis* from the Early Arenig of South Wales (Cope 1996*b*) where the modiolopsids *Cosmogoniophorina* and *Parallelodus* also occur. Middle Ordovician modiolopsids are recorded possibly from Spain (Babin & Gutiérrez-Marco 1991, pp. 123–124), from the Middle and Upper Ordovician of Argentina (Sánchez 1990, 1999*b*) and the Upper Ordovician of Laurentia (Pojeta 1971), Kazakhstania (Khalfin 1958) and Siberia (Krasilova 1970, 1979). The genus *Corallidomus* appears in the Upper Ordovician; this genus is noteworthy because it is the first boring bivalve (as first reported by Whitfield 1893) and lived in crypts excavated into colonial corals or bryozoans; the same mode of life was followed by *Semicorallidomus* Isberg, 1934.

Colpomyids normally have rather amorphous bulbous-shaped teeth, referred to as articulating devices by Pojeta and Gilbert-Tomlinson (1977 p. 27), that are not mounted on a hinge-plate. The earliest forms are from the Late Tremadoc of the Amadeus Basin of Australia, *Colpantyx* and *Xestoconcha* (Pojeta & Gilbert-Tomlinson 1977); the latter genus also occurs in the Early Arenig of South Wales (Cope 1996*b*). Colpomyids are known from the Upper Ordovician of Laurentia (Pojeta 1978), Siberia (Krasilova 1979) and western Gondwana (Sánchez 1999*b*) and survived into the Silurian.

The Modiolodontidae are modiolopsoids that have a small number of cardinal teeth in each valve. They are known from the Middle and Upper Ordovician and younger rocks. One of the earliest is the *Modioliodon* recorded from the Llanvirn of Avalonia by Cope (1999).

Anomalodesmata

The earliest generally accepted anomalodesmatan is *Arenigomya* from the Early Arenig of South Wales (Cope 1996*b*). *Arenigomya* shows already several of the principal characters of the group, including a finely pustulate shell ornament and an edentulous hinge that sometimes carries subumbonal articulating devices (see Cope 1996*b*, pl. 7, figs 12–14; text-fig. 7). These characters are evident in the genera *Cuneamya* and *Rhytimya* (the latter also possesses a posterior gape), both of which are known from the Late Ordovician of the United States, Canada and Siberia; *Cuneamya* is also recorded from Kazakhstania (Khalfin 1958). *Cuneamya* appears earliest, however, in the Middle Ordovician of Argentina (Sánchez 1990).

Another Ordovician group may also be anomalodesmatans: this is the elongate orthonotians which were elevated to subclass level by Pojeta (1978). Of the Ordovician genera Pojeta (1978) included in the Orthonotia, *Psiloconcha* was later transferred to the Solemyoida (Pojeta 1988) and the elongate *Cymatonota*, *Solenomorpha* and *Prothyris* would now be regarded by most workers as anomalodesmatans. *Breviorthonota* Krasilova, 1979, from the Upper Ordovician of Siberia, also belongs here. The earliest possible *Cymatonota* was recorded from the Early Arenig of the Montagne Noire by Babin (1982*a*) but as discussed above, there is a possibility that this specimen may be a solemyid. The orthonotids were included in the Anomalodesmata by Cox *et al.* (1969).

Waller (1990) suggested that the ligamental nymphs and outer prismatic shell layer of the modiomorphoids suggested that they were linked to the anomalodesmatans and this supported Carter & Aller's (1975) suggestion that the calcified spicules of the modiomorphoid periostracum indicated that their affinities lay with the anomalodesmatans. Fang & Morris (1997) showed that the modiomorphoideans as defined by Cox *et al.* (1969) included both the Devonian modiomorphoids and the Ordovician–Silurian modiolopsoids, a group with which they had frequently been confused, but which lacked the ligamental nymphs. The latter are now regarded as pteriomorphians and are discussed above.

Summary and conclusions

Although fairly widely dispersed during the Early and Mid-Cambrian, it is suspected that during the Late Cambrian bivalves may have become restricted to Gondwana, but they are so far unknown with certainty from that Epoch.

In the Tremadoc Epoch of the Ordovician Period bivalves are known from Gondwanan high latitudes (Morocco, Montagne Noire), median latitudes (Argentina) and low latitudes (Australia). The most diverse fauna so far discovered is that of the Amadeus Basin in Australia (Pojeta & Gilbert-Tomlinson 1977) which already contains pteriomorphians that are not known with certainty from higher latitudes in rocks of that age. The Early Ordovician initial diversification of the bivalves was restricted to Gondwana (including Avalonia at this time) and by the Mid-Arenig all the principal clades of bivalves had evolved. Some forms were decidedly high-latitude in distribution, whilst others were just as clearly confined to low latitudes; most areas had some endemic genera. Comparison between the high latitude of the Montagne Noire and the lower latitude of Avalonia for the Early Arenig (Fig. 3a) shows a far greater diversity in Avalonian latitudes. Avalonia has at least 18 genera belonging to all the major bivalve groups, with great diversity of pteriomorphians and to a lesser extent heteroconchs, whereas nuculoids and heteroconchs are equally important in the less diverse faunas of higher latitudes. Even though Avalonian was not far north of the Gondwanan margin in Early Arenig times, the difference in pteriomorphian diversity is pronounced. It is a pity that there are no known low-latitude faunas of this age with which to make further comparisons. It thus seems that climatic gradients played some part in controlling bivalve diversity in the Early Ordovician, but the fact that a few genera occur at both high and median latitudes suggests that such gradients may not have been particularly strong at this time. In some cases the gradual migration of a species across Gondwana can be tracked through time, whilst in other cases the same genus appears at widely separated areas, or latitudes, simultaneously.

Figure 4a shows abundance of bivalve species belonging to each major clade for the faunas of South Wales and the Montagne Noire; comparison of Figure 3a with Figure 4a shows some surprising differences. Most notable for Wales is the fact that the diverse pteriomorphians make up only 9% of individuals, with the fauna totally dominated by heteroconchs (76.6%). For the Montagne Noire the fauna is again dominated by heteroconchs (56.6%) whilst the pteriomorphians make up only 0.4% of the bivalve population there.

Early Ordovician bivalves seem particularly controlled by facies, occurring virtually exclusively in siliciclastic sediments and being most abundant in very shallow-water silty muds (Cope & Babin 1999). Even the deeper-water facies, such as the muds of the Montagne Noire, were certainly deposited in water less than 50 m deep (Babin 2000). These Early Ordovician bivalves were predominantly infaunal, although there may have been some semi-infaunal forms.

In the Mid-Ordovician bivalves managed to migrate from Gondwana and reached both Baltica and the Laurentian margin in early Mid-Ordovician times. Once established here they spread to reach all continental shelves by the earliest part of the Late Ordovician. Mid-Ordovician faunas are thus, again, virtually

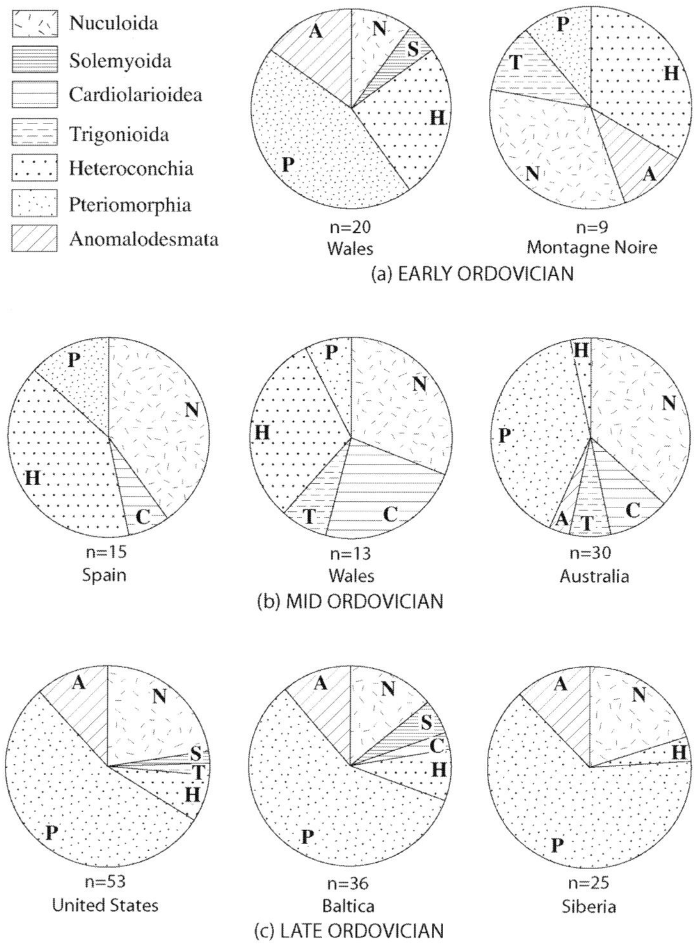

Fig. 3. Pie diagrams to illustrate the faunal composition and diversity of bivalve faunas in the Ordovician. (**a**) Early Ordovician. Comparison between the numbers of species comprising faunas of the Llangynog Inlier, Wales (Cope 1996*b*) and the near-polar Montagne Noire (Babin 1982*a*). Note the far greater abundance of pteriomorphians in Wales. (**b**) Mid-Ordovician. Comparison between the numbers of species comprising the faunas of high-latitude Spain (Babin & Gutiérrez-Marco 1991), mid-latitude Avalonia (Cope 1999) and low-latitude Australia (Pojeta & Gilbert-Tomlinson 1977). Note the increased importance of pteriomorphians at low latitudes whilst the heteroconchs are much more diverse at median to high latitudes. (**c**) Late Ordovician. Comparison of three low-latitude bivalve faunas to show faunal composition and generic diversity: Laurentia (data summarized from various sources), Baltica (primarily after Isberg 1934) and Siberia (Krasilova 1970, 1979).

exclusively Gondwanan and are dominated by two groups, the nuculoids and the heteroconchs, although there are a few pteriomorphs and even fewer anomalodesmatans. The preference of bivalves for shallow-water siliciclastic facies persisted through Mid-Ordovician times and it is noteworthy that one of the earliest Laurentian faunas, from Minnesota, and probably of Early Caradoc age, was in a sandstone facies (Sardeson 1896).

Comparison of Mid-Ordovician faunas from high, median and low latitudes (Fig. 3b) reveals a real difference in both bivalve diversity and faunal composition. The number of species at equatorial latitude is twice that of the highest latitude. Surprisingly, the percentage of nuculoid species in the faunas is remarkably similar across the latitudes, but the diversity is clearly greater at low latitudes. For high-latitude Gondwana (Spain) there are six nuculoid

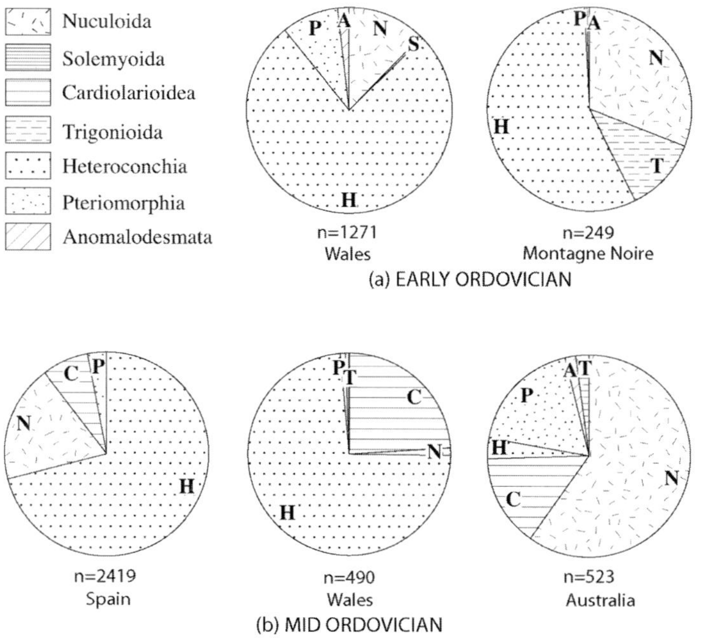

Fig. 4. Pie diagrams to illustrate abundance of bivalve individuals composing the faunas of which the species diversity is shown in Figure 3a, b. Precise numbers are available in the sources quoted in Figure 3a, b for most species, but for a few species numbers have been estimated from published information. (**a**) Compare with Figure 3a. Note the dominance of the heteroconchs in both faunas and the small percentage of the diverse pteriomorphians in the Welsh fauna. (**b**) Compare with Figure 3b. Note the dominance of the heteroconchs at high and median latitudes and the dominance of the nuculoids at low latitude.

species (40% of the fauna), for Avalonia (Wales and the Welsh Borderland) four species (31%) and for low-latitude Gondwana (Australia) 11 species (36.7%). The most pronounced latitudinal differences lie with the pteriomorphs and the heteroconchs. For the former, Spain has two species, one of which is a dubious record (13.3%), Wales one (7.5%), whilst Australia has 12 (40%). For the latter, Spain has six (40%), Wales four (31%) and Australia one (3.3%). Thus the preference of the pteriomorphian groups for low latitudes and heteroconchs for high latitudes is clearly apparent by Mid-Ordovician times. The apparently increased diversity of the high-latitude faunas of Spain when compared to the median latitudes of Avalonia probably reflects nothing more than the fact that the faunas from Spain were collected from 87 localities, whereas those from Wales and the Welsh Borderland were from nine localities.

Proportions of each bivalve clade based on abundance of individual bivalves is shown in Figure 4b. Here again there are some differences when compared with Figure 3b. Perhaps the most surprising is that nuculoids are far more abundant at low latitudes than at high latitudes: in the equatorial latitudes of Australia they account for no less than 59.8% of individuals. Comparative figures for Wales and Spain are 1.2% and 18.6% respectively. Pteriomorphians are shown to be far more abundant at low than at high latitudes and heteroconchs are confirmed as essentially preferring the median and high latitudes, displaying a dramatically reduced significance in the Australian faunas (3.3%) when compared to either Spain or Wales, where heteroconchs make up close to three-quarters of individuals in the bivalve faunas.

In the Late Ordovician a second major radiation of bivalves began; this was linked to the development of low-latitude carbonate platforms, particularly in Baltica and Laurentia, but a smaller expansion of pteriomorph groups also occurred in Kazakhstania and Siberia. This major radiation was principally amongst the semi-infaunal and epifaunal groups, producing a plethora of new genera of cyrtodontids, ambonychiids and byssally attached modiolopsoids, together with pterineids. The anomalodesmatans, which are primarily (but not exclusively) infaunal, were also more diverse at low latitudes, as too were the protobranch groups. The rapid diversification of bivalves in the Late Ordovician produced a large number of species as well as genera, but there has been little revisionary work on the faunas. For this reason, Figure 3c shows the number of genera in the Late Ordovician in three low-latitude areas. Insufficient is known of higher-latitude faunas to make meaningful comparisons.

The low-latitude shelves contained a high proportion of Late Ordovician endemic genera. Cope & Babin (1999) noted that out of 53 genera from the Laurentian Late Ordovician, over 40% were endemic. However, some low-latitude continental shelves were evidently in sufficient proximity to each other to allow interchange of genera. For example the strongly ribbed *Paraphtonia* Khalfin, 1958 from Kazakhstan also occurs in Siberia (Krasilova 1979) and genera such as *Cyrtodonta* are close to cosmopolitan. In stark contrast to these rich low-latitude faunas, the Late Ordovician faunas of high latitudes are quite impoverished and entirely dominated primarily by nuculoids; there are a few heteroconchs, but very few pteriomorphs or anomalodesmatans. Although there has been a lack of recent work on high-latitude Late Ordovician faunas, it appears that rates of endemism are lower, with Gondwana at about 30% and Avalonia at about 14% (Cope & Babin 1999).

These factors strongly suggest that latitudinal temperature differences became more pronounced in the Late Ordovician, heralding the Late Ashgill glaciation. This glaciation produced a major eustatic sea-level fall that exposed the low-latitude carbonate platforms. The resulting extinction event had a profound effect on bivalve stocks. The greatest losses were amongst the epifaunal and semi-infaunal groups, with major extinctions amongst the cyrtodontids, ambonychiids and pterineids. Other groups to be affected were the modiolopsids, amongst which the extinction of the coral-boring genera *Corallidomus* and *Semicorallidomus* ended a mode of life that was not to re-evolve until the Late Jurassic. On the other hand, at higher latitudes, the nuculoid-dominated communities appeared to have suffered least by the sea level fall, the infaunal forms adapting to the new sea-levels with little apparent difficulty; several genera survived apparently little changed into the Silurian.

There has been very little recent work published on Silurian bivalve faunas; this is a pity as the bivalves have the potential to provide much information about the recovery from the latest-Ordovician mass extinction. One fauna that has been studied is that of the Silurian of Wales and the Welsh Borderland (Ratter 1999), which is largely unpublished. Ratter's studies show that in the Llandovery Series these mid-latitude Avalonian faunas, predominantly from siliciclastic facies, are dominated by nuculoids, which make up half of the total of 18 recorded species. The remainder of the bivalve fauna consists, in order of diminishing importance, of anomalodesmatans, heteroconchs and trigonioids; there are no pteriomorphians. Diversity levels appear much lower in the Early Silurian and it seems that the Ashgill extinctions essentially reduced the bivalves to something like their Mid-Ordovician levels of diversity.

Thanks are due to L. C. Norton and L. E. Popov, both of the Geology Department, National Museum of Wales, for assistance in drafting of figures 1, 3 and 4, and figure 2, respectively.

References

BABIN, C. 1966. *Mollusques bivalves et céphalopodes du Paléozoïque armoricain*. Imprimerie Commerciale et Administrative, Brest.

BABIN, C. 1977. Étude comparée des genres *Babinka* Barrande et *Coxiconcha* Babin (Mollusques bivalves de l'Ordovicien) Intérêt phylogénétique. *Geobios*, **10**, 51–79.

BABIN, C. 1982*a*. Mollusqes bivalves et rostroconches. *In*: BABIN, C., COURTESSOLE, R., MELOU, M., PILLET, J., VIZCAINO, D. & YOCHELSON, E. L. (eds) *Brachiopodes (articulés) et mollusques (bivalves, rostroconches, monoplacophores, gastropodes) de l'Ordovicien inférieur (Trémadocien-Arenigien) de la Montagne Noire (France méridionale)*. Mémoire de la Société des Études Scientifiques de l'Aude, Sival, Carcassonne, 37–49, pls 8–12.

BABIN, C. 1982*b*. *Tromelinodonta* nov. gen., Bivalve lyrodesmatide (Mollusca) de l'Arenigien (Ordovicien inférieur). *Geobios*, **15**, 423–427.

BABIN, C. 1987. *Pseudarca*, Mollusque Bivalve singulier de l'Ordovicien supérieur du Massif armoricain. *Annales de la Société Géologique du Nord*, **105**, 253–258.

BABIN, C. 1993. Rôle des plates-formes gondwaniennes dans la diversification des mollusques bivalves durant l'Ordovicien. *Bulletin de la Société géologique de France*, **164**, 141–153.

BABIN, C. 2000. Ordovician to Devonian diversification of the Bivalvia. *American Malacological Bulletin*, **15**, 167–178.

BABIN, C. & BRANISA, L. 1987. *Ribeiria*, *Peelerophon* y otros moluscos del Ordovícico de Bolivia. *IV Congreso Latinamérico de Paleontología, Santa Cruz*, **1**, 119–129.

BABIN, C. & DESTOMBES, J. 1992 (dated 1990). Les Mollusques Bivalves et Rostroconches ordoviciens de l'Anti-Atlas marocain: Intérêt paléogéographique de leur inventaire. *Géologie Méditerranéenne*, **17**, 243–261.

BABIN, C. & GUTTIÉRREZ-MARCO, J. C. 1985. Un nouveau cycloconchide (Mollusca, Bivalvia) du Llanvirn inférieur (Ordovicien) des Monts de Tolède (Espagne). *Geobios*, **18**, 609–614.

BABIN, C. & GUTTIÉRREZ-MARCO, J. C. 1991. Middle Ordovician bivalves from Spain and their phyletic and palaeogeographic significance. *Palaeontology*, **34**, 109–147.

BARRANDE, J. 1881. *Système silurien du centre de la Bohême. 1ère partie. Recherches paléontologiques. Vol VI: Classe des Mollusques. Ordre des Acéphalés.* Bellman, Prague and Paris.

BARROIS, C. 1891. Mémoire sur la faune du grès armoricain. *Annales de la Société Géologique du Nord*, **19**, 134–351, pls 1–5.

BERG-MADSEN, V. 1987. *Tuarangia* from Bornholm (Denmark) and similarities in Baltoscandian and Australasian late Middle Cambrian faunas. *Alcheringa*, **11**, 245–259, pl. 7.

BRANSON, E. E. 1909. The fauna of the residuary Auburn Chert of Lincoln County, Missouri. *Transactions of the Academy of Science of St Louis*, **18**, 39–52.

CARTER, J. G. & ALLER, R. C. 1975. Calcification of the bivalve periostracum. *Lethaia*, **8**, 315–320.

CARTER, J. G. & SEED, R. 1998. Thermal potentiation and mineralogical evolution in *Mytilus* (Mollusca; Bivalvia). *In*: JOHNSTON, P. A. & HAGGART, J. W. (eds) *Bivalves: An Eon of Evolution — Paleobiological Studies Honoring Norman D. Newell.* University of Calgary Press, Calgary, 87–117.

CARTER, J. G., CAMPBELL, D. C. & CAMPBELL, M. R. 2000. Cladistic perspectives on early bivalve evolution. *In*: HARPER, E. M., TAYLOR, J. D. & CRAME, J. A. (eds) *The Evolutionary Biology of the Bivalvia.* Geological Society, London, Special Publications, **177**, 47–79.

CARTER, R. M. 1971. Revision of Arenig Bivalvia from Ramsey Island, Pembrokeshire. *Palaeontology*, **14**, 250–261.

COPE, J. C. W. 1995 (Dated 1996). The early evolution of the Bivalvia. *In*: TAYLOR, J. D. (ed.) *Origin and Evolutionary Radiation of the Mollusca.* Oxford University Press, Oxford, 361–370.

COPE, J. C. W. 1996*a*. 8. Bivalves. *In*: HARPER, D. A. T. & OWEN, A. W. (eds) *Fossils of the Upper Ordovician.* Palaeontological Association, Field Guides to Fossils, **7**, 95–115.

COPE, J. C. W. 1996*b*. Early Ordovician (Arenig) bivalves from the Llangynog Inlier, South Wales. *Palaeontology*, **39**, 979–1025.

COPE, J. C. W. 1997*a*. Affinities of the early Ordovician bivalve *Catamarcaia* Sánchez and Babin,1993 and its role in bivalve evolution. *Geobios Mémoire Spécial*, **20**, 126–131.

COPE, J. C. W. 1997*b*. The early phylogeny of the class Bivalvia. *Palaeontology*, **40**, 713–746.

COPE, J. C. W. 1999. Middle Ordovician bivalves from mid-Wales and the Welsh Borderland. *Palaeontology*, **42**, 467–499 .

COPE, J. C. W. 2000. A new look at early bivalve phylogeny. *In*: HARPER, E. M, TAYLOR, J. D. & CRAME, J. A. (eds) *The Evolutionary Biology of the Bivalvia.* Geological Society, London, Special Publications, **177**, 81–95.

COPE, J. C. W. & BABIN, C. 1999. Diversification of bivalves in the Ordovician. *Geobios*, **32**, 175–185.

COX, L. R., NEWELL, N. D., BOYD, D. W., *et al.*, 1969–1971. *In*: MOORE, R. C. (ed.) *Treatise on Invertebrate Paleontology. Part N. Mollusca 6, Bivalvia,* vols 1–3. Geological Society of America and University of Kansas Press, Boulder, Colorado and Lawrence, Kansas.

DEAN, W. T. & MONOD, O. 1970. The Lower Palaeozoic stratigraphy and faunas of the Taurus Mountains near Beysehir, Turkey. I. Stratigraphy. *Bulletin of the British Museum (Natural History), Geology Series*, **19**, 413–426.

DESPARMET, R., TERMIER, G. & TERMIER, H. 1971. Sur un bivalve protobranche anté-arénigien trouvé au nord de Wardak (Afghanistan). *Geobios*, **4**, 143–150.

DZIK, J. 1994. Evolution of 'small shelly fossils' assemblages. *Acta Palaeontologica Polonica*, **39**, 247–313.

EICHWALD, E. 1856. Beitrag zur geographischen Verbreitung der fossilen Thiere Russlands Alte Period. *Bulletin de la Société des Naturalistes de Moscou*, **17**, 247–339.

FANG, ZONG-JIE & MORRIS, N. J. 1997. The genus *Pseudosanguinolites* and some modioliform bivalves (mainly Palaeozoic). *Palaeoworld*, **7**, 50–74.

FISCHER, P. H. 1880–1887. *Manual de conchyliologie et de paléontologie conchyliologique.* F. Savy, Paris.

FOERSTE, A. F. 1910. Preliminary notes on Cincinnatian and Lexington fossils of Ohio, Indiana, Kentucky and Tennessee. *Bulletin of the Science Laboratories of Denison University*, **16**, 17–87.

FORTEY, R. A. & MORRIS, S. F. 1982. The Ordovician trilobite *Neseuretus* from Saudia Arabia and the palaeogeography of the *Neseuretus* fauna related to Gondwanaland in the earlier Ordovician. *Bulletin of the British Museum (Natural History), Geology Series*, **36**, 63–75.

GEYER, G. & STRENG, M. 1998. Middle Cambrian pelecypods from the Anti-Atlas, Morocco. *Revista Española di Paleontología*, **no. extra Homenajae al Prof. Gonzalo Vidal**, 83–96.

GROBBEN, K. 1894. Zur Kenntniss der Morphologie, Der Verwandschaftsverhältnisse und des Systems der Mollusken. *Sitzungsberichte der Kaiserlichen Akademie der Wissenschaften, Mathematisch-Naturwissenschaftlichen Classe*, **103**, 61–86.

GUO, FUXIANG. 1988. New genera of fossil bivalves from Yunnan. *Yunnan Geology*, **7**, 112–144 (in Chinese with English abstract).

GUTTIÉRREZ-MARCO, J. C. & BABIN, C. 1999. *Lyrodesma* et autres mollusques bivalves des Quartzites Botella (Ordovicien Moyen) de la Zone Centre-ibérique (Espagne). *Revista Española di Paleontología*, **no. extra Homenajae al Prof. J. Truyols**, 229–238.

HALL, J. 1847. Palaeontology. *Geological Survey of New York*, **1**, 1–338.

HARRINGTON, H. J. 1938. Sobre las faunas del Ordoviciano inferior del Norte Argentino. *Revista del Museo de La Plata, Nueva Serie*, **1**, 109–289, 14 pls.

HARRISON, W. B. & HARRISON, L. K. 1975. A Maquoketa-like molluscan community in the Brassfield Formation (Early Silurian) of Adams County, Ohio. *Bulletins of American Paleontology*, **67**, 197–231.

HAVLIČEK, V. & KŘÍŽ, J. 1978. Middle Cambrian *Lamellodonta simplex* Vogel: 'bivalve' turned brachiopod *Trematobolus simplex* (Vogel). *Journal of Paleontology*, **52**, 972–975.

HERTWIG, C. W. T. R. 1895. *Lehrbuch der Zoologie.* Gustav Fischer, Jena.

HICKS, H. 1873. On the Tremadoc rocks in the neighbourhood of St. David's, South Wales and their fossil contents. *Quarterly Journal of the Geological Society of London*, **29**, 39–52, pls 3–5.

HINZ-SCHALLREUTER, I. 1995. Muscheln (Pelecypoda) aus dem Mittelkambrium von Bornholm. *Geschiebekunde aktuell*, **11**, 71–84.

HINZ-SCHALLREUTER, I. 2000. Middle Cambrian Bivalvia from Bornholm and a review of Cambrian bivalved Mollusca. *Revista Española de Micropaleontología*, **32**, 225–242.

ISBERG, O. 1934. *Studien über Lamellibranchiaten des Leptaenakalkes in Dalarna.* Håkan Ohlssons Buchdruckerei, Lund.

JELL, P A. 1980. Earliest known pelecypod on Earth. A new Early Cambrian genus from South Australia. *Alcheringa*, **4**, 233–239.

JOHNSTON, P. A. 1996. Middle Ordovician (Caradoc) pelecypods from the Advance Formation, British Columbia. *Bulletin of the Geological Survey of Canada*, **491**, 113–129.

JOHNSTON, P. A. & ZHANG RENJIE. 1998. Re-interpretation of *Sinodora* Pojeta and Zhang, 1984, an unusual bivalve (Trigonioida) from the Devonian of China. *In*: JOHNSTON, P. A. & HAGGART, J. W. (eds) *Bivalves: An Eon of Evolution—Paleobiological Studies Honoring Norman D. Newell.* University of Calgary Press, Calgary. 361–375.

KHALFIN, L. L. 1958. Plastinchatozhabernye mollyuski ordovika Chu-Iliyskikh gor. *Trudy Geologicheskogo Instituta Akademii Nauk SSSR*, **9**, 130–136 (in Russian).

KOROBKOV, I. A. 1954. *Sprvochnik i metodicheskoe rukovodstro po tretichnum Mollyuskam Plastinchatozhabernye.* Nauchno-tecnoï Isledov, Leningradskoi Otdelenie (in Russian).

KRASILOVA, I. N. 1970. Droustvortchatie Molliouski Srednego i Verkhnego Ordovika Khr. Sette-Daban. *In*: ROSMAN, K. S., IVANOVA, V. A., KRASILOVA, I. N. & MODZALEVSKAIA, E. A. (eds) Biostratigrafiya verkhnego ordovika Severo-Vostoka SSSR. *Trudy Geologicheskogo Instituta Akademii Nauk SSSR*, **205**, 168–187, pls 26–27 (in Russian).

KRASILOVA, I. N. 1976. Noyi rod *Siberoctenia* (Bivalvia) iz ordovika Sibisrskoi Plarformy. *Paleontologicheskii Zhurnal*, **10**, 144–147 (in Russian).

KRASILOVA, I. N. 1979. *In*: MENNER, V. V. (ed.) Fauna ordovika sredneï Sibiry. *Trudy Geologicheskogo Instituta Akademii Nauk SSSR*, **330** (in Russian).

KŘÍŽ, J. 1995. *Coxiconcha* Babin, 1966 from the Llanvirn of the Prague Basin (Bivalvia, Ordovician, Bohemia) and the function of some "accessoric" muscles in Recent and fossil Bivalvia. *Věstník Českého geologického ústavu*, **70**, 45–50.

LAMONT, A. 1946. Lamellibranchs from the Lower Drummuck Group, (Ashgillian), Girvan, Scotland. *Cement, Lime and Gravel*, **20**, 364–366.

LILJEDAHL, L. 1984. Silurian silicified bivalves from Gotland. *Sveriges Geologiska Undersökning, Series C*, **804**, 1–82.

MCALESTER, A. L. 1965. Systematics, affinities and life habits of *Babinka*, a transitional Ordovician lucinoid bivalve. *Palaeontology*, **8**, 231–246.

MCALESTER, A. L. 1966. Evolutionary and systematic implications of a transitional Ordovician lucinoid bivalve. *Malacologia*, **3**, 433–439.

MCKINNON, D. I. 1982. *Tuarangia paparua* n. gen. and n. sp. A late Middle Cambrian pelecypod from New Zealand. *Journal of Paleontology*, **56**, 589–598.

MILLER, S. A. 1874. Monograph of the Lamellibranchiata of the Cincinnati group. *Cincinnati Quarterly Journal of Science*, **1**, 211–231.

MILLER, S. A. 1877. *The American Palaeozoic Fossils, a Catalogue of the Genera and Species.* Published by the author, Cincinnati.

MORRIS, N. J. 1978. The infaunal descendants of the Cycloconchidae: an outline of the evolutionary history of the Heteroconchia, Superfamilies Cycloconchacea to Chamacea. *Philosophical Transactions of the Royal Society of London, Series B*, **284**, 259–275.

MORRIS, N. J. & FORTEY, R. A. 1976. The significance of *Tironucula* gen. nov. to the study of bivalve evolution. *Journal of Paleontology*, **50**, 701–709.

NEUMAN, R. B., BRUTON, D. L. & POJETA, J. 1997. Fossils from the Ordovician 'Upper Hovin Group' (Caradoc–Ashgill), Trondheim Region, Norway. *Norges geologiske undersøkelse Bulletin*, **432**, 25–58.

NEUMAYR, M. 1884. Zur Morphologie der Bivalvenschlosses. *Kaiserlich-Königlichen Akademie der Wissenschaften, Wien. Mathematische-Naturwissenschaftliche Klasse, Sitzungberichte, Abteilung 1*, **88**, 385–418.

NEWELL, N. D. 1965. Classification of the Bivalvia. *American Museum Novitates*, **2206**.

NEWELL, N. D. & BOYD, D. W. 1975. Parallel evolution in early trigoniacean bivalves. *Bulletin of the American Museum of Natural History*, **154**, 53–162.

PELSENEER, P. 1889. Sur la classification phylogénétique des pélécypodes. *Bulletin Scientifique de la France et de la Belgique*, **20**, 27–52.

PFAB, L. 1934. Revision der Taxodonta des böhmischen Silurs. *Palaeontographica, Abteilung A*, **80**, 195–253, pls 2–4.

PHILLIPS, J. 1848. *In*: PHILLIPS, J. & SALTER, J. W. *Palaeontological appendix to Professor John Phillips' memoir on the Malvern Hills compared with the Palaeozoic districts of Abberley etc.* Memoir of the Geological Survey of Great Britain, **2**, 331–386, pls 4–30.

POJETA, J. 1962. The pelecypod genus *Byssonychia* as it occurs in the Cincinnatian at Cincinnati, Ohio. *Palaeontographica Americana*, **4**(30), 170–216.

POJETA, J. 1966. North American Ambonychiidae. *Palaeontographica Americana*, **5**(36), 131–238.

POJETA, J. 1971. *Review of Ordovician Pelecypods*. United States Geological Survey, Professional Papers, **695**.

POJETA, J. 1973. *Fordilla troyensis* Barrande: the oldest known pelecypod. *Science*, **180**, 866–868.

POJETA, J. 1975. *Fordilla troyensis* Barrande and early pelecypod phylogeny. *Bulletins of American Paleontology*, **67**(287), 363–384.

POJETA, J. 1978. The origin and early taxonomic diversification of pelecypods. *Philosophical Transactions of the Royal Society of London, Series B*, **284**, 225–246.

POJETA, J. 1980. Molluscan phylogeny. *Tulane Studies in Geology and Paleontology*, **16**, 55–80.

POJETA, J. 1988. The origin and Paleozoic diversification of solemyoid bivalves. *Memoir of the New Mexico Bureau of Mines and Mineral Resources*, **44**, 201–271.

POJETA, J. 1997. *Claudeonychia babini* nov. gen. et nov. sp. of Ordovician ambonychiid pelecypods from Cincinnati, Ohio, USA. *Geobios Mémoire Spécial*, **20**, 433–436.

POJETA, J. 2000. Cambrian Pelecypoda (Mollusca). *American Malacological Bulletin*, **15**, 157–166.

POJETA, J. & GILBERT-TOMLINSON, J. 1977. Australian Ordovician pelecypod molluscs. *Bulletin of the Bureau of Mineral Resources, Geology and Geophysics*, **174**, 1–64, pls 1–29.

POJETA, J. & RUNNEGAR, B. 1985. The early evolution of Diasome Molluscs. *In*: TRUEMAN, E. R. & CLARKE, M. R. (eds) *The Mollusca, Volume 10, Evolution*. Academic Press, New York, 295–336.

POJETA, J. & ZHANG, RENJIE. 1986. Systematic paleontology of the Devonian pelecypods of Guangzi and Michigan. *In*: POJETA, J. (ed.) *Devonian Rocks and Lower and Middle Devonian Pelecypods of Guangzi, China and the Traverse Group, Michigan*. United States Geological Survey, Professional Papers, **1394**.

RATTER, V. A. 1999. *Silurian Bivalvia from Wales and the Welsh Borderland*. PhD thesis, University of Wales.

RATTER, V. A. & COPE, J. C. W. 1998. New neotaxodont bivalves from the Silurian of South Wales and their phylogenetic significance. *Palaeontology*, **41**, 975–991.

REED, F. R. C. 1946. Notes on some lamellibranchs from Quarrel Hill, Girvan. *Geological Magazine*, **83**, 201–205, pl. 14.

SALTER, J. W. 1859. Fossils from the base of the Trenton Group. *In*: *Figures and Descriptions of Canadian Organic Remains*. Geological Survey of Canada, Decade 1, 1–47.

SÁNCHEZ, T. M. 1986. La fauna de bivalvos en la Formación Santa Gertrudis (Ordovícico) de la Provincia de Salta (Argentina). *Ameghiniana*, **23**, 131–139.

SÁNCHEZ, T. M. 1990. Bivalvos del Ordovícico medio-tardio de la Precordillera de San Juan (Argentina). *Ameghiniana*, **27**, 251–261.

SÁNCHEZ, T. M. 1996. Un nuevo género de Tironuculidae (Bivalvia, Palaeotaxodonta) en el Arenigiano del Noroeste argentino. *Geobios*, **29**, 683–690.

SÁNCHEZ, T. M. 1997*a*. Natasiinae, a new subfamily of the Tironuculidae (Bivalvia, Palaeotaxodonta) from the Acoite Formation (Early–Middle Arenig), Northwestern Argentine. *Geobios Mémoire Spécial*, **20**, 471–475.

SÁNCHEZ, T. M. 1997*b*. Additional Mollusca (Bivalvia and Rostroconchia) from the Suri Formation, Early Ordovician (Arenig), western Argentina. *Journal of Paleontology*, **71**, 1046–1054.

SÁNCHEZ, T. M. 1999*a*. Late Ordovician (Early Caradoc) bivalves from the Sierra de Villicum (Argentine Precordillera). *Journal of Paleontology*, **73**, 66–76.

SÁNCHEZ, T. M. 1999*b*. Caradoc bivalves from the Argentine Precordillera: a local radiation-extinction event. *Geobios*, **32**, 333–340.

SÁNCHEZ, T. M. & BABIN, C. 1993. Un insolite bivalve, *Catamarcaia* n.g., de l'Arenig (Ordovicien inférieur) d'Argentine. *Comptes rendus de l'Académie des Sciences, Paris, Série II*, **316**, 265–271.

SÁNCHEZ, T. M. & BABIN, C. 1994. Los géneros *Redonia* y *Catamarcaia* (Mollusca, Bivalvia) de la Formación Suri (Ordovícico Temprano, Oeste de Argentina) y su interés Paleobiogeográfico. *Revista Española de Paleontología*, **9**, 81–90.

SÁNCHEZ, T. M. & SUÁREZ-SORUCO, R. 1996. Primer registro de lyrodesmatidos (Bivalvia, Heteroconchia) en América del sur (Formación San Benito) Ordovícico superior de Bolivia. *Memorias del XII Congreso Geológico de Bolivia, Tarija, Bolivia*, 217–219.

SARDESON, F. W. 1896. The Saint Peter Sandstone. *Bulletin of the Minnesota Academy of Natural Sciences*, **4**, 64–88.

SARDESON, F. W. 1939*a*. Early pelecypod *Vanuxemia* in Minnesota. *Pan-American Geologist*, **71**, 283–293.

SARDESON, F. W. 1939*b*. Early pelecypod *Cyrtodonta* in Minnesota. *Pan-American Geologist*, **71**, 337–346.

SHARPE, D. 1853. Description of the new species of Zoophyta and Mollusca. Appendix B. *In*: RIBEIRO, C., SHARPE, D. & JONES, T. R. On the Carboniferous and Silurian formations in the neighbourhood of Bussaco in Portugal. *Quarterly Journal of the Geological Society of London*, **9**, 146–158, pls 7–9.

SHERGOLD, J. H., GORTER, J. D., NICOLL, R. S. & HAINES, P. W. 1991. Stratigraphy of the Pacoota Sandstone (Cambrian–Ordovician), Amadeus Basin, N. T. *Bulletin of the Bureau of Mineral Resources, Geology and Geophysics*, **237**, 1–14.

SOOT-RYEN, H. 1964. Nuculoid pelecypods from the Silurian of Gotland. *Arkiv för Mineralogi och Geologi, Kungliga Svenska Vetenskapsakademien, Series 3*, **28**, 489–519.

SOOT-RYEN, H. 1969. A new species of *Babinka* (Bivalvia) from the Lower Ordovician of Öland, Sweden. *Palaeontology*, **12**, 173–177, pls 1–5.

SOOT-RYEN, H. & SOOT-RYEN, T. 1960. The Middle Ordovician of the Oslo Region, Norway. 11, Pelecypoda. *Norsk Geologisk Tidsskrift*, **40**, 81–122, pls I–IV.

STANLEY, S. M. 1977. Co-adaptation in the Trigoniidae, a remarkable family of burrowing Bivalves. *Palaeontology*, **20**, 869–899.

STAROBOGATOV, YA. L. 1971. Systematic position of the Ordovician bivalve *Babinka*. *Paleontological Journal*, **5**, 384–387 (English translation by American Geological Institute published 1972).

STEINMANN, G. & HOEK, H. 1912. Das Silur und Cambrium des Hochlandes von Bolivia und ihre Fauna. *Neues Jahrbuch für Mineralogie, Geologie und Paläontologie,Beilige-Band*, **34**, 176–252.

STUBBLEFIELD, C. J. 1938. The types and figured specimens in Phillips and Salter's palaeontological appendix to John Phillips' memoir on 'The Malvern Hills compared with the Palaeozoic districts of Abberley, etc.'. *Summary of Progress of the Geological Survey of Great Britain and the Museum of Practical Geology*, **1936**, 27–51.

TAYLOR, J. D. & GLOVER, E. A. 2000. Functional anatomy, chemosymbiosis and evolution of the Lucinidae. *In*: HARPER, E. M, TAYLOR, J. D. & CRAME, J. A. (eds) *The Evolutionary Biology of the Bivalvia.* Geological Society, London, Special Publications, **177**, 207–225.

THORAL, M. 1935. *Contribution à l'étude paléontologique de l'Ordovicien inférieur de la Montagne Noire et révision sommaire de la faune cambrienne de la Montagne Noire*. Imprimerie Charité, Montpellier.

TONI, R. T. 1975. Upper Ordovician bivalves from the Oslo Region, Norway. *Norsk Geologisk Tidsskrift*, **55**, 135–156.

TORSVIK, T. H. 1998. Palaeozoic palaeogeography: A North Atlantic viewpoint. *GFF*, **120**, 109–118.

TROMELIN, G. LE G. DE & LEBESCONTE P. 1875. *Note sur quelques fossiles des grès Siluriens de Saint-Germain-sur-Ile, la Bouxière, Champeau*. Éditions Cottonec, Quimper.

TROMELIN, G. LE G. DE & LEBESCONTE P. 1876. Essai d'un catalogue raisonné des fossiles siluriens des départements de Maine-et-Loire, de la Loire-Inférieure et du Morbihan, avec des observations sur les terrains Paléozoïques de l'Ouest de la France. *Comptes rendus du Congrès de l'Association Française pour l'Avancement des Sciences, Nantes*, **1875**, 601–661.

TROMELIN, G. LE G. DE & LEBESCONTE P. 1877. Étude sur les terrains Paléozoïques de la Basse-Normandie. *Comptes rendus du Congrès de l'Association Française pour l'Avancement des Sciences, Le Havre*, **1876**, 493–501.

TUNNICLIFF, S. P. 1982. A revision of late Ordovician bivalves from Pomeroy, Co. Tyrone, Ireland. *Palaeontology*, **25**, 43–88.

TUNNICLIFF, S. P. 1987. Caradocian bivalve molluscs from Wales. *Palaeontology*, **30**, 677–690.

ULRICH, E. O. 1879. New genera and species of fossils from the Lower Silurian about Cincinnati. *Journal of the Cincinnati Society of Natural History*, **2**, 1–22.

ULRICH, E. O. 1892. New Lower Silurian Lamellibranchiata chiefly from Minnesota rocks. *Geological and Natural History Survey of Minnesota*, **19**, 211–248.

ULRICH, E. O. 1894. The Lower Silurian Lamellibranchiata of Minnesota. *In*: *Final Report of the Geological and Natural History Survey of Minnesota*, Volume **3**, 475-628 (published and distributed under separate cover prior to publication of the entire Volume 3 in 1897).

VAN DE POEL, L. 1959. Faune malacologique du Hervien. *Institut Royal des Sciences Naturelles de Belgique, Bulletin*, **35**(15) 1–26, **35**(16) 1–28.

WALLER, T. R. 1990. The evolution of ligament systems in the Bivalvia. *In*: MORTON, B. (ed.) *The Bivalvia — Proceedings of a Symposium in Honour of Sir Charles Maurice Yonge*. Hong Kong University Press, Hong Kong, 49–71.

WALLER, T. R. 1998. Origin of the molluscan class Bivalvia and a phylogeny of major groups. *In*: JOHNSTON, P. A. & HAGGART, J. W. (eds) *Bivalves: An Eon of Evolution — Paleobiological Studies Honoring Norman D. Newell.* University of Calgary Press, Calgary, 1–45.

WHITFIELD, R. P. 1893. Contributions to the paleontology of Ohio. *Ohio Division of the Geological Survey*, **7**(2), 407–494, pl. 13.

Phylogeny of the Reedocalymeninae (Trilobita): implications for Early Ordovician biogeography of Gondwana

SAMUEL T. TURVEY

Department of Earth Sciences, Oxford University, Parks Road, Oxford OX1 3PR, United Kingdom (e-mail: sam_turvey@hotmail.com*)*

Abstract: The shallow-water *Neseuretus* Association has been recognized by many authors as an important indicator of the former extent of Gondwana during the Early Ordovician. Phylogenetic relationships of the Reedocalymeninae (Arenig–?Early Silurian) were investigated using cladistic analysis, incorporating 22 species of *Neseuretus* as well as representatives of all other reedocalymenine genera. The area cladogram derived from this analysis contains four subclades each containing areas that span much of the palaeogeographic extent of Gondwana, which are interpreted as representing separate biogeographic events during the evolution of the subfamily. As the majority of taxa included in the analysis are associated with shallow shelf facies, this suggests that no significant environmental barriers existed across the continent during the Early Ordovician. Consideration of area relationships both for different subclades within the area cladogram, and within a consensus area cladogram, also supports the idea of a faunal cline between eastern and western Gondwana during this time interval. Further cladistic analysis of different trilobite taxa can be used to test these ideas.

The palaeogeography of the Early Palaeozoic is poorly understood relative to that of more recent geological history, and faunal data have been widely used to determine continental configurations during this time period. Biogeographic sensitivity varies between different taxa: whereas some groups (e.g. graptolites) commonly cannot be used to detect palaeogeographic boundaries, trilobites have been recognized as being able to indicate biogeographic patterns which are congruent with abiotic geological information such as palaeomagnetism, geodynamic features and climatically sensitive sediments (Fortey & Mellish 1992).

Faunal provinciality was very high during the Early Ordovician (Arenig–Llanvirn), as a result of both continental and climatic differentiation (Cocks & Fortey 1988, 1990), and geographic separation is regarded as a more important control on faunal composition than biofacies during this interval (Fortey & Mellish 1992). The distributions of different trilobite taxa can be used to differentiate Laurentia, Baltica and Gondwana, the three main continental regions which existed during the Ordovician (Zhou & Dean 1989), although some faunal admixture between regions at similar latitudes on the different continents can also be detected, such as the presence of Baltican genera in carbonates at Sobova, south central Turkey (Dean 1973, 1975).

Whereas Laurentian and Baltican trilobite faunas show little intracontinental provinciality during the Early Ordovician, considerable faunal differences have long been detected between different geographically separated Gondwanan cratonic faunas during this interval (e.g. Stubblefield 1939). For example, western and eastern regions of the palaeocontinent are characterized by different trilobite groups, and have been recognized respectively by several authors as the '*Selenopeltis* Province' or 'calymenacean–dalmanitacean Province' and the '*Asaphopsis* Province' (Whittington & Hughes 1972; Cocks & Fortey 1990). Gondwana, the largest of the Early Ordovician palaeocontinents, stretched from polar to equatorial latitudes, extending from the South Pole to more than 30°N (Cocks 2001), so unlike Laurentia or Baltica it represented a multiclimate region (Spjeldnæs 1961, 1981; Cocks & Fortey 1988, 1990). As temperature is regarded as a first-order control on faunal distribution (Fortey & Mellish 1992), the corresponding well-developed climatic gradient across the palaeocontinent can account for much of this faunal variation, with Palmer (1972) and Whittington (1973) being among the first to consider that the distribution of Early Ordovician trilobites around Gondwana may be related to latitudinal belts around a North African South Pole. Geographic separation must also have determined patterns of trilobite provinciality and endemicity to some extent, both across the continents forming the core of Gondwana, and between these continents and the numerous small

From: CRAME, J. A. & OWEN, A. W. (eds) 2002. *Palaeobiogeography and Biodiversity Change: the Ordovician and Mesozoic–Cenozoic Radiations*. Geological Society, London, Special Publications, **194**, 53–68.
0305-8719/02/$15.00

tectonic plates and terranes distributed along its borders or adjacent to it (termed 'peri-Gondwana'). However, the characterization and geographic location of these smaller tectonic units during the Early Ordovician is still strongly debated, even for geographic regions such as western Europe which have been exposed to considerable study, and few authors agree on the number of oceans, tectonic plates and terranes involved (Scotese & McKerrow 1990; Paris 1998; Cocks 2001).

Different authors have used trilobite data to propose differing biogeographic patterns for Early Ordovician Gondwana and peri-Gondwana. Area relationships across the palaeocontinent have largely been investigated on the basis of overall faunal similarity, using both quantitative (e.g. Whittington & Hughes 1972; Fortey & Mellish 1992) and qualitative approaches. Cladistic biogeographic analysis of Early Ordovician taxa has tended instead to consider the relationships between Laurentia, Baltica and Gondwana (e.g. Peers 1997). South China and Australia have been regarded as representing a distinct biogeographic region separate from central Europe during this time period by some authors (Palmer 1972; Cocks & Fortey 1988, 1990; Peers 1997). An alternative division of Gondwana into a Sino-European Region and a South American-Australian Region has also been advocated (Li 1994). Other authors have emphasized the distinction between SE Asian and European regions. Zhou & Dean (1989) suggested that trilobite faunas from all parts of Gondwanan and peri-Gondwanan east Asia may belong to a single faunal province, in turn divisible into sub-provinces, with further research into SE Asian Early Ordovician trilobites highlighting faunal similarities between South China, Tarim and Indo-China (Zhou *et al.* 1998*a*, *b*). A distinct Northern Gondwanan Province, consisting of North Africa, the Middle East, and western and central Europe exclusive of Avalonia, has also been recognized and subdivided into different domains partly on the basis of trilobite data (Paris 1998). There seems to be no faunal evidence for the presence of oceanographic or other environmental barriers separating different areas of Ordovician Gondwana (Zhou & Dean 1989). Spjeldnæs (1961, 1981) and Cocks & Fortey (1988, 1990) have suggested that the relationships between different Gondwanan cratonic trilobite faunas are best understood in terms of a geographic and climatic cline across the entire palaeocontinent. Gondwanan faunas can thus be interpreted as being gradational across the large palaeocontinent, with intermediate faunal regions variously allied by different authors to both eastern and western biogeographic provinces representing 'mixed' faunas.

The *Neseuretus* Association

Biofacies has been recognized as an important determinant of the quality of the biogeographic signal provided by different trilobite faunas. The generalized palaeogeographic model proposed by Fortey & Owens (1978) suggests that, for a series of different trilobite associations occurring in different water depths along a continent-edge profile, the associations occurring nearest inshore should display the highest degree of endemicity; the environmental barrier separating them from neighbouring geographic regions is higher than for deeper-water associations and so is more likely to act as a barrier to gene flow and induce speciation. According to this model, shallow-water trilobite associations can thus potentially provide the highest resolution biogeographic information. However, shallow-water trilobite associations can only be of use in biogeographic analysis if the genera or subfamilies they contain are geographically widespread rather than localized.

The trilobite association expected to provide the best biogeographic information according to these criteria is the *Neseuretus* Association, first identified by the term '*Calymene Tristani*-Stufe' (Born 1918; see also Hammann 1983) and also referred to variously as '*Neseuretus* shales', the '*Neseuretus* Community' or the '*Neseuretus* Fauna'. This association, described by Fortey & Morris (1982), is generally found in coarse- to fine-grained clastics such as decalcified ironstones, mudrocks and the Grès Armoricain or Armorican Quartzite, which extends from Brittany to Iberia and also probably occurs in eastern Newfoundland (the 'Armoricain Grit' of Van Ingen 1914). The reedocalymenine genus *Neseuretus* is the dominant faunal component. It occurs either on its own, or associated with a sparse, low-diversity trilobite fauna, which includes genera such as *Kerfornella*, *Plaesiacomia*, *Eohomalonotus*, *Iberocoryphe*, *Crozonaspis*, *Taihungshania*, *Ogyginus*, *Merlinia* and *Liomegalaspides* in different parts of the association's geographic range (Fortey & Owens 1978; Fortey & Morris 1982; El-Khayal & Romano 1985; Rabano 1990; Zhou *et al.* 1998*a*).

The *Neseuretus* Association has the widest geographic distribution of any Early Ordovician Gondwanan trilobite association, occurring in Avalonia, southern and western Europe, North Africa, the Middle East, Indo-China, South

Fig. 1. Early Ordovician (Arenig) palaeogeography, showing the Arenig–basal Caradoc geographic distribution of the 22 *Neseuretus* species included in the cladistic analysis. Key: 1, South China (Yangtze Platform): *N. concavus* (Arenig), *N. elegans* (Arenig-Llanvirn), *N. intermedius* (Arenig), *N. planus* (Arenig), *N. shensiensis* (Arenig); 2, Indo-China: *N. elegans* (Arenig-Llanvirn), *N. turveyi* (Llanvirn); 3, Arabia: *N. tristani* (Llanvirn); 4, South Turkey: *N. sexangulus* (Arenig); 5, Libya: *N. tristani* (Llanvirn); 6, France (Armorica, Montagne Noire); *N. arenosus* (Arenig), *N. tristani* (Llanvirn); 7, Spain (Iberia): *N. avus* (Llanvirn), *N. henkei* (Llanvirn), *N. leonensis* (?Arenig/Llanvirn), *N. tristani* (Llanvirn-basal Caradoc); 8, Anti-Atlas, Morocco: *N. attenuatus* (Llanvirn), *N. tristani* (Llanvirn); 9, Eastern Avalonia: *N. caerhunensis* (Arenig), *N. monensis* (Arenig), *N. murchisoni* (Arenig), *N. parvifrons* (Arenig), *N. ramseyensis* (Arenig); 10, Western Avalonia: *N. vaningeni* (Arenig); 11: South America: *N. chaschuilensis* (Arenig, Argentina), *N. lipanensis* (Arenig-Llanvirn, Argentina), *N. sanlucasensis* (Llanvirn, Bolivia). Map generated by David Lees (Natural History Museum, London).

China and the Central Andean basin of South America, and is regarded as one of the more reliable indicators of the former extent of Gondwana (although species assigned to *Neseuretus* or its junior synonym *Synhomalonotus* from various Central Asian regions, some of which were listed by Fortey & Morris (1982) and Rabano (1990) as valid representatives of the genus, represent different reedocalymenine genera) (Fig. 1). As early as 1937, Kobayashi noted that the genus (as *Synhomalonotus*) was characteristic of the 'Euro-Meridional' Province in the Early Ordovician, and later established the term '*Asaphopsis–Taihungshania–Neseuretus* Fauna' to describe the Early and Middle Ordovician southern faunal province (Kobayashi 1976, 1987). The association is interpreted as typically occurring in shallow inner-shelf environments (Fortey & Morris 1982). However, various species of *Neseuretus* have also been recorded in shallow outer-shelf deposits across much of the geographic range of the genus, in faunas composed of slightly deeper-water trilobite taxa such as *Colpocoryphe*, *Salterocoryphe*, *Placoparia*, *Zeliszkella*, *Ectillaenus* and *Hanchungolithus*, indicating that the genus had a wider palaeobathymetric tolerance than has previously been suggested (Dean 1966; Yin & Lee 1978; Courtessole *et al.* 1981; Beckly 1989; Henry 1989; Rabano 1990; Zhou *et al.* 1998*a*). The *Neseuretus* Association was originally regarded as an indicator of cold water, circumpolar shelf seas (Fortey & Morris 1982). This interpretation has been revised due to the occurrence of the association in Early Ordovician tropical regions such as the Yangtze Platform of the South China Plate; *Neseuretus* is now regarded as a eurytopic genus with exceptional latitudinal tolerance, adapted to general inshore clastic environments (Cocks & Fortey 1988).

Neseuretus has traditionally been regarded as basal within the Reedocalymeninae, representing the ancestral stock from which other genera evolved (e.g. Whittington 1966*b*; Dean 1975). Dean (1967*a*, *b*) and Zhou & Dean (1989) interpreted *Neseuretus* as having originated in southern and western Europe, continuing to evolve in its region of origin but also dispersing eastwards around the Gondwanan margins to arrive in South China during the Llanvirn. However, these ideas on faunal migration have been biased by differing local preservation in different geographic regions of the appropriate shallow-water clastic sediments to which *Neseuretus* was restricted; for example, appropriate transgressive facies are only known from the lower Arenig in south Wales (Fortey & Owens 1987). They have also been based on inaccurate stratigraphic correlation between different Gondwanan regions; for example, although the South Chinese *Neseuretus* species were originally interpreted as occurring in Llanvirn deposits (Lu 1975), these are now considered to have been deposited during the Arenig (Mu 1974; Chen *et al.* 1995; Zhou *et al.* 1998*a*). Further problems for biogeographic interpretation have arisen as a result of some authors regarding *Synhomalonotus* as a separate genus from *Neseuretus* (e.g. Moore 1959; Ross 1975). *Neseuretus* stocks in different Gondwanan regions have alternatively been interpreted as representing separate lineages, which may have been environmentally separated because of the transience of the *Neseuretus* Association in a marginal setting (Beckly 1989). Fortey & Owens (1987) considered that individual *Neseuretus* species probably ranged widely over Gondwana, as well as observing, following Hammann (1983), that different species were associated with particular facies, suggesting ecological differentiation; however, so far only *N. tristani* (Brongniart *in* Desmarest, 1817), the geologically youngest species of *Neseuretus*, has been recorded from several Ordovician palaeoplates.

Phylogenetic analysis

Although *Neseuretus* is regarded as an important genus both biogeographically and for understanding the evolution of the Calymenidae, no quantitative phylogenetic analysis has previously been conducted to determine the pattern of relationships between different *Neseuretus* species. This is in part because of the large number of species currently recognized within the genus. Whereas the seven other genera assigned to the Reedocalymeninae (*Calymenella*, *Calymenesun*, *Neseuretinus*, *Pradoella*, *Reedocalymene*, *Sarrabesia* and *Vietnamia*) contain relatively few species, over 60 different species or subspecies have at various times been established for either *Neseuretus* or *Synhomalonotus*, many of which are poorly known and may represent invalid taxa. Representatives of the Reedocalymeninae are shown in Figure 2.

Cladistic analysis was conducted on the Reedocalymeninae, incorporating 22 of the better-known species of *Neseuretus* and a representative species from each of the other reedocalymenine genera. Type material was studied where possible, although additional material for many taxa was also considered, and type species were included for most reedocalymenine genera. *Reedocalymene expansa* Yi, 1957 was used instead of *R. unicornis* (Reed, 1917), as the type species is poorly known and the genus was revised comprehensively on the basis of extensive new material of *R. expansa* by Peng *et al.* (2000). *Calymenella preboisseli* Beckly, 1989 was included on stratigraphic grounds instead of *C. boisseli* Bergeron, 1890, as this species represents the only Arenig representative of an otherwise Late Ordovician genus. *Calymenesun* was coded using *C. granulosa* Lu, 1975 instead of *C. tingi* (Sun, 1931), as well-preserved material representing this species was available for analysis. *Bavarilla hofensis* (Barrande, 1868), *Pharostomina oepiki* Sdzuy, 1955 and *Protocalymene mcallisteri* Ross, 1967 were selected as outgroup taxa to determine character polarity within the Reedocalymeninae, as they have been interpreted by previous authors as either closely related to the Calymenidae or representing basal calymenids, or in the case of *Protocalymene* (= 'aff. *Calymenidius* sp. indet.' of Whittington 1965) as a possible calymenid (Whittington 1966b; Fortey 1983; Hammann 1983; D.J. Siveter pers. comm. 2001).

Twenty-five exoskeletal characters were coded for the analysis. As most *Neseuretus* species are only known from holaspid cranidia and pygidia, all of the characters used were taken from these two exoskeletal regions. This prevented the cladistic data matrix from containing large amounts of missing values, which could lower the resolution of the analysis by producing large numbers of equally parsimonious trees, or produce spurious theories of character evolution (Kitching *et al.* 1998). Certain characters which have been used by

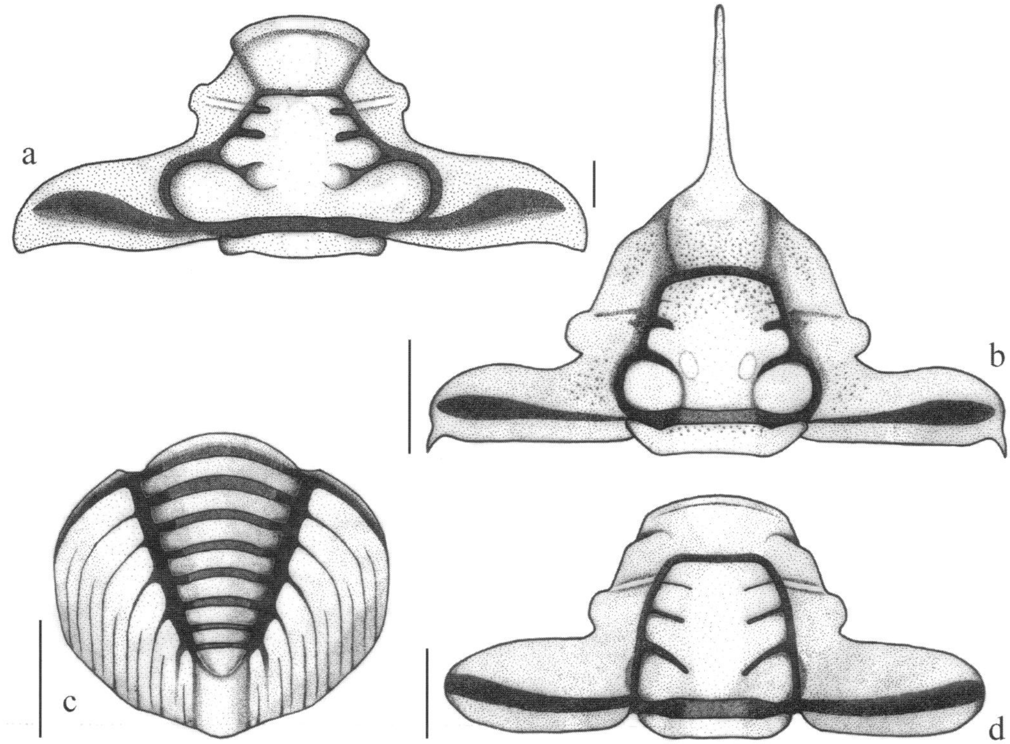

Fig. 2. Representatives of the Reedocalymeninae. (**a**) *Vietnamia douvillei*. (**b**) *Calymenesun granulosa*. (**c**, **d**) *Neseuretus elegans*. Scale bars = 5 mm. Note that *N. elegans* displays an incipient anterior furrow (character 14, state 1), whilst *V. douvillei* and *C. granulosa* both display complete anterior furrows (state 2). The slight fixigenal depressions opposite the 1p lobes in *N. elegans* represent paraglabellar areas rather than distinct abaxial curvature of the axial furrows.

previous authors to discuss aspects of calymenid phylogeny and evolution, notably those relating to the hypostome and rostral plate (e.g. Whittington 1966*a*; Hammann 1983; Fortey 1990; Peng *et al.* 2000), were thus excluded from the analysis.

During determination of character states, attempts were made to prevent preservational variation from artificially influencing cladogram topology, especially in characters regarded as particularly vulnerable to such non-biological variation (e.g. by combining 'absent' and 'weakly defined' under the same character state for characters relating to furrow strength). Variation between internal and external moulds was also recognized. Poorly preserved material was generally not analysed, although the deformed type material for *N. ramseyensis* Hicks, 1873 was studied in addition to the well-preserved specimens figured by Fortey & Owens (1987). Quantitative characters were coded informally, with discontinuities in the distribution of character states between different taxa recognized and used to define character state boundaries. The data matrix is presented in Table 1.

(1) Midpoint of palpebral lobe (exsag.) as proportion of distance from posterior margin of occipital ring to anterior margin of glabella: 0 = 0.83 or greater, 1 = 0.70 to less than 0.83, 2 = below 0.70 (variation between different taxa included in the analysis ranges from 0.6 to 0.9, with several taxa coded for state 1 displaying some variation in palpebral lobe position).
(2) Length of preglabellar area as proportion of total cranidial length (excluding 'pre-cranidial' projection if present): 0 = less than 0.3, 1 = 0.3 or greater (variation within ingroup taxa ranges from 0.2 to 0.4).
(3) Length of 1s furrow (tr., measured from

Table 1. *Data matrix for character state distributions used in phylogenetic analysis*

	1 2
Species	1 2 3 4 5 6 7 8 9 0 1 2 3 4 5 6 7 8 9 0 1 2 3 4 5
Bavarilla hofensis (Barrande, 1868)	1 0 1 0 2 1 0 0 0 0 0 1 0 0 1 1 0 0 0 0 1 0 0 0 0
Calymenella preboisseli Beckly, 1989	2 1 1 0 0 0 0 0 0 1 0 0 0 1 0 0 1 B 0 0 1 0 0 1 0
Calymenesun granulosa Lu, 1975	2 1 0 0 1 1 1 1 0 0 1 1 0 2 0 1 1 1 0 0 2 2 1 0 0
Neseuretinus turcicus (Dean, 1967a)	1 1 1 0 1 1 1 1 1 0 1 0 0 2 1 1 2 2 0 0 0 1 0 1 0
Neseuretus arenosus Dean, 1966	1 0 1 0 ? ? 0 0 0 0 0 0 0 1 0 0 1 1 0 0 0 0 0 1 0
N. attenuatus (Gigout, 1951)	1 0 1 0 0 0 0 0 0 0 0 0 1 1 0 0 ? ? ? ? ? ? ? ? ?
N. avus Hammann, 1977	A 0 1 0 0 0 1 0 1 0 0 1 1 2 0 0 1 B 0 0 0 0 0 1 0
N. chaschuilensis Vaccari & Waisfeld, 1994	1 0 1 0 ? ? ? 0 0 0 0 1 ? ? 0 0 1 0 0 1 0 0 1 ? 0
N. caerhunensis Beckly, 1989	2 1 1 0 0 0 0 0 0 0 0 1 0 1 0 0 1 2 ? 1 ? 1 0 1 0
N. concavus Lu, 1975	0 1 1 0 0 0 0 0 0 0 0 0 1 2 1 0 1 B 0 0 0 0 1 1 0
N. elegans Lee *in* Yin & Lee, 1978	1 0 1 0 0 0 1 0 0 0 0 0 0 1 0 0 1 1 0 0 0 0 0 1 0
N. henkei Hammann, 1977	0 1 1 0 0 0 1 0 0 0 0 1 0 2 0 0 1 1 0 0 0 0 0 1 0
N. intermedius Lu, 1975	0 A 1 1 ? ? 1 1 1 0 1 0 0 2 0 0 2 1 0 0 0 0 1 1 0
N. leonensis Hammann, 1983	1 0 1 1 0 ? 0 0 0 0 0 1 0 2 0 0 ? 0 ? ? 0 0 ? ? 0
N. lipanensis Waisfeld, 1997	0 1 1 1 0 0 1 0 0 0 0 1 0 2 0 0 1 1 0 0 0 0 1 1 0
N. monensis (Shirley, 1936)	2 1 1 0 0 0 0 0 0 0 0 1 0 1 0 0 1 1 0 ? 0 0 0 0 0
N. murchisoni (Salter, 1865a)	1 A 1 0 0 0 0 0 0 0 0 1 0 B 0 0 1 0 0 ? 0 0 0 1 0
N. parvifrons (M'Coy *in* Sedgwick & M'Coy, 1851)	1 1 1 0 0 0 A 0 0 0 0 1 0 2 0 0 1 A 1 1 0 0 0 1 0
N. planus Lu, 1975	0 0 1 1 ? ? 1 1 1 0 1 0 0 2 0 0 2 0 0 0 0 0 0 1 0
N. ramseyensis Hicks, 1873	1 0 1 0 0 0 0 0 A 0 1 0 1 0 0 1 2 1 0 0 1 0 0 1
N. sanlucasensis Přibyl & Vaněk, 1980	0 0 1 ? ? ? 1 0 1 0 0 1 0 2 0 0 2 1 0 ? 0 0 0 1 0
N. sexangulus Dean, 1971	1 0 1 1 ? 0 0 0 0 0 0 0 0 2 0 0 1 ? ? ? ? ? ? ? ?
N. shensiensis (Lu, 1957)	0 0 1 0 0 0 0 0 0 0 0 0 0 2 0 0 1 1 0 0 1 0 1 0 0
N. tristani (Brongniart *in* Desmarest, 1817)	1 0 1 0 0 0 1 0 0 0 0 1 1 2 0 0 1 1 0 0 0 0 0 1 0
N. turveyi Zhou, Dean & Luo *in* Zhou *et al.*, 2001	1 0 1 0 0 0 1 0 0 0 0 0 0 2 0 0 1 1 0 0 0 0 0 1 0
N. vaningeni Dean *in* Dean & Martin, 1978	1 0 1 1 0 0 0 0 0 0 0 1 0 2 0 0 1 0 1 0 0 0 1 1 0
Pharostomina oepiki Sdzuy, 1955	1 0 1 0 2 1 0 0 1 0 0 0 0 0 1 0 2 0 0 ? 1 0 0 0 0
Pradoella pradoi Hammann, 1977	1 0 1 1 0 0 0 0 0 1 0 1 0 2 0 0 1 B 0 0 1 0 0 1 0
Protocalymene mcallisteri Ross, 1967	2 0 0 0 1 1 0 1 0 0 0 0 0 0 1 0 2 0 0 0 0 0 0 1 0
Reedocalymene expansa Yi, 1957	2 0 1 0 1 1 1 0 0 0 0 0 0 2 0 1 1 1 0 0 0 0 1 1 0
Sarrabesia teichmuelleri Hammann & Leone, 1997	0 0 1 0 1 1 2 1 1 0 0 0 0 2 0 0 1 2 0 0 2 1 0 0 1
Vietnamia douvillei (Mansuy, 1908)	0 1 1 0 ? ? 2 1 1 0 0 1 0 2 0 0 1 2 0 0 2 1 0 0 1

Characters and states are as described in the text. Character numbers are listed at the top of the table. Missing data are indicated with a question mark. Character states listed as 'A' and 'B' indicate multistate coding, where 'A' = (0&1) and 'B' = (1&2).

base of 2p lobe) as proportion of width of unfurrowed median lobe of glabella (tr., along same transverse line as that of furrow width): 0 = 0.40 or less width of median lobe, 1 = above 0.40 width of median lobe (such that combined width of both 1s furrows is approximately equal to or greater than width of median lobe).

(4) Position of anterior margin of postocular cheek at fixigenal midlength (tr.): 0 = approximately opposite anterior margin of 1p lobe, 1 = opposite or anterior to 2p lobe.

(5) Facial suture: 0 = gonatoparian, 1 = proparian, 2 = opisthoparian.

(6) Genal spine: 0 = absent, 1 = present.

(7) Shape of posterior border furrow: 0 = equal width along fixigenal margin ('ptychoparioid condition'), 1 = expanding abaxially, 2 = sigmoidal.

(8) Axial furrow adjacent to 1p lobe: 0 = straight, 1 = curved outwards (abaxially convex).

(9) Glabellar outline: 0 = subrectangular, 1 = tapering strongly forwards ('truncated cone'/subtrapezoidal).

(10) Strength of glabellar furrows: 0 = strongly defined, 1 = absent or weakly defined.

(11) 1s furrow (tr.): 0 = does not reach occipital furrow, 1 = reaches occipital furrow.

(12) Strength of intermediate lobe (swelling at adaxial base of 2p lobe): 0 = absent or weakly defined, 1 = strongly defined.

(13) Shape of posterior margin of 4p lobe: 0 = lateral margin at similar abaxial position to anterior margin of 3p lobe, 1 = indented adaxially relative to 3p lobe.

(14) Anterior furrow (furrow running from anterior margin to distal portion of preglabellar furrow, forming lateral boundary to preglabellar area, *sensu* Whittard 1960; see Fig. 2): 0 = absent, 1 = incipient, only expressed anterolaterally ('fixigenal inflation' or 'fixigenal overhang' condition), 2 = extends posteriorly to preglabellar furrow (may superficially appear to be a posteriorly positioned anterior border furrow if true anterior border furrow is absent, e.g. Dean & Martin 1978, pl. 4, figs 1, 4, 7, pl. 5, fig. 1).
(15) Anterior border furrow: 0 = absent or weakly defined, 1 = strongly defined.
(16) Anterior border (dorsal view): 0 = transversely straight to gently anteriorly convex, 1 = strongly convex and pointed, possibly forming distinct anterior 'precranidial' projection.
(17) Surface tubercles (size and density): 0 = no tuberculation, 1 = light tuberculation, 2 = heavy tuberculation.
(18) Number of pygidial axial rings: 0 = up to and including five, 1 = six or seven, 2 = eight or more (the majority of *Neseuretus* species have five to seven axial rings, and several are described as having 'six or seven', although the number of axial rings varies from two to ten between different taxa included in the analysis).
(19) Dorsal outline of terminal piece: 0 = continuous with anterior part of pygidial axis, 1 = anterior region laterally expanded ('swollen').
(20) Lateral outline of terminal piece: 0 = continuous with anterior part of pygidial axis, 1 = inflated.
(21) Postaxial pygidial morphology: 0 = weakly defined terminal piece, with narrower postaxial ridge extending to margin ('*Neseuretus* condition'), 1 = posterior margin of terminal piece defined by distinct posterior border furrow, with no postaxial ridge present, 2 = posterior border furrow absent, narrow terminal piece fused with prominent postaxial ridge to form a continuous elongate structure which extends posteriorly to the pygidial margin ('*Sarrabesia*/*Vietnamia* condition').
(22) Pygidial axis anterior to terminal piece: 0 = evenly tapered, 1 = funnel-shaped (tapering to a point slightly anterior to the terminal piece, posterior to which the axial furrows become parallel), 2 = swollen (axial furrows abaxially convex anterior to a parallel-sided posterior section).
(23) Pygidial lateral pleural furrow/vincular furrow (= 'cincture' of Dean & Zhou 1988): 0 = absent, 1 = present.
(24) Pygidial interpleural furrows: 0 = absent or only present very close to the pygidial margin, 1 = clearly defined across much of pleural fields.
(25) Pygidial rib morphology: 0 = flattened, 1 = ridge-like.

Cladistic analysis was conducted on PAUP*4.0b4a (Swofford 2001), using a heuristic search with 50 random addition sequence replicates. Of the eight multistate characters, four (5, 7, 21 and 22) were treated as unordered and four (1, 14, 17 and 18) as ordered. Multistate taxa were coded as polymorphisms. All characters were unweighted. Searching was performed twice, once using accelerated transformation optimization (ACCTRAN) and a second time using delayed transformation optimization (DELTRAN).

Within these parameters, four minimal length cladograms of 106 steps, occupying a single island, were recovered (Consistency Index = 0.4151, Retention Index = 0.5867). ACCTRAN and DELTRAN optimization resulted in identical sets of cladograms, as did subsequent reweighting of the dataset. The strict consensus of these cladograms is shown in Figure 3a (a semistrict consensus has the same topology). Support for individual nodes within the consensus cladogram was assessed with bootstrap analysis (Felsenstein 1985) using PAUP*4.0b4a; bootstrap values are displayed on the cladogram in Figure 3a. Removing the three taxa with five or more missing values (*Neseuretus attenuatus*, *N. chaschuilensis* and *N. sexangulus*) from the character matrix and rerunning the analysis does not improve resolution in the *Neseuretus* subclade, recovering five different cladograms of 103 steps (Consistency Index = 0.4272, Retention Index = 0.5845) (Fig. 3b), and so the consensus cladogram in Figure 3a is here used as the basis for further analysis.

The consensus cladogram will provide the basis for a systematic revision of the Reedocalymeninae in a separate study, and it is used here as the basis for biogeographic analysis. The reliability of the consensus cladogram obviously underpins both any taxonomic or biogeographic assumptions based on it. As several of the nodes are only weakly supported by bootstrap analysis, some caution should be employed when evaluating biogeographic ideas based on the topology of the resultant area cladogram. A qualitative but straightforward method of investigating the validity of the cladogram is a consideration of the degree of congruence

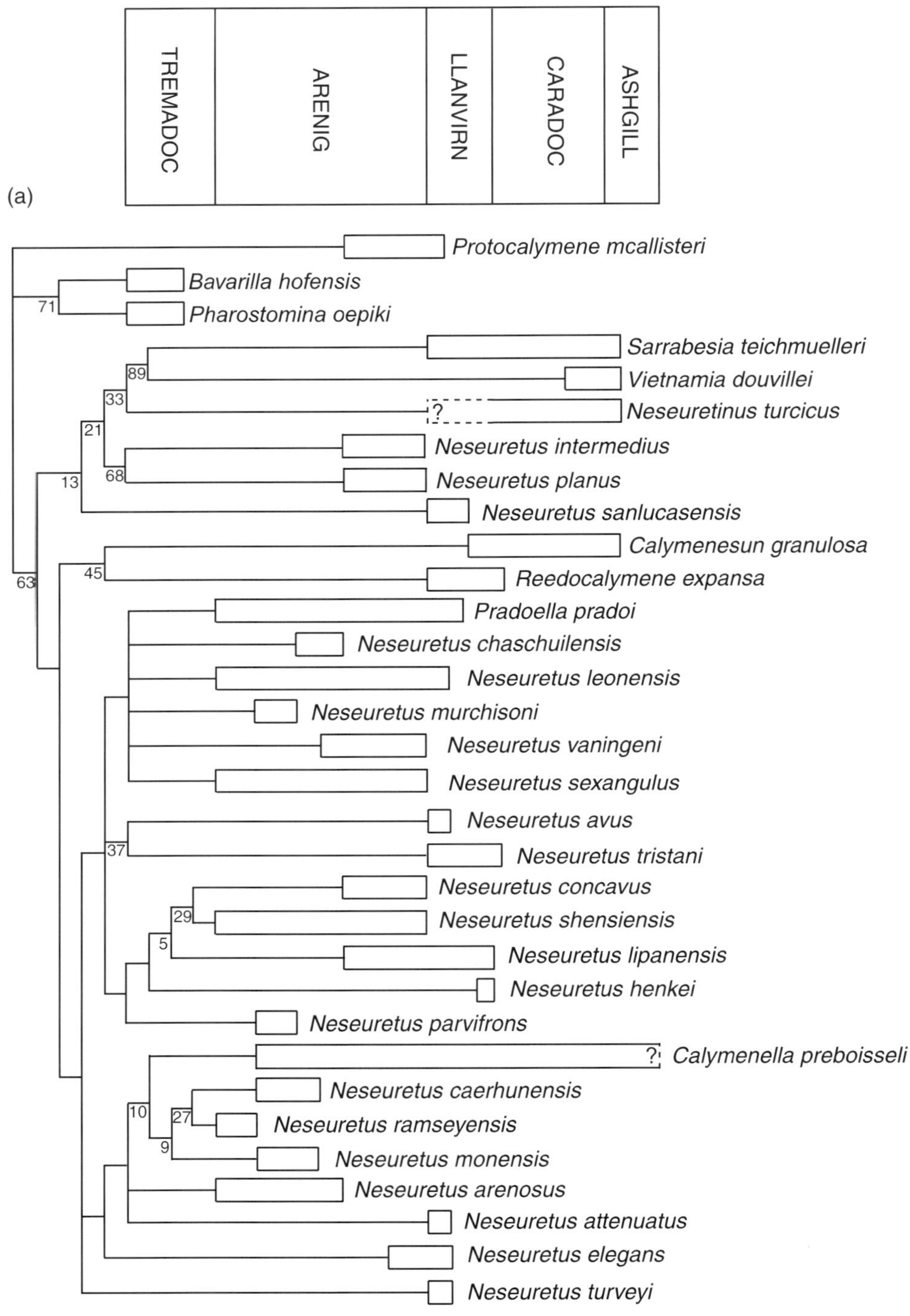

Fig. 3. (**a**) Strict consensus cladogram for the Reedocalymeninae. Numbers in small type below nodes represent bootstrap percentages, for all nodes with relative frequencies equal to or greater than 5%. Stratigraphic ranges are displayed for each taxon; these represent combined ranges for all representatives of the genus for non-*Neseuretus* reedocalymenine genera. Ranges may represent overestimates of true stratigraphic range, due to poor stratigraphic information available for some taxa. Stratigraphic ranges of British taxa have been correlated to the revised British Ordovician stratigraphy of Fortey *et al.* (2000).

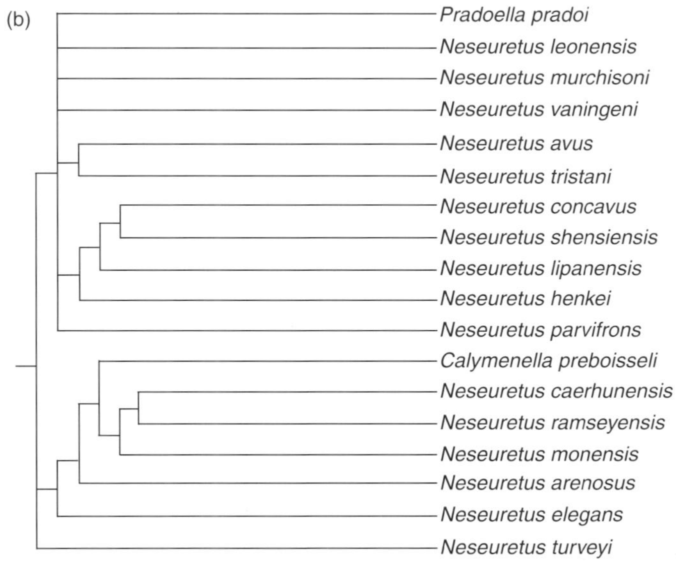

Fig. 3. (**b**) *Neseuretus* subclade from second strict consensus cladogram generated by removing *N. attenuatus*, *N. chaschuilensis* and *N. sexangulus* from the dataset.

between the consensus tree and the known stratigraphic ranges of the ingroup taxa. As *Neseuretus* represents a derived member of the Reedocalymeninae in the consensus tree, all of the major subclades are constrained to have evolved by the Arenig. This necessitates ghost ranges during the Arenig for *N. sanlucasensis* and the *Reedocalymene–Calymenesun* subclade only, providing fairly good correlation at series level, although the stratigraphic occurrence of *Neseuretinus*, *Sarrabesia* and *Vietnamia* remains poorly understood for many Asian localities and little confidence can be placed on stratigraphic ranges within this subclade (e.g. Dean 1967*b*; Pillet & de Lapparent 1969; Hammann & Leone 1997; Zhou *et al.* 1998*a*).

Cladistic biogeography

The consensus cladogram in Figure 3a was converted into an area cladogram for biogeographic analysis by substituting the names of the different ingroup taxa with the Early Ordovician geographic area(s) from which they have been recorded (Fig. 4). The area cladogram is simplified, as area redundancy has been corrected. For the non-*Neseuretus* reedocalymenine taxa, the combined biogeographic area occupied by all representatives of the genus recognized herein was recorded on the area cladogram.

Area delimitation in some cases requires explanation. France+Spain are regarded as representing a single biogeographic area, following the strong faunal similarities between these regions noted by previous authors (e.g. Paris 1998). Şengör (1984, 1987) proposed that Turkey, the Sibumasu and Indo-China terranes and areas of Central Asia represented a peri-Gondwanan continent called Cimmeria. However, Turkey represents a complex region, with northern and southern Turkey now interpreted as having occupied separate palaeogeographic positions during the Ordovician. Southern Turkey, source of the Turkish material of *Neseuretus* and *Neseuretinus*, was tentatively assigned to a high-latitude position on the margin of Gondwana at the eastern end of the present-day Mediterranean (Dean *et al.* 2000; Cocks 2001), the eastern region of the North Gondwanan Province of Paris (1998). Following these authors, southern Turkey is here provisionally interpreted as having been separate from Cimmeria during the Ordovician, and is grouped with Sardinia, another eastern Mediterranean component of Paris' (1998) North Gondwanan Province. As noted above, Early Ordovician faunas from South China, Tarim and Indo-China also share strong similarities (e.g. Zhou & Dean 1989), with the two South Chinese *Neseuretus* species *N. elegans* Yin *in* Yin & Lee, 1978 and *N. tungtzuensis* Sheng, 1958 also recorded from Indo-China by Zhou *et al.* (1998*a*,

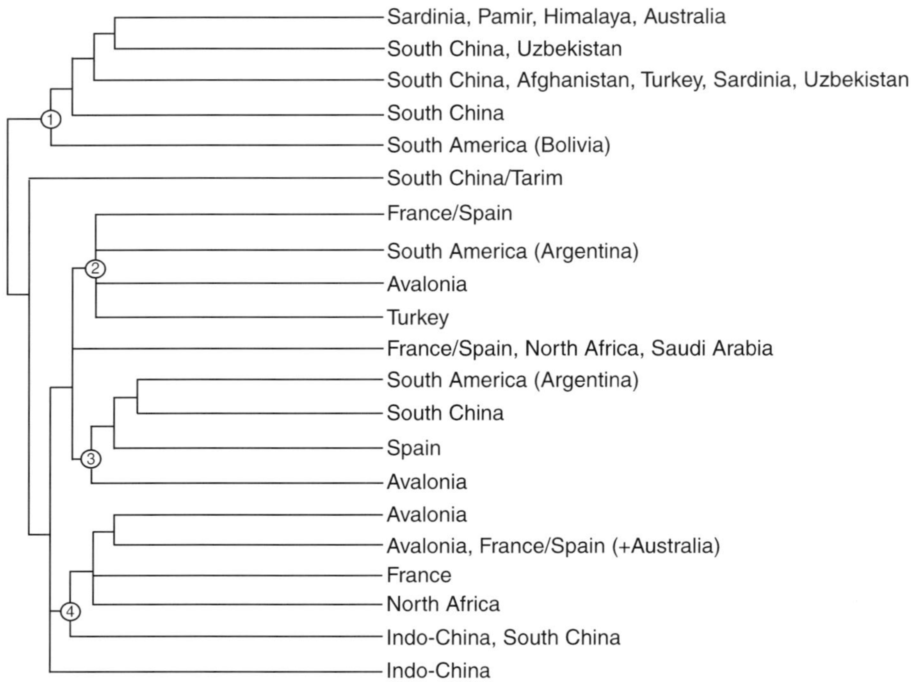

Fig. 4. Area cladogram derived from the strict consensus cladogram in Figure 3a. The nodes numbered 1 to 4 are discussed in the text.

2001). These three regions are here interpreted as representing a single Early Ordovician SE Asian palaeogeographic area. A restricted interpretation of Cimmeria is thus employed in this analysis.

Area assignment was generally straightforward, with the exception of some Asian and Australian taxa. The Himalayan species *Calymene nivalis* Salter, 1865*b* was interpreted as representing *Neseuretus* by Dean (1967*b*, 1975), Fortey & Morris (1982), Morris & Fortey (1985) and Rabano (1990); examination of the type material in the Natural History Museum, London, suggests that this species should instead be referred to *Sarrabesia*. '*Synchomalonotus*' *sulcatus* Kolobova *in* Sokolov & Yolkin, 1978, from the Ashgill of Uzbekistan, previously reassigned to *Sarrabesia* by Hammann & Leone (1997), is here interpreted as a species of *Vietnamia*. *Calymenesun longinasuta* Dean & Zhou, 1988, from the early Ashgill of the Zap Valley, SE Turkey, is extremely similar to *Neseuretinus turcicus*, and is certainly congeneric. Several different authors (Banks 1988; Corbett & Banks 1974; Legg 1976) have recorded the presence of reedocalymenine taxa in Australia, but the taxonomic position of these specimens has remained unclear. Edgecombe *et al.* (1999) tentatively assigned some of this material to *Sarrabesia*; although other Australian material may well represent different reedocalymenine taxa, only *Sarrabesia* is considered to occur in Australia during the Ordovician in this analysis.

The consensus cladogram indicates that *Neseuretus* as currently understood represents both a paraphyletic and a polyphyletic genus. Whereas *Neseuretus sensu stricto* (i.e. *Neseuretus* exclusive of *N. sanlucasensis*, *N. intermedius* and *N. planus*) is known to have occurred in shallow inner-shelf conditions amenable to high-resolution biogeographic analysis, and *Calymenella* and *Pradoella* occupied similar shallow-water conditions during the Early Ordovician (e.g. Beckly 1989; Rabano 1990), the ecological requirements of Middle Ordovician Asian reedocalymenine taxa, and the species previously assigned to *Neseuretus* which occur at the base of the *Neseuretinus–Sarrabesia–Vietnamia* subclade, are less well understood or are known to have differed. *Neseuretinus turcicus* is recorded from offshore shallow shelf conditions in the Portixeddu and Tuviois formations, Sardinia (Leone *et al.* 1991, as ?*Calymenesun* sp.; Hammann & Leone 1997), and *Sarrabesia teichmuelleri* occurs in coarse-grained sandstones of the Punta Serpeddi Formation, Sardinia, which are also indicative of shelf facies (Hammann & Leone 1997). *N. longinasuta* occurs alongside a varied trilobite assemblage in shales and mudstones of the Şort Tepe Formation in Turkey, suggestive of similar environmental conditions (Dean & Zhou 1988). Both *Neseuretinus birmanicus* and *Vietnamia douvillei* are known from siltstones from the upper Na Mo Formation in Vietnam; the formation's depositional environment is poorly understood, but both lithofacies and biofacies comparisons suggest that it represents shelf facies (Tong-Dzuy Thanh, pers. comm. 2001). '*Neseuretus*' *intermedius* occurs in outer shelf facies on the Yangtze Platform, in association with characteristically shelf-slope taxa such as raphiophorids and *Nileus* (Wang *et al.* 1987, as *N. xiadongensis*). *Reedocalymene expansa* is abundant in shales from the Miaopo Formation of the Yangtze Platform, which has been interpreted as representing a deep-water aulacogenic environment by Chen & Qiu (1986). *Calymenesun* also occurs in deep-water facies (e.g. Tripp *et al.* 1989).

However, all of these other reedocalymenine taxa still show relatively restricted geographic ranges, which suggests that their incorporation into cladistic biogeographic analysis will not lower the resolution of the pattern of area relationships (although see the discussion on the possible geographic range of *Neseuretinus* in Hammann & Leone 1997, p. 114). Similarly, although most of these taxa occurred during the Caradoc or Ashgill, a time period with different continental configurations to the Arenig–Llanvirn interval (Scotese & McKerrow 1990), the consensus cladogram indicates that most of the divergences involving these lineages are stratigraphically constrained to have occurred during the Early Ordovician. All of the reedocalymenine taxa included in cladistic analysis are therefore also included in cladistic biogeographic analysis. However, the occurrence of *Calymenella* in Australia during the Early Silurian (A. Sandford, pers. comm. 2001) is excluded from this analysis, as this would appear to represent a range expansion which occurred after the time period under consideration here.

The area cladogram in Figure 4 contains four subclades containing areas that span much of the palaeogeographic extent of Gondwana (nodes 1–4 in Fig. 4). This is interpreted as indicating that several separate, independent biogeographic events occurred during the evolution of the Reedocalymeninae, which can thus be compared with one another to reach an understanding of Gondwanan area relationships. Node 2 consists of an unresolved polytomy in the area cladogram, and so can provide no information on the relationships between the four different areas (France/Spain, South America, Avalonia and Turkey) included within this subclade. The other three nodes are either well-resolved (nodes 1, 4) or adequately resolved (node 3). Nodes 1 and 4 display the easternmost Gondwanan areas at the base of the subclade, with progressively more western Gondwanan areas tending to be nested progressively further within the subclade. At node 1, the far eastern Gondwanan region of South America represents the sister group to Central Asian and eastern Mediterranean regions and South China; at node 4, the western Gondwanan regions of Avalonia, France, Spain and North Africa are more closely related to one another than any is to the more easterly situated SE Asian regions of Indo-China and South China. Node 3 shows the opposite pattern of area relationships, with South America and South China as sister areas, together sister to Spain, with the westernmost region of Avalonia as the sister area to the rest of the subclade.

The ordering of successively nested areas within each of the resolved nodes matches previous ideas on the relative geographic positions of Gondwanan and peri-Gondwanan tectonic units in the Early Ordovician. These three sets of area relationships can therefore be interpreted as supporting the idea of a faunal cline mirroring the climatic cline across Gondwana during this time interval. However, the three resolved subclades appear to represent biogeographic shifts in opposite directions across Gondwana. The opposing nested patterns of area relationships thus represent non-congruent biogeographic events, and contain paralogous areas which would conflict with duplications of themselves to generate ambiguous data if combined as a consensus cladogram, reducing the information available from the area cladogram in Figure 3a. As it is imperative to limit paralogy in area cladistics in order to detect biogeographic congruence (Ebach 1999), combining all three into a single consensus area cladogram in an attempt to provide further information on Early Ordovician Gondwanan area relationships is thus to be avoided.

The subclades contained within nodes 1 and 4 appear to represent faunal shifts in the same direction across Gondwana (Fig. 5a, b). As these two subclades share the area of SE Asia in

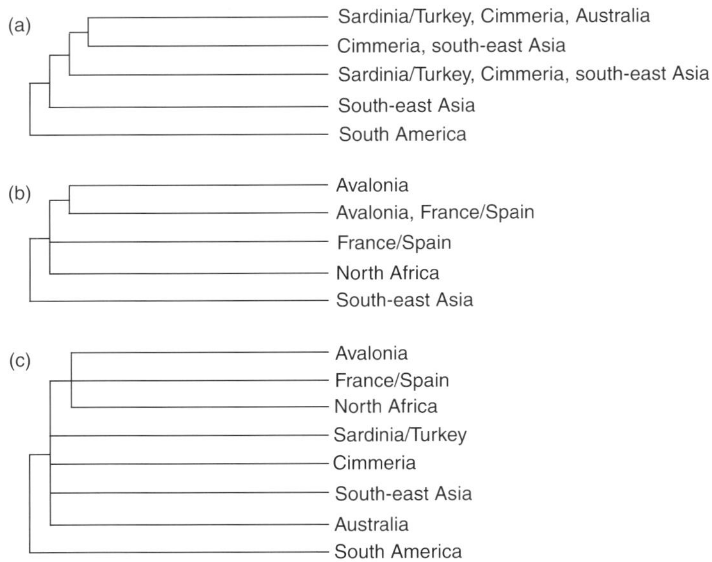

Fig. 5. (**a**, **b**) Area cladograms for nodes 1 and 4 of original area cladogram in Figure 4, with revised area boundaries. (**c**) Consensus area cladogram of nodes 1 and 4, showing complete pattern of area relationships of Gondwanan and peri-Gondwanan regions.

common, they can be combined to produce a consensus cladogram, which represents the relationships between different Gondwanan areas indicated by the congruent westward faunal shifts within nodes 1 and 4 (Fig. 5c). This consensus cladogram shows that the faunal cline extended across the entire region of Gondwana for which Early Ordovician deposits are available. South America and Avalonia+western Europe+North Africa respectively represent the eastern and western limits of the cline, with no evidence for a faunal connection linking these two areas directly across the Early Ordovician southern polar region.

Conclusions and future research

The area cladograms generated from the cladistic analysis of the Reedocalymeninae support certain previously held ideas on Gondwanan biogeography. The presence of multiple subclades within the Reedocalymeninae, each containing a wide range of both eastern and western Gondwanan areas, suggests that no significant environmental barriers existed across the palaeocontinent even for shallow inner-shelf marine taxa. The pattern of area relationships both within these subclades and as evidenced by the consensus cladogram in Figure 5c are also consistent with the idea of a faunal cline occurring across the palaeocontinent during the Early Ordovician. No reedocalymenine taxa have been recorded from the Tremadoc, but some *Neseuretus* species are known from the base of the Arenig, suggesting that this pattern of area relationships refers approximately to the interval between these two Ordovician series.

This cladistic biogeographic analysis offers a preliminary hypothesis on the relationships between different Gondwanan and peri-Gondwanan geographic areas for the Early Ordovician. However, the inter-relationships between the various Asian areas included in the analysis are unresolved and require further study, and some of the nodes used to generate the area cladograms in Figures 4 and 5 are only weakly supported by bootstrap analysis. The hypothesis thus requires testing against area cladograms generated with other Early Ordovician taxa.

Cladistic analysis of several other Early Ordovician trilobite groups could be used to assess support for this pattern of area relationships. Of the shallow-water taxa which sometimes co-occur with *Neseuretus* in the *Neseuretus* Association, several genera (e.g. *Kerfornella*, *Plaesiacomia*, *Eohomalonotus*, *Iberocoryphe*, *Crozonaspis* and *Taihungshania*) are only known from restricted regions of Gondwana, and so their phylogenies cannot be used to generate comparable area cladograms. Other shallow-water trilobite groups do occur over much of the same geographic region as

Neseuretus during the Early Ordovician. Asaphids are a common component of the *Neseuretus* Association across the entire Gondwanan region; however, the relationships within the family are currently poorly understood, and thorough taxonomic revision is required before they can be used to test the hypothesis of area relationships generated with the Reedocalymeninae.

Early Ordovician trilobite taxa potentially able to provide a test for this hypothesis include the dikelokephalinid *Hungioides*, which ranges from western Europe to South America (Fortey & Peel 1983), and two different groups of trinucleids which occur in slightly deeper shelf conditions. The closely related hanchungolithine trinucleid genera *Hanchungolithus* and *Ningkianolithus*, regarded by Li (1994) as occurring only in South China, have in fact also been recorded in Avalonia, France, the Middle East, Indo-China and possibly Spain (Dean 1966; El-Khayal & Romano 1985; Beckly 1989; Rabano 1990), although South American material originally assigned to *Hanchungolithus* (Hughes *et al.* 1975, as *Ichangolithus*) has recently been transferred to the Trinucleinae (Baldis & Pöthe de Baldis 1995). A second trinucleid group comprises the genera *Lordshillia*, *Anebolithus*, *Famatinolithus* and *Incaia*, which appear to be closely related to each other within the Trinucleinae, and occur in Avalonia, South China, New Zealand and South America (Hughes *et al.* 1975). Future cladistic research on some of these groups can hopefully provide further information on Early Ordovician area relationships across Gondwana and peri-Gondwana.

This project was greatly helped by the encouragement and assistance of D.J. Siveter, who engaged the author with considerable discussion on the evolution of the Calymenidae, and provided access to his substantial reference collection. Discussion with T.G. Waters generated further ideas on appropriate methods of biogeographic analysis. P. Janvier and Tong-Dzuy Thanh provided useful information on *Vietnamia douvillei*, and casts of most of Mansuy's type material of this species were provided by P. Racheboeuf. The Early Ordovician map used in Figure 1 was generated by D. Lees, and M.D. Sutton provided welcome technical assistance. W.T. Dean, A.W. Owen and M. Romano reviewed the manuscript and made several useful suggestions for its improvement. Funding for the project was provided by a Natural Environment Research Council D.Phil studentship.

References

BALDIS, B. & PÖTHE DE BALDIS, E. D. 1995. Trilobites Ordovicicos de la Formacion Las Aguaditas (San Juan, Argentina) y consideraciones estratigraficas. *Boletin de la Academia Nacional de Ciencas, Corboda*, **60**, 409–435.

BANKS, M. R. 1988. The base of the Silurian System in Tasmania. *In*: COCKS, L. R. M. & RICKARDS, R. B. (eds) A global analysis of the Ordovician-Silurian boundary. *Bulletin of the British Museum (Natural History), Geology*, **43**, 191–194.

BARRANDE, J. 1868. Silurische Fauna aus der Umgebung von Hof in Bayern. *Neues Jahrbüch für Mineralogie*, **6/7**, 641–696.

BECKLY, A. J. 1989. A new Arenig trilobite fauna from the Bangor area, North Wales. *Bulletin of the British Museum (Natural History), Geology*, **45**, 1–20.

BERGERON, J. 1890. Sur une forme nouvelle de trilobite de la famille des Calymenidae [genre *Calymenella*]. *Bulletin de la Societé Géologique de France*, **18**, 365–371.

BORN, A. 1918. Die *Calymene Tristani*-Stufe (mittleres Unter-silur) bei Almadén, ihre Fauna, Gliederung und Verbreitung. *Abhandlungen der senckenbergischen naturforschenden Gesellschaft*, **36**, 309–358.

CHEN XU & QIU JUNYU. 1986. Ordovician palaeoenvironmental reconstruction of Yichang area, W. Hubei. *Journal of Stratigraphy*, **10**, 1–15.

CHEN XU, RONG JIAYU, WANG XIAOFENG, WANG ZHIHAO, ZHANG YUANDONG & ZHANG RENBIN. 1995. *Correlation of the Ordovician rocks of China*. International Union of Geological Sciences, Publication **31**.

COCKS, L. R. M. 2001. Ordovician and Silurian global geography. *Journal of the Geological Society, London*, **158**, 197–210.

COCKS, L. R. M. & FORTEY, R. A. 1988. Lower Palaeozoic facies and faunas around Gondwana. *In*: AUDLEY-CHARLES, M. G. & HALLAM, A. (eds) *Gondwana and Tethys*. Geological Society, London, Special Publications, **37**, 183–200.

COCKS, L. R. M. & FORTEY, R. A. 1990. Biogeography of Ordovician and Silurian faunas. *In*: MCKERROW, W. S. & SCOTESE, C. R. (eds) *Palaeozoic Palaeogeography and Biogeography*. Geological Society, London, Memoirs, **12**, 97–104.

CORBETT, K. D. & BANKS, M. R. 1974. Ordovician stratigraphy of the Florentine Synclinorium, south-west Tasmania. *Papers and Proceedings of the Royal Society of Tasmania*, **55**, 1–55.

COURTESSOLE, R., PILLET, J. & VIZCAÏNO, D. 1981. Nouvelles donnees sur la biostratigraphie de l'Ordovicien inferieur de la Montagne Noire. Révision des Taihungshaniidae, de *Megistaspis* (*Ekeraspis*) et d'*Asaphopsoides* (Trilobites). *Mémoire de la Société des Etudes Scientifiques de l'Aude*. Carcassonne, France.

DEAN, W. T. 1966. The Lower Ordovician stratigraphy and trilobites of the Landeyran valley and the neighbouring district of the Montagne Noire, south-western France. *Bulletin of the British Museum (Natural History), Geology*, **12**, 247–353.

DEAN, W. T. 1967*a*. The correlation and trilobite fauna of the Bedinan Formation (Ordovician) in south-eastern Turkey. *Bulletin of the British Museum (Natural History), Geology*, **15**, 83–123.

DEAN, W. T. 1967*b*. The distribution of Ordovician shelly faunas in the Tethyan region. In: ADAMS, C. G. & AGER, D. V. (eds) *Aspects of Tethyan Biogeography*. Systematics Association, Publication **7**, 11–44.

DEAN, W. T. 1971. The Lower Palaeozoic stratigraphy and faunas of the Taurus Mountains near Beyşehir, Turkey. II. The trilobites of the Seydişehir Formation (Ordovician). *Bulletin of the British Museum (Natural History), Geology*, **20**, 3–24.

DEAN, W. T. 1973. The Lower Palaeozoic stratigraphy and faunas of the Taurus Mountains near Beyşehir, Turkey. III. The trilobites of the Sobova Formation (Lower Ordovician). *Bulletin of the British Museum (Natural History), Geology*, **24**, 279–348.

DEAN, W. T. 1975. Cambrian and Ordovician correlation and trilobite distribution in Turkey. *Fossils and Strata*, **4**, 353–373.

DEAN, W. T. & MARTIN, F. 1978. Lower Ordovician acritarchs and trilobites from Bell Island, eastern Newfoundland. *Geological Survey of Canada Bulletin*, **284**, 1–35.

DEAN, W. T. & ZHOU ZHIYI. 1988. Upper Ordovician trilobites from the Zap Valley, south-east Turkey. *Palaeontology*, **31**, 621–649.

DEAN, W. T., MONOD, O., RICKARDS, R. B., OSMAN DEMIR & BULTYNCK, P. 2000. Lower Palaeozoic stratigraphy and palaeontology, Karadere-Zirze area, Pontus Mountains, northern Turkey. *Geological Magazine*, **137**, 555–582.

DESMAREST, A. G. 1817. Crustacés Fossiles. *In*: BIOT, J. B. *et al.* (eds) *Nouveau Dictionnaire d'Histoire Naturelle*, **8** [Cor–Cun]. Deterville, Paris, 495–519.

EBACH, M. C. 1999. Paralogy and the centre of origin concept. *Cladistics*, **15**, 387–391.

EDGECOMBE, G. D., BANKS, M. R. & BANKS, D. M. 1999. Upper Ordovician Phacopida (Trilobita) from Tasmania. *Alcheringa*, **23**, 235–257.

EL-KHAYAL, A. A. & ROMANO, M. 1985. Lower Ordovician trilobites from the Hanadir Shale of Saudi Arabia. *Palaeontology*, **28**, 401–412.

FELSENSTEIN, J. 1985. Confidence limits on phylogenies: an approach using the bootstrap. *Evolution*, **39**, 783–791.

FORTEY, R. A. 1983. Cambrian–Ordovician trilobites from the boundary beds in western Newfoundland and their phylogenetic significance. *Special Papers in Palaeontology*, **30**, 179–211.

FORTEY, R. A. 1990. Ontogeny, hypostome attachment and trilobite classification. *Palaeontology*, **33**, 529–576.

FORTEY, R. A. & MELLISH, C. J. T. 1992. Are some fossils better than others for inferring palaeogeography? The early Ordovician of the North Atlantic as an example. *Terra Nova*, **4**, 210–216.

FORTEY, R. A. & MORRIS, S. F. 1982. The Ordovician trilobite *Neseuretus* from Saudi Arabia, and the palaeogeography of the *Neseuretus* fauna related to Gondwanaland in the earlier Ordovician. *Bulletin of the British Museum (Natural History), Geology*, **36**, 63–75.

FORTEY, R. A. & OWENS, R. M. 1978. Early Ordovician Arenig) stratigraphy and faunas of the Carmarthen district, south-west Wales. *Bulletin of the British Museum (Natural History), Geology*, **30**, 225–294.

FORTEY, R. A. & OWENS, R. M. 1987. The Arenig Series in South Wales. *Bulletin of the British Museum (Natural History), Geology*, **41**, 69–307.

FORTEY, R. A. & PEEL, J. S. 1983. The anomalous bathyurid trilobite *Ceratopeltis* and its homeomorphs. *Special Papers in Palaeontology*, **30**, 51–57.

FORTEY, R. A., HARPER, D. A. T., INGHAM, J. K., OWEN, A. W., PARKES, M. A., RUSHTON, A. W. A. & WOODCOCK, N. H. 2000. A revised correlation of Ordovician rocks in the British Isles. *Geological Society Special Report*, **24**, 1–83.

GIGOUT, M. 1951. Études géologiques sur la Méséta Marocaine Occidentale (arrière Pays de Casablanca, Mazagan et Safi). *Travaux de l'Institut scientifique chérifien*, **3**, 1–507.

HAMMANN, W. 1977. Neue Calymenacea (Trilobita) aus dem Ordovizium von Spanien. *Senckenbergiana Lethaea*, **58**, 91–97.

HAMMANN, W. 1983. Calymenacea (Trilobita) aus dem Ordovizium von Spanien; ihre Biostratigraphie, Ökologie und Systematik. *Abhandlungen der senckenbergischen naturforschenden Gesellschaft*, **542**, 1–177.

HAMMANN, W. & LEONE, F. 1997. Trilobites of the post-Sardic (Upper Ordovician) sequence of southern Sardinia. Part I. *Beringeria*, **20**, 1–217.

HENRY, J-L. 1989. Paléoenvironnements et dynamique de faunes de trilobites dans l'Ordovicien (Llanvirn supérieur – Caradoc basal) du massif Armoricain (France). *Palaeogeography, Palaeoclimatology, Palaeoecology*, **73**, 139–153.

HICKS, H. 1873. On the Tremadoc rocks in the neighbourhood of St. David's, Pembrokeshire, with special reference to those of the Arenig and Llandeilo groups and their fossil contents. *Quarterly Journal of the Geological Society of London*, **29**, 39–52.

HUGHES, C. P., INGHAM, J. K. & ADDISON, R. 1975. The morphology, classification and evolution of the Trinucleidae (Trilobita). *Philosophical Transactions of the Royal Society of London. B. Biological Sciences*, **272**(920), 537–607.

KITCHEN, I. J., FOREY, P. L., HUMPHRIES, C. J. & WILLIAMS, D. M. 1998. *Cladistics. The Theory and Practice of Parsimony Analysis* (second edition). Oxford University Press, Oxford.

KOBAYASHI, T. 1937. The Cambro-Ordovician shelly faunas of South America. *Journal of the Faculty of Science, Imperial University of Tokyo, Section II (Geology, Mineralogy, Geography, Seismology)*, **4**, 369–522.

KOBAYASHI, T. 1976. On the Ordovician Trilobite Provinces. *Proceedings of the Japanese Academy*, **52**, 232–235.

KOBAYASHI, T. 1987. The Cambrian and Ordovician systems in the south side of Eurasia and the origin of the Tethys Sea. *Journal of Geography*, **96**, 39–368.

LEGG, D. P. 1976. Ordovician trilobites and graptolites from the Canning Basin, western Australia. *Geologica et Palaeontologica*, **10**, 1–58.

LEONE, F., HAMMANN, W., LASKE, R., SERPAGLI, E. & VILLAS, E. 1991. Lithostratigraphic units and biostratigraphy of the post-sardic Ordovician sequence in south-west Sardinia. *Bollettino della Società Paleontologica Italiana*, **30**, 201–235.

LI ZHIMING. 1994. Ordovician. *In*: YIN HONGFU (ed.) *The Palaeobiogeography of China*. Clarendon Press, Oxford, 64–87.

LU YENHAO. 1957. Trilobita. *In*: *Index fossils of China. Invertebrates (3)*. Science Press, Beijing (in Chinese).

LU YENHAO. 1975. Ordovician trilobite faunas of central and southwestern China. *Palaeontologica Sinica, New Series B*, **11**, 1–463.

MANSUY, H. 1908. Contribution à la Carte Géologique de l'Indo-Chine. Paléontologie. *Goùvernement Général de l'Indo-Chine, Direction Générale des Travaux Publics, Service des Mines*. Imprimerie d'Extrême-Orient, Hanoi-Haiphong.

MOORE, R.C. (ed.) 1959. *Treatise on Invertebrate Paleontology. Part O. Arthropoda 1*. University of Kansas Press and Geological Society of America, Lawrence, Kansas.

MORRIS, S. F. & FORTEY, R. A. 1985. *Catalogue of the Type and Figured Specimens of Trilobita in the British Museum (Natural History)*. British Museum (Natural History), London.

MU ENZHI. 1974. Evolution, classification and distribution of Graptoloidea and graptodendroids. *Scientia Sinica*, **17**, 227–238.

PALMER, A. R. 1972. Problems of Cambrian biogeography. *24th International Geological Congress*, **7**, 310–315.

PARIS, F. 1998. Early Palaeozoic palaeobiogeography of northern Gondwana regions. *Acta Universitatis Carolinae – Geologica*, **42**, 473–483.

PEERS, S. 1997. *The taxonomy, phylogeny and palaeobiogeography of the trilobite families Pliomeridae and Encrinuridae – reconstructing the Ordovician world using evolving lineages*. PhD thesis, Glasgow University.

PENG SHANCHI, LIN TIANRUI & LI YUE. 2000. Notes on the genus *Reedocalymene* Kobayashi, 1951 (Trilobita, Ordovician). *Acta Palaeontologica Sinica*, **39**, 63–75.

PILLET, J. & DE LAPPARENT, F. 1969. Description des trilobites ordoviciens, siluriens et dévoniens d'Afghanistan. *Annales de la Société géologique du Nord*, **84**, 323–333.

PŘIBYL, A. & VANĚK, J. 1980. Ordovician trilobites of Bolivia. *Rozpravy Československé Akademie Věd*, **90**, 1–90.

RABANO, I. 1990. *Trilobites del Ordovícico Medio del sector meridional de la zona Centroibérica española*. Publicaciones especiales del boletín geológico y minero, Instituto Tecnológico Geominero de España, Madrid.

REED, F. R. C. 1917. Ordovician and Silurian fossils from Yun-Nan. *Memoirs of the Geological Survey of India, Palaeontologica Indica, New Series*, **6**, 1–69.

ROSS, R. J. Jr. 1967. Some Middle Ordovician brachiopods and trilobites from the Basin Ranges, western United States. *Geological Survey Professional Paper*, **523-D**, 1–43.

ROSS, R. J. Jr. 1975. Early Paleozoic trilobites, sedimentary facies, lithospheric plates, and ocean currents. *Fossils and Strata*, **4**, 307–329.

SALTER, J. W. 1865*a*. A Monograph of the British Trilobites from the Cambrian, Silurian and Devonian Formations. *Palaeontographical Society Monograph*, **2**, 81–128.

SALTER, J. W. 1865*b*. *In*: SALTER, J. W. & BLANDFORD, H. F. *Palaeontology of Niti in the Northern Himalaya: being descriptions and figures of the Palaeozoic and Secondary fossils collected by Colonel Richard Strachey, R. E.* Calcutta.

SCOTESE, C. R. & MCKERROW, W. S. 1990. Revised world maps and introduction. *In*: MCKERROW, W. S. & SCOTESE, C. R. (eds) *Palaeozoic Palaeogeography and Biogeography*. Geological Society, London, Memoirs **12**, 1–21.

SDZUY, K. 1955. Die Fauna der Leimitz-Schiefer (Tremadoc). *Abhandlungen der senckenbergischen naturforschenden Gesellschaft*, **492**, 1–74.

SEDGWICK, A. & M'COY, F. 1851. *A synopsis of the classification of the British Palaeozoic rocks, with a systematic description of the British Palaeozoic fossils in the geological museum of the University of Cambridge. Pt. 1*. Cambridge.

ŞENGÖR, A. M. C. 1984. *The Cimmeride orogenic system and the tectonics of Eurasia*. Geological Society of America, Special Paper, **195**.

ŞENGÖR, A. M. C. 1987. Tectonics of the Tethysides: orogenic collage development in a collisional setting. *Annual Reviews of Earth and Planetary Science*, **15**, 213–244.

SHENG, S. F. 1958. The Ordovician trilobites of southwestern China. *Acta Palaeontologica Sinica*, **6**, 169–204.

SHIRLEY, J. 1936. Some British trilobites of the family Calymenidae. *Quarterly Journal of the Geological Society of London*, **92**, 384–422.

SOKOLOV, B. S. & YOLKIN, E. A. (eds). 1978. Pogranichnye sloi ordovika i silura Altye-Sayanskoy oblasti i Tyen-Shanya. *Transactions of the Academy of Sciences of the USSR, Siberian Branch, Institute of Geology and Geophysics*, **397**, 1–221.

SPJEDNÆS, N. 1961. Ordovician climatic zones. *Norsk Geologisk Tidsskrift*, **41**, 45–77.

SPJEDNÆS, N. 1981. Lower Palaeozoic palaeoclimatology. *In*: HOLLAND, C. H. (ed.) *Lower Palaeozoic Rocks of the World. Volume 3. Lower Palaeozoic of the Middle East, Eastern and Southern Africa, and Antarctica.* John Wiley and Sons, Chichester, 199–256.

STUBBLEFIELD, C. J. 1939. Some aspects of the distribution and migration of trilobites in the British Lower Palaeozoic faunas. *Geological Magazine*, **76**, 49–72.

SUN, Y. C. 1931. Ordovician trilobites of central and southern China. *Palaeontologica Sinica, Series B*, **7**, 1–47.

SWOFFORD, D. L. 2001. *PAUP*: Phylogenetic Analysis Using Parsimony (*and other methods). Version 4.0b4a*. Sinauer Associates, Sunderland, MA.

TRIPP, R. P., ZHOU ZHIYI & PAN ZHENQIN. 1989. Trilobites from the Upper Ordovician Tangtou Formation, Jiangsu Province, China. *Transactions of the Royal Society of Edinburgh: Earth Sciences*, **80**, 25–68.

VACCARI, N. E. & WAISFELD, B. G. 1994. Nuevos trilobites de la Formacion Suri (Ordovicico temprano) en la region de Chaschuil, Provincia de Catamarca, implicancias bioestratigraficas. *Ameghiniana*, **31**, 73–86.

VAN INGEN, G. 1914. Cambrian and Ordovician faunas of south-eastern Newfoundland (Abstract). *Bulletin of the Geological Society of America*, **25**, 138.

WAISFELD, B. G. 1997. Trilobites calymenáceos de la Formacíon Acoite (Arenigiano) en el contrafuerte occidental de la Cordillera Oriental Argentina. *Ameghiniana*, **34**, 333–343.

WANG XIAOFENG, XIANG LIWEN, NI SHIZHAO, *ET AL.* 1987. *Biostratigraphy of the Yangtze Gorges Area (2). Early Palaeozoic Era.* Geological Publishing House, Beijing.

WHITTARD, W. F. 1960. The Ordovician trilobites of the Shelve Inlier, west Shropshire. *Palaeontographical Society Monograph*, **4**, 117–162.

WHITTINGTON, H. B. 1965. Trilobites of the Ordovician Table Head Formation, western Newfoundland. *Bulletin of the Museum of Comparative Zoology, Harvard University*, **132**, 275–442.

WHITTINGTON, H. B. 1966*a*. Trilobites of the Henllan Ash, Arenig Series, Merioneth. *Bulletin of the British Museum (Natural History), Geology*, **11**, 489–505.

WHITTINGTON, H. B. 1966*b*. Phylogeny and distribution of Ordovician trilobites. *Journal of Paleontology*, **40**, 696–737.

WHITTINGTON, H. B. 1973. Ordovician trilobites. *In*: HALLAM, A. (ed.) *Atlas of Palaeobiogeography*. Elsevier Science Publications, Oxford, 13–18.

WHITTINGTON, H. B. & HUGHES, C. P. 1972. Ordovician geography and faunal provinces deduced from trilobite distribution. *Philosophical Transactions of the Royal Society of London. B. Biological Sciences*, **263**(850), 235–278.

YI YONGEN. 1957. The Caradocian trilobite fauna from the Yangtze-Gorges. *Acta Palaeontologica Sinica*, **5**, 527–559.

YIN GONGZHENG & LEE SHANJI. 1978. Trilobita. *In*: *Atlas of Palaeontology of Southwest China, Guizhou (1)*. Geological Publishing House, Beijing (in Chinese).

ZHOU ZHIYI & DEAN, W. T. 1989. Trilobite evidence for Gondwanaland in east Asia during the Ordovician. *Journal of Southeast Asian Earth Sciences*, **3**, 131–140.

ZHOU ZHIYI, DEAN, W. T. & LUO HUILIN. 1998*a*. Early Ordovician trilobites from Dali, west Yunnan, China, and their palaeogeographical significance. *Palaeontology*, **41**, 429–460.

ZHOU ZHIYI, DEAN, W. T., YUAN WENWEI & ZHOU TIANRONG. 1998*b*. Ordovician trilobites from the Dawangou Formation, Kalpin, Xinjiang, northwest China. *Palaeontology*, **41**, 693–735.

ZHOU ZHIYI, LUO HUILIN, ZHOU ZHIQIANG & YUAN WENWEI. 2001. Palaeontological constrains on the extent of the Ordovician Indo-China Terrane in western Yunnan. *Acta Palaeontologica Sinica*, **40**, 310–317.

The spatial and temporal diversification of Early Palaeozoic vertebrates

M. PAUL SMITH, PHILIP C. J. DONOGHUE & IVAN J. SANSOM

School of Earth Sciences, University of Birmingham, Edgbaston, Birmingham B15 2TT, UK (e-mail: m.p.smith@bham.ac.uk; p.c.j.donoghue@bham.ac.uk; i.j.sansom@bham.ac.uk*)*

Abstract: Recent discoveries have dramatically altered traditional views of the stratigraphic distribution and phylogeny of Early Palaeozoic vertebrates and permit a reappraisal of biogeographic patterns and processes over the first 120 million years of vertebrate evolution. Stratigraphic calibration of the phylogenetic trees indicates that most of the pre-Silurian record can be inferred only through ghost ranges. Assessment of the available data suggests that this is due to a shift in ecological niches after the latest Ordovician extinction event and a broadening of geographical range following the amalgamation of Euramerica during the early Silurian. Two major patterns are apparent in the biogeographic data. Firstly, the majority of jawless fishes with dermoskeletal, plated 'armour' were highly endemic during Cambrian–Ordovician time, with arandaspids restricted to Gondwana, galeaspids to China, and anatolepids, astraspids and, possibly, heterostracans confined to Laurentia. These Laurentian groups began to disperse to other continental blocks as the 'Old Red Sandstone continent' amalgamated through a series of tectonic collisions. The second major pattern, in contrast, encompasses a number of microsquamous and naked, jawed and jawless primitive vertebrates such as conodonts, thelodonts, placoderms, chondrichthyans and acanthodians, which dispersed rapidly and crossed oceanic barriers to attain cosmopolitan distributions, although many have Laurentian origins. A clear difference in dispersal potential exists between these two types of fishes. Overall, the development of biogeographic patterns in Early Palaeozoic vertebrates involved a complex interaction of dispersal, vicariance and tectonic convergence.

Elliott *et al.* (1991) reviewed pre-Silurian vertebrates, concluding that only six Ordovician genera could be recognized unequivocally, and that none of the reports of Cambrian taxa were sustainable on available evidence. The oldest vertebrates were considered to be *Arandaspis* and the poorly known taxon *Porophoraspis* from the early Llanvirn Stairway Sandstone of central Australia (Ritchie & Gilbert-Tomlinson 1977). In addition, *Sacabambaspis* had been described from the Caradoc of Bolivia by Gagnier *et al.* (1986) and was considered to be closely related to *Arandaspis* (Gagnier *et al.* 1986; Elliott *et al.* 1991). The most diverse vertebrate fauna was thought to be that from the Harding Sandstone (Caradoc) of Colorado, USA, which had been the first locality to yield pre-Silurian vertebrates. Walcott (1892) described three species of fishes from the Harding Sandstone: *Astraspis desiderata*, *Eriptychius americanus* and *Dictyorhabdus priscus*. The last is not now considered to be a vertebrate, although its affinities remain obscure (Sansom *et al.* 2001). A third vertebrate taxon was, however, known from the unit (Denison 1967; Smith 1991) and was subsequently described as *Skiichthys halsteadi* by Smith & Sansom (1997). Of these taxa, only *Astraspis*, *Arandaspis* and *Sacabambaspis* were reliably understood in terms of their anatomy and, in phylogenetic terms, Elliott *et al.* (1991) considered *Astraspis* to be the most derived, with *Arandaspis* + *Sacabambaspis* as a sister taxon.

The biogeography of *Astraspis*, *Arandaspis* and *Sacabambaspis* was examined by Elliott *et al.* (1991) in the light of both dispersal and vicariance models. Their dispersal model interpreted the spatial and temporal data rather literally in biogeographic terms, invoking a dispersal route from the older, Australian, locality across Gondwana to Bolivia and thence across an oceanic barrier to Laurentia, amounting to what might be termed an 'out of Gondwana' model. It was recognized that the final trans-oceanic migration was an obstacle to a dispersal model given the apparent shallow-water ecological specialism of these fishes, but it was also noted that a vicariant model was not supported, and that both were in conflict with the palaeocontinental reconstructions of Scotese (1986) and Scotese & McKerrow (1990).

Since the review of Elliott *et al.* (1991), there has been a dramatic increase in the quality of the fossil record of Early Palaeozoic vertebrates which has arisen both through the discovery of

From: Crame, J. A. & Owen, A. W. (eds) 2002. *Palaeobiogeography and Biodiversity Change: the Ordovician and Mesozoic–Cenozoic Radiations*. Geological Society, London, Special Publications, **194**, 69–83.
0305-8719/02/$15.00

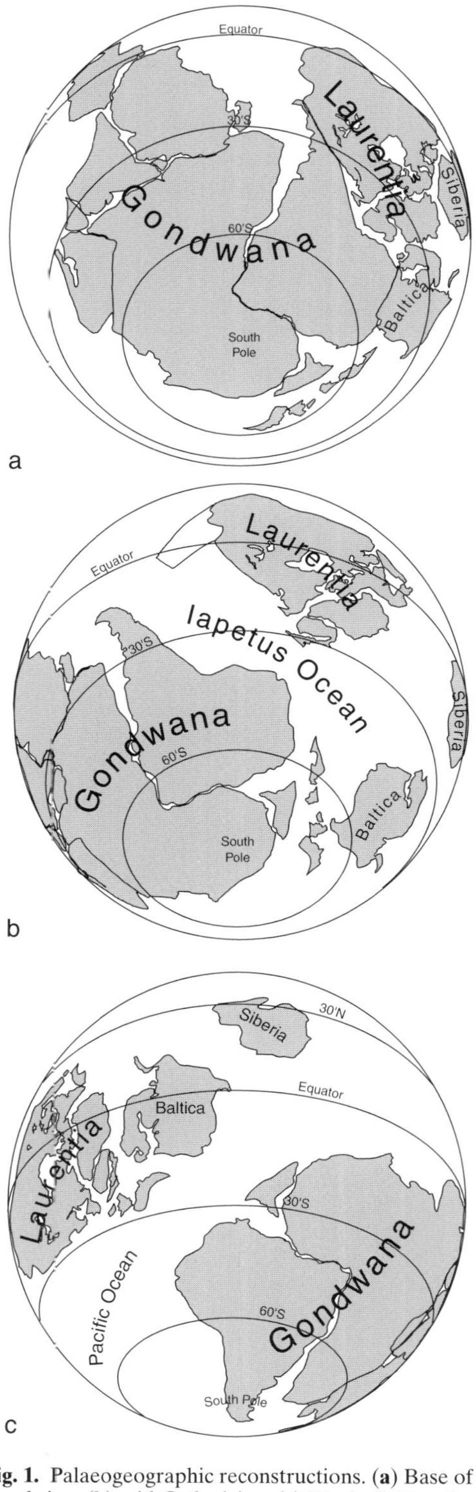

Fig. 1. Palaeogeographic reconstructions. (**a**) Base of Cambrian; (**b**) mid-Ordovician; (**c**) Wenlock. Based on Dalziel (1997).

new faunas and from reappraisals of affinity in well-known faunas. These advances have included the discovery of probable Early Cambrian vertebrates in the Chengjiang Lagerstätte (Shu *et al.* 1999*a*), the recognition of possible vertebrates in the Middle Cambrian Burgess Shale Lagerstätte of British Columbia (Simonetta & Insom 1993; Smith *et al.* 2001), the confirmation of the Late Cambrian–Early Ordovician sclerite *Anatolepis* as a vertebrate (Smith *et al.* 1996, 2001) and the discovery of unanticipated diversity in the Harding Sandstone of Colorado (Smith *et al.* 1995; Sansom *et al.* 1995, 1996, 2001). Perhaps the most significant discovery in terms of the biodiversity and geographical range of early vertebrates has been the addition of conodonts to the clade. The presence of preserved soft tissues had already suggested to Aldridge *et al.* (1986) that the affinities of this highly diverse group lay with the vertebrates, and a wealth of new data has now been advanced in support of the hypothesis. These include additional soft tissue finds (Smith *et al.* 1987; Aldridge *et al.* 1993; Gabbott *et al.* 1995), the recognition of vertebrate hard tissue synapomorphies in conodont elements (Sansom *et al.* 1992, 1994; Sansom 1996; Donoghue 1998; Donoghue & Aldridge 2001) and detailed cladistic analysis (Donoghue *et al.* 2000). The inclusion of conodonts as the earliest vertebrates to possess a mineralized skeleton, more derived than either hagfishes or lampreys, increases the generic and specific biodiversity of Cambro-Ordovician vertebrates by two orders of magnitude.

Does the substantial amount of new data acquired subsequent to the review of Elliott *et al.* (1991) permit more detailed and better supported models of Early Palaeozoic vertebrate biogeography?

Early Palaeozoic palaeogeography

The determination of biogeographic trends in Early Palaeozoic vertebrates is highly dependent on the accuracy, and choice, of palaeocontinental reconstructions. For Cambrian–Ordovician vertebrates, this is particularly true of the relative positions of Laurentia, Australia and South America, whereas constraints on the timing of collisions that assembled the 'Old Red Sandstone (ORS) continent' are critical for interpretations of Silurian–Devonian biogeography. To avoid circularity, we have used palaeocontinental reconstructions constructed from palaeomagnetic and tectonic data, and have eschewed those that already incorporate faunal data (e.g. Scotese & McKerrow 1990).

Traditionally, the Iapetan margin of southern Laurentia has been reconstructed as the conjugate margin of NW Africa prior to the opening of Iapetus, following the early reconstructions of Wilson (1966) and an implicit acceptance of an 'accordian style' opening and closure model (Dalziel 1997). However, the evidence for pre-Iapetan juxtaposition of the Laurentian and West African cratons is not strong and evidence has accumulated in favour of a radical alternative which arose out of the 'SWEAT' hypothesis of Moores (1991). Independently, Dalziel (1991) and Hoffman (1991) suggested that, prior to the opening of Iapetus and the final break-up of the Neoproterozoic supercontinent Rodinia, Laurentia was located between East Antarctica–Australia (East Gondwana) and South America (West Gondwana) (Fig. 1a). Following rifting, Laurentia was considered to have followed a clockwise rotating path around the conjoined South American and West African cratons (Dalziel 1991, 1997). Strong supporting evidence comes from the similarities between the fauna, stratigraphy and geological structure of SW USA and the Argentinean Precordillera (Dalla Salda *et al.* 1992; Astini *et al.* 1995; Thomas & Astini 1996) and there is now general agreement that the latter terrane originated as part of the Laurentian craton, located within the Ouachita embayment. Opinion differs, however, on the way in which the terrane was decoupled from Laurentia and attached to South America. Dalziel (1997) suggested that it originated as a promontory of Falkland–Malvinas Plateau type whereas Astini *et al.* (1995) favoured a rifted microcontinent that detached from Laurentia, drifted across Iapetus and collided with the South American margin. This has some relevance to early vertebrate biogeography, as discussed below, but in both cases the conjugate nature of the SE Laurentian and South American margins is implicit. The reconstructions of Dalziel (1997) are used as the basis for our assessments of Early Palaeozoic palaeobiogeography (Fig. 1a–c), although we recognize that not all details of these are universally accepted.

With regard to the assembly of the ORS continent, evidence has mounted to the point where a reasonably well-constrained and consistent model is available. The earliest collisions are of island arcs, with successive terrane accretion events (see van Staal *et al.* (1998) and Armstrong & Owen (2001) for reviews) serving as a prelude to the assembly of Avalonia, Baltica and Laurentia in mid-Silurian time. On the NE margin of Laurentia, early arc collisions are absent, and the first major assembly event was the collision with Baltica (Elevold *et al.* 2000). On the basis of palaeomagnetic data and kinematic indicators, it has long been recognized that this collision was oblique to the margins, with a sinistral component. Despite this obliquity, the collision is remarkably synchronous along the entire 2000 km length of this sector of the Caledonides. In NW Scotland, mylonites in the Moine thrust zone were dated by Freeman *et al.* (1998) as 437–430 Ma (Llandovery) and brittle deformation extends from 430 to 408 Ma (latest Llandovery–earliest Devonian). In comparison, prograde metamorphism in East Greenland has been dated at 435–423 Ma with anatexis at 430–422 Ma (Elvevold *et al.* 2000; Hartz *et al.* 2001) and in eastern North Greenland, late thrusts truncate earliest Wenlock (425 Ma) foreland basin turbidites (Higgins *et al.* 2001).

The closure of the Tornquist Sea, between Avalonia and Baltica, probably began in the latest Ordovician by dextral 'soft' amalgamation of the two plates (McKerrow *et al.* 1991, 2000; Torsvik *et al.* 1996). Palaeomagnetic data confirm that the sector of Iapetus between Avalonia and Laurentia was also closed, within the limits of palaeomagnetic resolution, by the Wenlock (Mac Niocaill 2000). The location of Siberia during this interval is more open to question. There is good evidence that Siberia was geographically inverted with respect to modern coordinates during the Early Palaeozoic (Torsvik *et al.* 1995, 1996; Smethurst *et al.* 1998) and drifted northwards, apparently in tandem with Euramerica (= Laurentia + Baltica + Avalonia) (Smethurst *et al.* 1998). Rotation of Siberia and collision with Baltica occurred between 360 Ma (earliest Carboniferous) and 250 Ma (end-Permian) (Smethurst *et al.* 1998), and McKerrow *et al.* (1991) interpreted Siberia as having an almost juxtaposed relationship with Euramerica during Silurian–Early Devonian time. There is no evidence for collision but there may not have been a significant oceanic barrier separating Novaya Zemlya and the northern Siberian margin. Cocks (2001, p. 199) noted that shelly taxa were mainly endemic at species level, with some endemic genera, indicating that the palaeocontinent was 'not particularly isolated during much of the period'. An additional, and key, component of Siberia from the standpoint of vertebrate biogeography is the terrane of Tuva (Young 1991, 1993), which had amalgamated with Siberia by the early Silurian (Bachtadse *et al.* 2000) but does contain late Silurian and Devonian endemic vertebrates (Afanassieva & Janvier 1985; Young 1993; Janvier 1996*a*).

The stratigraphic and biogeographic record

Cambrian vertebrates

Although molecular clock estimates imply an extensive pre-Cambrian history to the vertebrate clade (Kumar & Hedges 1998), at present there are no palaeontological data to support this conclusion. The Cambrian record begins with the Chengjiang Lagerstätte (basal Atdabanian, 530 Ma) which has yielded a number of taxa that have been interpreted as vertebrates. Shu *et al.* (1999*a*) described two unmineralized vertebrate taxa, *Haikouichthys* and *Myllokunmingia*, from the deposit, which they considered to be closely related to lampreys. Other Chengjiang taxa have also been interpreted as vertebrates. *Yunnanozoon* and *Haikouella* were considered to be stem group vertebrates by Chen *et al.* (1999) and Holland & Chen (2001), but this is not well-supported by the preserved anatomical evidence. *Xidazoon* was likened to *Pipiscius*, a jawless vertebrate from the Carboniferous Mazon Creek Lagerstätte of Illinois, USA, by Shu *et al.* (1999*b*), but newly discovered material of the latter indicates that it is not a vertebrate (Conway Morris, pers. comm. 2000).

A more equivocal vertebrate record is present in the Burgess Shale Lagerstätte of British Columbia, Canada. Simonetta & Insom (1993) described an unnamed taxon (Smithsonian Institution specimen USNM 198612) and likened the preserved structures in the head region to the cranial cartilages and sensory organ supports of extant lampreys. Testing of this intriguing suggestion probably awaits the discovery of additional material.

The earliest evidence of biomineralizing vertebrates is present in the mid-Late Cambrian where the phosphatic sclerite *Anatolepis* (Bockelie & Fortey 1976; Repetski 1978; Smith *et al.* 1996) and euconodonts have an almost simultaneous first occurrence (Smith *et al.* 2001). The vertebrate character of earliest euconodont hard tissues has been confirmed in a number of studies (e.g. Sansom *et al.* 1992; Donoghue 1998). If a phylogenetic link with paraconodonts, initially proposed by Bengtson (1976), is demonstrated, then the origin of biomineralizing vertebrates would be extended back to the Middle Cambrian. Euconodonts are cosmopolitan and their first appearance is globally synchronous, at least within the limits of biostratigraphic resolution. In contrast, *Anatolepis* is entirely confined to the margins of Laurentia. The affinities of *Anatolepis* have been the subject of some debate, but the presence of dentine indicates that it is a vertebrate (Smith *et al.* 1996) whilst the presence of dermoskeletal 'armour' (rather than visceroskeletal 'teeth') suggests that it is more derived than conodonts. For ease of reference, groups of jawless fish more derived than conodonts are referred to in this paper as 'ostracoderms', a paraphyletic but nonetheless useful term in this context.

Ordovician vertebrates

In contrast to the six genera, and seven species, recognized by Elliott *et al.* (1991), the total biodiversity of Ordovician vertebrates is now known to number several hundred species, of which the vast majority are conodonts. However, the recognized diversity of non-conodont vertebrates has also increased significantly. Analysis of the microvertebrate fauna of the Harding Sandstone of Colorado and a number of roughly coeval clastic units in North America has revealed that, in contrast to the three taxa recognized in earlier studies, at least 20 species are present. More remarkably, this fauna contains a large number of first appearances of major clades, including astraspids, thelodonts and chondrichthyans (Sansom *et al.* 2001). The age of these units is thus critical to accurate constraint of the ghost ranges of vertebrate taxa. Conodonts are present in all units examined and indicate that the Harding Sandstone is of *undatus* Chronozone age in the graphically correlated stratigraphic framework of Sweet (1984), and that related units with similar vertebrate faunas span the *compressa–undatus* Chronozone interval (mid-Mohawkian, mid-Caradoc). The chondrichthyan fauna includes mongolepids, which are here considered to be stem-chondrichthyans based on the shared presence of neck canals in the scales, but the absence of other chondrichthyan synapomorphies. In addition to Laurentia, thelodonts are also found in Upper Ordovician localities in Baltica (Timan-Pechora; Karatajūtė-Talimaa 1997) and peri-Gondwanan terranes (Austria; Cocks 2000; Bogolepova pers. comm.).

The application of micropalaeontological methodologies in the Amadeus Basin of central Australia has also revealed more diverse faunas than previously suspected. Young (1997) described microvertebrate faunas ranging in age from early Arenig to early Caradoc. Some of the sclerites are of doubtful affinity, but *Apedolepis* (Llanvirn) is a probable vertebrate (Young 1997; Sansom *et al.* 2001) and the range-base of *Porophoraspis* has been extended from the Llanvirn to the lower Arenig. *Pirchanchaspis*

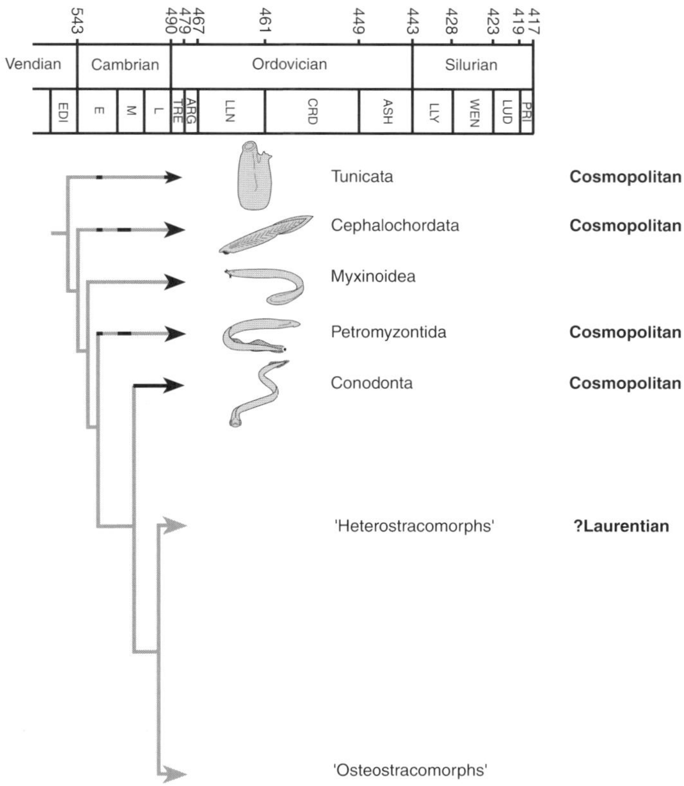

Fig. 2. Stratigraphic distribution and phylogenetic relationships of Cambrian vertebrates, with the biogeographic distribution of each group during this interval indicated. Black bars indicate observed occurrences, grey lines indicate inferred ghost ranges. Phylogenetic tree based on Donoghue *et al.* (2000) and Donoghue & Smith (2001); stratigraphic data from sources in Smith *et al.* (2001); absolute ages of series and system boundaries from sources in Cooper (1999). ARG, Arenig; ASH, Ashgill; CRD, Caradoc; E, Early; EDI, Ediacaran; L, Late; LLN, Llanvirn; LLY, Llandovery; LUD, Ludlow; M, Middle; PRI, Pridoli; TRE, Tremadoc; WEN, Wenlock.

rinconensis, recorded from Bolivia by Erdtmann *et al.* (2000), may represent an additional Gondwanan species.

Euconodonts underwent an explosive radiation at the beginning of the Ordovician, increasing rapidly to a standing diversity of 30–40 genera that exploited a wide range of ecological niches (cf. Aldridge 1988; Sweet 1988). The effect of the latest Ordovician extinction event is dramatic in conodonts, with standing diversity reduced to no more than 10 genera.

Silurian vertebrates

All the major groups of lower vertebrates are represented in the Silurian, which marks a dramatic increase in the quality and quantity of the non-conodont vertebrate record (extensively reviewed by Blieck & Janvier 1991; Janvier 1996*a*). Conodonts recovered rapidly from their minimum diversity of 10 genera at the base of the Silurian, to reach levels of approximately 20 at the Llandovery–Wenlock boundary. Thereafter, conodonts declined to a standing diversity of approximately 10–15 genera for the remainder of the Silurian (cf. Aldridge 1988; Sweet 1988).

The phylogeny of primitive vertebrates

Donoghue *et al.* (2000) performed a phylogenetic analysis of chordate interrelationships incorporating all of the major groups of extinct and extant jawless vertebrates, including conodonts. The three equally most parsimonious

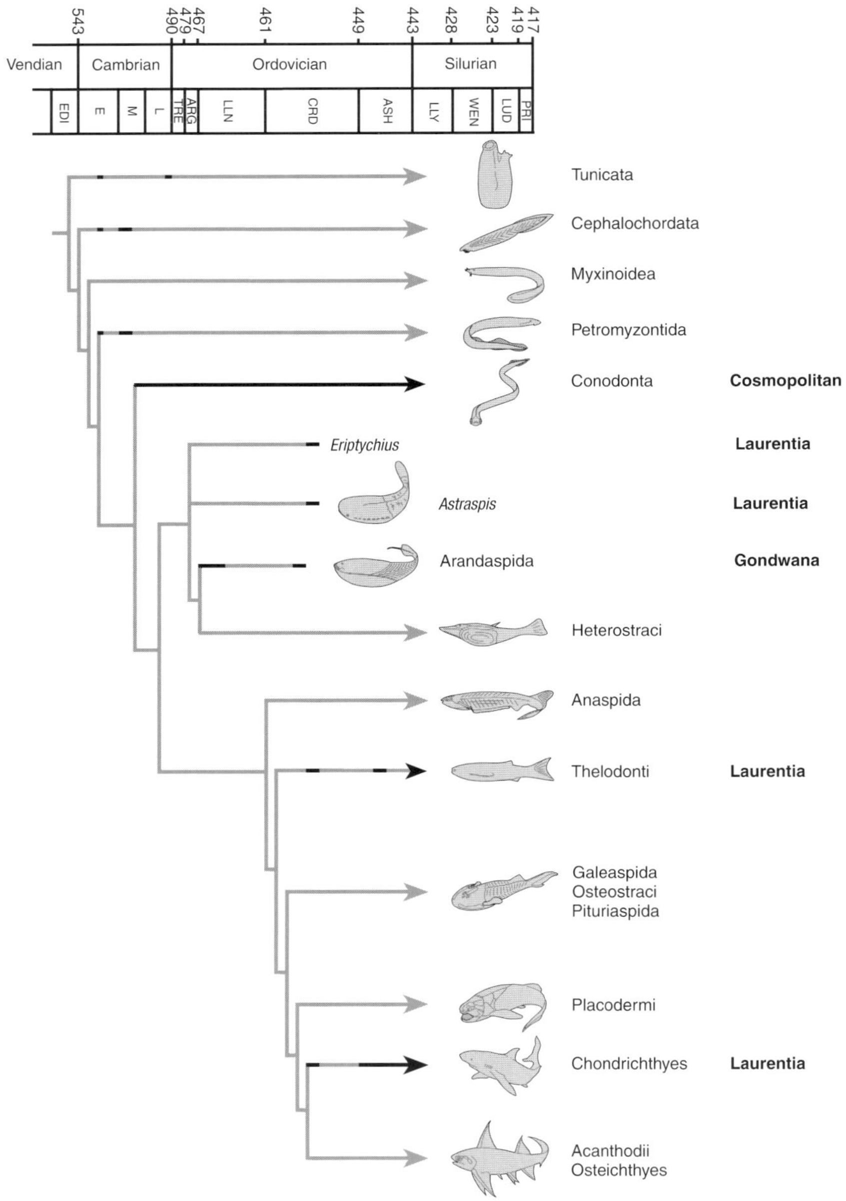

Fig. 3. Stratigraphic distribution and phylogenetic relationships of Cambrian–Ordovician vertebrates, with the biogeographic distribution of each group during the Ordovician indicated. Black bars indicate observed occurrences, lines indicates inferred ghost ranges. Phylogenetic tree based on Donoghue *et al.* (2000) and Donoghue & Smith (2001); stratigraphic data from sources in Sansom *et al.* (2001) and Smith *et al.* (2001); absolute ages of series and system boundaries from sources in Cooper (1999). Series abbreviations as in Figure 2.

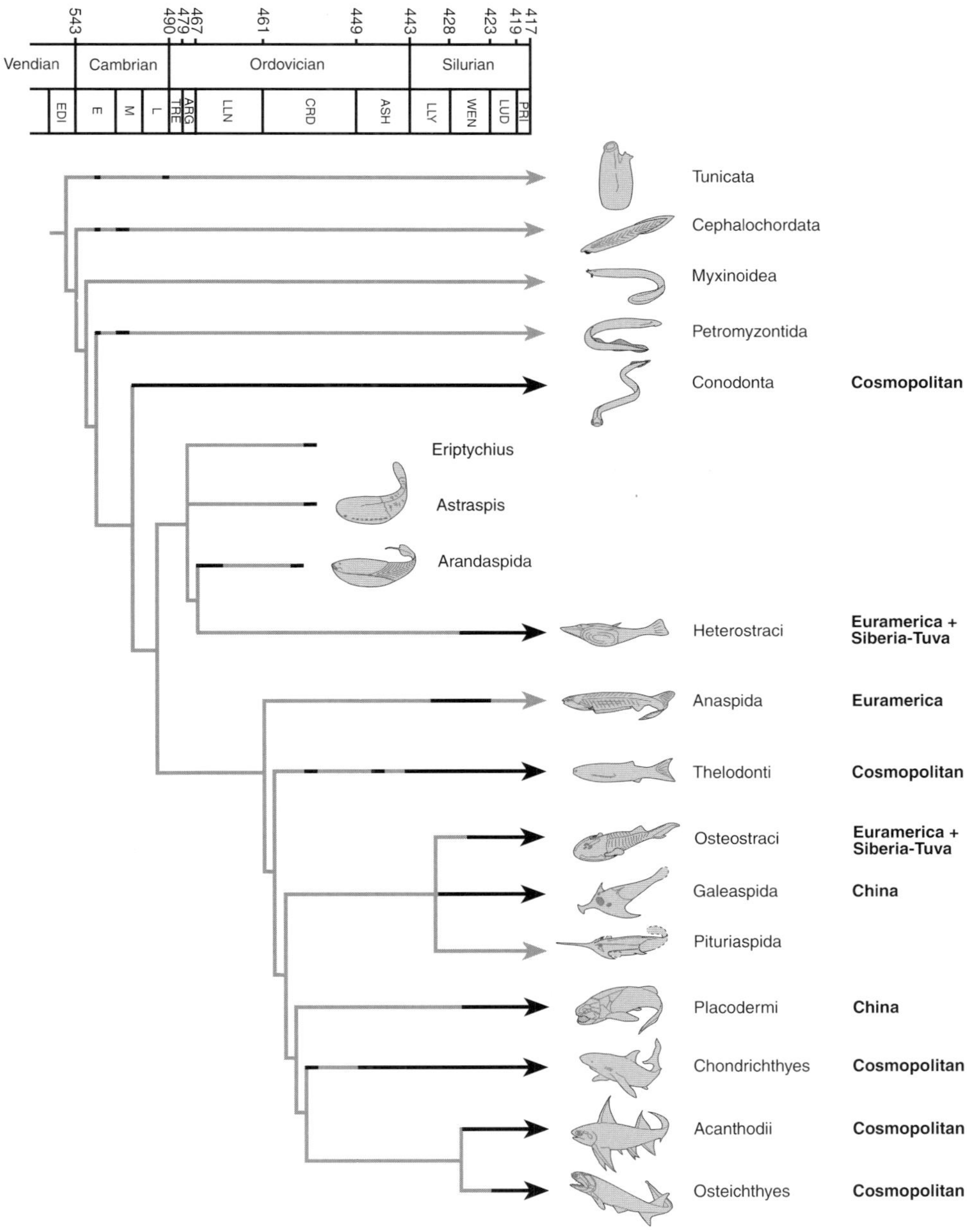

Fig. 4. Stratigraphic distribution and phylogenetic relationships of Cambrian–Silurian vertebrates, with the biogeographic distribution of each group during the Silurian indicated. Black bars indicate observed occurrences, lines indicates inferred ghost ranges. Phylogenetic tree based on Donoghue *et al.* (2000) and Donoghue & Smith (2001); stratigraphic data from sources in Blieck & Janvier (1991), Janvier (1996*a*), Sansom *et al.* (2001), Smith *et al.* (2001) and other sources discussed in the text; absolute ages of series and system boundaries from sources in Cooper (1999). Series abbreviations as in Figure 2.

trees differed only in the relationship of osteostracans, galeaspids and pituriaspids, which together constitute the sister group to jawed vertebrates. Myxinoids and petromyzontids lie at the base of the craniate/vertebrate clade, with conodonts resolved as the sister group of all other jawless and jawed vertebrates. The relationships of derived jawless vertebrates were further resolved by Donoghue & Smith (2001) who argued that the majority of thelodonts comprise a monophyletic group and constitute the sister taxon to osteostracans, pituriaspids, galeaspids and jawed vertebrates (Figs 2–4).

Ghost ranges and their implications

Calibrating cladograms against stratigraphic range data enables the completeness of the fossil record to be assessed as it provides a means of inferring the existence of unsampled taxa. Such an approach was first adopted by Gauthier *et al.* (1988), and the concept of 'ghost lineages' or 'ghost ranges' was subsequently developed by Norell (1992), amongst others. As useful as ghost ranges are, they are no panacea and require a number of assumptions upon which to base the inference of an unsampled (or unsampleable) fossil record. The most basic assumption requires that the cladogram is at least a reasonable reflection of the true tree of relationships. The second requires that the taxa represented in the cladogram must be monophyletic, since the inclusion of paraphyletic taxa leads to spurious inferences of ghost lineages (see e.g. Wagner 1998).

On the first count, we note that although the tree used is only relatively weakly supported at various nodes (Donoghue *et al.* 2000; Donoghue & Smith 2001), a degree of confidence is provided by the consistency of the signal arrived at following experimentation with the dataset (Donoghue *et al.* 2000) and its congruence with analyses of independent datasets (e.g. Forey & Janvier 1993, 1994; Forey 1995; Janvier 1996*b*). There is also good evidence to support the assumption that the operational taxa are monophyletic given that synapomorphies for each of these groups have been identified through character analysis (see e.g. Janvier 1996*a*) and numerical cladistic analysis of lower rank taxa corroborates this assumption (Donoghue & Smith 2001).

On the basis of these assumptions, it is possible to interpret the stratigraphically calibrated trees presented in Figures 2–4. The trees indicate that although the fossil record of most major groups of lower vertebrates does not begin until the Silurian, all have ghost lineages that extend into the Ordovician, suggesting that the early evolutionary history of these groups is at best unsampled, or at worst unrepresented. In consequence, assessments of diversity change carried out at family level and above (e.g. Benton 1999) cannot be applied with any confidence to the analysis of early vertebrate evolution, since entire orders are missing from the Ordovician record.

Biogeographic trends – patterns and processes

Cambrian–Ordovician

Evidence for the biogeographic distribution of Early and Mid-Cambrian vertebrates is scant and insufficient to derive secure models, other than to conclude that, by the Mid-Cambrian, vertebrates were probably present in both China and Laurentia. It is not until the first appearance of biomineralized vertebrates, *Anatolepis* and euconodonts, in the Late Cambrian, that there are sufficient data to draw firm conclusions regarding biogeographic patterns. As a clade, euconodonts are cosmopolitan and have a synchronous first appearance within the limits of biostratigraphic resolution. At lower taxonomic levels, many euconodont taxa are also cosmopolitan (Miller 1984) although some endemic genera and species are present in the Cambrian, and provincialism develops in the Ordovician between high- and low-latitude faunas (Rasmussen 1998; Armstrong & Owen 2002).

In distinct contrast, the early 'ostracoderm' *Anatolepis* is exclusively circum-Laurentian in its distribution. Indeed, there are no known cosmopolitan 'ostracoderms' during the Cambrian–Ordovician interval, leading to highly regionalized distribution patterns. Many major clades, including astraspids, heterostracans, thelodonts and chondrichthyans have their earliest occurrences in Laurentia, together with a number of plesiomorphic taxa of uncertain affinity, some as yet unnamed (Sansom *et al.* 2001), that are known exclusively from microvertebrate assemblages. That such taxa are exclusively Laurentian in their distribution during the Ordovician is supported by their repeated occurrence in Upper Ordovician deposits throughout Laurentia (Sansom *et al.* 2001, unpublished data) and their absence from coeval microvertebrate assemblages from Gondwana and elsewhere (e.g. Young 1997), which does not seem to be an artefact of sampling (see below). The available dataset thus implies that the latest common ancestor of all 'ostracoderms' and jawed vertebrates was Laurentian, rather than Gondwanan (*contra* Elliott *et al.* 1991).

In this context, the phylogenetic and spatial relationship between the arandaspids and other 'ostracoderms' is puzzling, since the group is entirely restricted to Gondwana (Fig. 3). *Astraspis*, the sister taxon of the arandaspids + heterostracans, is present throughout Laurentia, but endemic to it. In turn, heterostracans have a first record in the Wenlock of Laurentia (Soehn & Wilson 1990) and throughout their stratigraphic range are restricted to Euramerica + Siberia–Tuva. Therefore, despite the strict endemicity exhibited by 'ostracoderm' groups, the presence of arandaspids in Gondwana requires an absence of barriers between Laurentia and Gondwana for at least some of the time interval between the Late Cambrian and earliest Ordovician (Fig. 1a, b), and this biogeographic pattern lends some support to Dalziel's (1997) model for tectonic interaction between Laurentia and South America during the Ordovician.

Most groups exhibit endemicity through to the Wenlock, although significant exceptions amongst the 'ostracoderms' and jawed vertebrates are thelodonts and mongolepid chondrichthyans. The oldest thelodonts are exclusively Laurentian (Sansom *et al.* 2001), but by the end of the Ordovician they had dispersed to Baltica (Timan-Pechora) and peri-Gondwana, and were present in Siberia and Tuva by the end of the Llandovery. A similar pattern is evident amongst the mongolepid chondrichthyans that have a first occurrence in Laurentia (Sansom *et al.* 2001) but which, by the end of the Llandovery, were present in North China, South China and Mongolia (Karatajūtė-Talimaa 1996; Sansom *et al.* 2000).

Silurian

The ghost lineages suggest that the origin of most of the major groups of jawed and jawless lower vertebrates lies within the Ordovician and, thus, that the dramatic change in vertebrate distribution during the Silurian is deceptive. There is, nevertheless, an increase in the biogeographic range, abundance and diversity of all groups during the Silurian.

Blieck & Janvier (1991) recognized four Silurian vertebrate provinces: Euramerica (Laurentia, Baltica, Avalonia plus Kara-Taimyr), Siberia, Tuva and China. Within the Euramerican Province, tectonic convergence in the form of successive terrane and continent–continent collisions led to the assembly of Euramerica. This tectonic setting, with an absence of deep oceanic barriers, enabled 'ostracoderms' to disperse across the range of terranes and continental blocks that made up the ORS continent. The assembly of Euramerica also coincides with the increased abundance of a range of vertebrate clades in shallow marine environments. The timing of dispersal correlates well with dates for the collision events derived from other sources (see above), since none of the elements of the dispersal predate the Wenlock. The assembly of the ORS continent and the dispersal of vertebrates are also coeval with the breakdown of endemism in invertebrate faunas (Hallam 1994).

As noted above, Tuva had docked with Siberia by the early Silurian (Bachtadse *et al.* 2000), but the exact position of Siberia–Tuva and its relationship to Euramerica during the later Silurian and Devonian is the subject of some equivocation (cf. Torsvik *et al.* 1995; Cocks 2001). However, stratigraphic data indicate that osteostracans and heterostracans spread from Euramerica to Siberia–Tuva, suggesting the absence of deep water barriers at that time. Once this initial dispersal had taken place, relatively high levels of endemicity were then maintained between the two blocks during the late Silurian and Devonian, with amphiaspid heterostracans and tannuaspid osteostracans being unique to Siberia and Tuva respectively (Blieck & Janvier 1991; Young 1991, 1993). The presence of the distinctive endemic tannuaspid fauna within the Tuva part of the Siberia–Tuva block (Blieck & Janvier 1991; Janvier 1996*a*) correlates with the development of the endemic *Tuvaella* brachiopod fauna, and may be related to the high-latitude position of Tuva (Cocks 2001).

The Silurian biogeography of vertebrates in China and Vietnam contrasts markedly with that of Euramerica. Osteostracans and heterostracans are absent and faunas are dominated by endemic galeaspids from the Llandovery onwards, together with thelodonts and, from the Wenlock, acanthodians and placoderms (Wang 1995; P'an *et al.* 1996; Thanh *et al.* 1997). The high degree of endemicity is consistent with substantial oceanic separation between China and Euramerica/Siberia, but the palaeogeographic origin of galeaspids is puzzling because their closest known relatives are exclusively Laurentian/Euramerican (Donoghue & Smith 2001). This posits the conclusion that the latest common ancestor of galeaspids and their nearest relatives was, again, Laurentian. The only derived 'ostracoderms' to have dispersed from Laurentia to China were the thelodonts, and this group may offer a clue to the origin of endemic galeaspids in China.

In Gondwana, 'ostracoderms' disappear from the record after the Caradoc, presumably as a result of glaciation (Elliott *et al.* 1991) and the only reliably documented Silurian vertebrates from Gondwana are conodonts, thelodonts and jawed vertebrates (Blieck & Janvier 1991). Indeed, some parts of Gondwana apparently remained unpopulated until the dispersal of jawed vertebrates during the Devonian. This is demonstrably not a sampling artefact, since conodont faunas from Australia and Laurentia are comparable, but 'ostracoderm' biodiversity is not. By the Early Devonian, an endemic placoderm province (the 'wuttagoonaspid–phyllolepid' province) is recognizable in East Gondwana which began to break down in the Late Devonian (Young 1991, 1993). West Gondwana, in contrast, is relatively depauperate in placoderms and has chondrichthyan–acanthodian-dominated faunas at that time (Young 1993).

The Silurian thus records a complex history of dispersal, vicariance and tectonic convergence. Acanthodians, thelodonts and chondrichthyans continue to be widely dispersed and almost certainly had genuine trans-oceanic dispersal capability (*contra* Blieck & Janvier 1991, p. 377). Heterostracans, anaspids and osteostracans also began to disperse, but only after the tectonic assembly of the ORS continent had removed oceanic barriers. In Siberia, a degree of vicariant endemism is observed after the original input of heterostracans and osteostracans from Euramerica, and in China a significant degree of endemism is also established following initial dispersal.

Where are all the Cambro-Ordovician vertebrates?

One of the most intriguing aspects of this new perspective on Cambro-Ordovician vertebrate biodiversity does not in itself stem from the discovery of vertebrate remains in rocks of this age but, rather, the discovery of significant gaps in the record in the form of ghost ranges. The improved stratigraphic constraints and better understanding of phylogenetic relationships provided by Cambro-Ordovician vertebrates reveal long ghost ranges which imply that most of the major groups of 'ostracoderms' and primitive jawed vertebrates have an evolutionary history that extends into the Ordovician. It is likely that many of the newly discovered Cambro-Ordovician microvertebrate remains of currently uncertain affinity will fulfil the prediction of these ghost lineages, but there are many more ghost lineages than there are tangible candidates. There are at least three possible, and non-mutually exclusive, explanations for the dramatic improvement in the quality of the vertebrate fossil record during the early–middle Silurian: (a) the increase in the taxonomic diversity and disparity of vertebrate fossil record accurately reflects an early Silurian cladogenic event (cf. Blieck & Janvier 1991) and the inferences of ghost lineages are entirely spurious; (b) Cambrian and Ordovician vertebrates are rare because there is a systematic bias against the preservation of the environments in which they lived; and/or (c) Cambro-Ordovician vertebrates were ecologically distinct from their middle Palaeozoic relatives and their fossil record is either undersampled or the preservation of fossil remains unlikely because of systematic bias.

The fossil record of Cambrian and Ordovician vertebrates (except for conodonts) is so poor that it is not possible to test comprehensively whether or not vertebrates were affected by the latest-Ordovician extinction event. Nevertheless, there is evidence that at least some vertebrate groups were affected. The fossil record of conodonts exhibits a dramatic drop in taxonomic diversity during this event such that the entire clade almost became extinct (Aldridge 1988; Armstrong 1995). However, this model has yet to be tested against phylogenetic trees and it is notable that not one of the major conodont groups became extinct. Furthermore, all of the Ordovician vertebrate remains discovered thus far can be identified either as members of clades surviving into the middle Palaeozoic, or as possible sister taxa to one or more of these clades. Thus, although it is likely that the vertebrate clade as a whole was affected by the extinction event locally, especially at low taxonomic level and in terms of numerical diversity, there is no evidence that any major vertebrate groups arose or met their demise at this time, a pattern mirrored in invertebrate faunas (Droser *et al.* 2000). The absence of extinction in major vertebrate groups indicates that the increase in the quality of the vertebrate record during the Silurian is not the result of major evolutionary turnover at high taxonomic level.

The second possibility, that the fossil record of early vertebrates is poor because of a systematic bias against the preservation of shallow-water nearshore lithofacies, is also doubtful. Such environments are both extensively preserved and crop out extensively throughout the

Ordovician of Laurentia in North America. This interval has been intensively sampled for conodonts, which are abundantly preserved throughout, but the 'ostracoderm' record is confined to specific intervals that seem to coincide with highstand events (Sansom *et al.* 2001). It is clear, therefore, that although shallow-water, nearshore environments are widely preserved, they contain no record of vertebrates, other than conodonts. Thus, it would appear that the third of the alternatives is worthy of further consideration and investigation, namely that Ordovician 'ostracoderms' were ecologically distinct from their middle Palaeozoic relatives and are preserved in very different facies which are either unsampled and/or systematically underrepresented in the rock record. If this is the case, the two end members that could serve as refugia for Ordovician vertebrates are freshwater and deep marine environments. The former may be excluded. Despite continuing appeals for the freshwater origin of vertebrates (Graffin 1992; Griffith 1994), no Cambro-Ordovician vertebrates are found in freshwater environments and Blieck & Janvier (1991) argued that most or all Silurian vertebrates were marine. To the contrary, *Anatolepis* occurs predominantly in outer-shelf, circum-Laurentian settings and the Harding Sandstone and Stairway Sandstone faunas may coincide with sea-level highstands (Sansom *et al.* 2001). This suggests that the missing Ordovician record may be found in deeper-water, outboard settings which, due to their position, tend to be more deformed and metamorphosed than cratonic interiors. Prospecting these areas for vertebrate faunas may provide additional insights into early vertebrate evolution.

What underlies endemism and cosmopolitanism during the Silurian?

There is a notable correlation between the biogeographic distribution of vertebrates during the Silurian and their anatomical design. For instance, thelodonts are the only monophyletic group of 'ostracoderms' to exhibit cosmopolitanism during the Silurian, the other groups being restricted to the margins of Laurentia and, later, Euramerica plus Siberia–Tuva (with the exception of galeaspids, which are restricted to China). In common with the thelodonts and conodonts, jawed vertebrates also exhibit cosmopolitanism during this interval. It is possible that the heavily armoured groups of 'ostracoderms' remained endemic to the ORS continent throughout the middle Palaeozoic because they were constrained against active swimming in deeper water and/or open ocean environments. Indeed, it has been argued by Gutmann (1967) and Gutmann & Bonik (1983) that, in the absence of a mineralized axial skeleton, the heavily armoured 'ostracoderms' (the *Panzerfische* of Gutmann) used their extensive dermal skeleton to brace musculature. The mineralized axial skeleton of jawed vertebrates (a synapomorphy of the clade) provides a biomechanically superior structure against which to brace musculature. As part of their 'new head' hypothesis, Gans & Northcutt (1983) noted that the origin of vertebral elements immediately precedes the advent of paired fins, as the final stage of their scenario in which vertebrate evolution was driven by a change in feeding strategies towards active predation. We suggest, therefore, that the heavily armoured dermal skeleton of most pre-jawed vertebrates was a limiting factor in their dispersal. Far from there being a direct competitive link between the 'ostracoderms' and their jawed relatives (see e.g. Purnell 2001), it is possible that the greater dispersal potential of the jawed vertebrates was a significant factor in the longevity of the group, while the 'ostracoderms' underwent a progressive decline to their demise in the Late Devonian. This greater dispersal potential would, in turn, enable the subsequent development of areas of endemism and concomitantly higher levels of global biodiversity which are encountered in Devonian jawed vertebrates.

Conclusions

We began with the 'out of Gondwana' vicariance model of Ordovician vertebrate spatial and temporal diversification (Elliott *et al.* 1991) and, with the incorporation of new and more complete datasets, conclude with a new model that incorporates a complex interaction of dispersal, vicariance, and tectonic convergence during the Early Palaeozoic. New datasets and the elucidation of existing datasets lead to the suggestion that much of the early evolutionary history of the major early vertebrate clades lies within the Ordovician, or even the Cambrian. Furthermore, it is likely that the record of these events is poor because the organisms were ecologically distinct from their middle Palaeozoic relatives, rather than because of a systematic bias in the preservation of environments typical of these counterparts.

We provide this assessment of the evolutionary palaeobiogeography of Early Palaeozoic vertebrates based on the data that is currently to hand. However, biogeographic analyses based upon historical evidence are, by their very nature, reliant upon incomplete datasets and it is likely that unexpected discoveries will lead to radical reinterpretation of the data presented herein. Nevertheless, we feel that the database has grown to an extent that we can begin to build testable models with the aim of stimulating the search for new data in both time intervals and palaeoenvironments in which the remains of the earliest vertebrates might otherwise be unexpected and therefore remain unsampled.

We are grateful to A. Owen and A. Crame for inviting us to contribute to the Lyell Meeting. The paper has been significantly improved by comments from H. Armstrong, T. Hallam and P. Janvier, and the work has been funded by NERC grants GR3/10272 and NER/B/S/2000/00284 and by NERC Post-Doctoral Research Fellowship GT5/99/ES/2.

References

AFANASSIEVA, O. & JANVIER, P. 1985. *Tannuaspis*, *Tuvaspis* and *Ilemoraspis*, endemic osteostracan genera from the Silurian and Devonian of Tuva and Khakassia (USSR). *Geobios*, **18**, 493–506.

ALDRIDGE, R. J. 1988. Extinction and survival in the Conodonta. *In:* LARWOOD, G.P. (ed.) *Extinction and Survival in the Fossil Record.* Systematics Association Special Volume, **34**, 231–256.

ALDRIDGE, R. J., BRIGGS, D. E. G., CLARKSON, E. N. K. & SMITH, M. P. 1986. The affinities of conodonts – new evidence from the Carboniferous of Edinburgh, Scotland. *Lethaia*, **19**, 279–291.

ALDRIDGE, R. J., BRIGGS, D. E. G., SMITH, M. P., CLARKSON, E. N. K. & CLARK, N. D. L. 1993. The anatomy of conodonts. *Philosophical Transactions of the Royal Society of London, Series B*, **340**, 405–421.

ARMSTRONG, H. A. 1995. High-resolution biostratigraphy (conodonts and graptolites) of the Upper Ordovician and Lower Silurian – evaluation of the Late Ordovician mass extinction. *Modern Geology*, **20**, 41–68.

ARMSTRONG, H. A. & OWEN, A. W. 2001. Terrane evolution of the paratectonic Caledonides of northern Britain. *Journal of the Geological Society, London*, **158**, 475–486.

ARMSTRONG, H. A. & OWEN, A. W. 2002. Euconodont diversity changes in a cooling and closing Iapetus Ocean. *In*: CRAME, J. A. & OWEN, A. W. (eds) *Palaeobiogeography and Biodiversity Change: the Ordovician and Mesozoic–Cenozoic Radiations*. Geological Society, London, Special Publications, **194**, 85–98.

ASTINI, R. A., BENEDETTO, J. L., & VACCARI, N. E. 1995. The early Paleozoic evolution of the Argentine Precordillera as a rifted, drifted and collided terrane – A geodynamic model. *Geological Society of America Bulletin*, **107**, 253–273.

BACHTADSE, V., PAVLOV, V. E., KAZANSKY, A. Y. & TAIT, J. 2000. Paleomagnetism of the Tuva Terrane (Russia) – evidence for low latitudes of the Siberian craton in Siluro-Devonian times. *Journal of Geophysical Research*, **105**(B6), 13,509–13,518.

BENGTSON, S. 1976. The structure of some Middle Cambrian conodonts, and the early evolution of conodont structure and function. *Lethaia*, **9**, 185–206.

BENTON, M. J. 1999. The history of life: databases in palaeontology. *In:* HARPER, D. A. T. (ed.) *Numerical Palaeobiology.* John Wiley & Sons, Chichester, 249–283.

BLIECK, A. & JANVIER, P. 1991. Silurian vertebrates. *Special Papers in Palaeontology*, **44**, 345–389.

BOCKELIE, T. & FORTEY, R. A. 1976. An early Ordovician vertebrate. *Nature*, **260**, 36–38.

CHEN, JUN-YUAN, HUANG, DI-YING & LI, CHIA-WEI. 1999. An early Cambrian craniate-like chordate. *Nature*, **402**, 518–522.

COCKS, L. R. M. 2000. The Early Palaeozoic geography of Europe. *Journal of the Geological Society, London*, **157**, 1–10.

COCKS, L. R. M. 2001. Ordovician and Silurian global geography. *Journal of the Geological Society, London*, **158**, 197–210.

COOPER, R. A. 1999. The Ordovician time scale – calibration of graptolite and conodont zones. *Acta Universitas Carolinae – Geologica*, **43**(1/2), 1–4.

DALLA SALDA, L. H., DALZIEL, I. W. D., CINGOLANI, C. A. & VARELA, R. 1992. Did the Taconic Appalachians continue into southern South America? *Geology*, **20**, 1059–1062.

DALZIEL, I. W. D. 1991. Pacific margins of Laurentia and East Antarctica–Australia as a conjugate rift pair: Evidence and implications for an Eocambrian supercontinent. *Geology*, **19**, 598–601.

DALZIEL, I. W. D. 1997. Neoproterozoic–Paleozoic geography and tectonics: Review, hypothesis, environmental speculation. *Geological Society of America Bulletin*, **109**, 16–42.

DENISON, R. H. 1967. Ordovician vertebrates from Western United States. *Fieldiana Geology*, **16**, 131–192.

DONOGHUE, P. C. J. 1998. Growth and patterning in the conodont skeleton. *Philosophical Transactions of the Royal Society of London, Series B*, **353**, 633–666.

DONOGHUE, P. C. J. & ALDRIDGE, R. J. 2001. Origin of a mineralised skeleton. *In:* AHLBERG, P. E. (ed.) *Major Events in Early Vertebrate Evolution: Palaeontology, Phylogeny, Genetics and Development.* Taylor & Francis, London, 85–105.

DONOGHUE, P. C. J. & SMITH, M. P. 2001. The anatomy of *Turinia pagei* (Powrie) and the phylogenetic status of the Thelodonti. *Transactions of the Royal Society of Edinburgh: Earth Sciences*, **92**, 15–37.

DONOGHUE, P. C. J., FOREY, P. L. & ALDRIDGE, R. J. 2000. Conodont affinity and chordate phylogeny. *Biological Reviews*, **75**, 191–251.

DROSER, M. L., BOTTJER, D. J., SHEEHAN, P. M. & MCGHEE, G. R. 2000. Decoupling of taxonomic and ecologic severity of Phanerozoic marine mass extinctions. *Geology*, **28**, 675–678.

ELLIOTT, D. K., BLIECK, A. R. M. & GAGNIER, P.-Y. 1991. Ordovician vertebrates. *Geological Survey of Canada, Paper*, **90–9**, 93–106.

ELVEVOLD, S., ESCHER, J. C., FREDERILSEN, K. S., *ET AL*. 2000. Tectonic architecture of the East Greenland Caledonides 72°–74°30' N. *Danmark og Grønlands Geologiske Undersøgelse Rapport*, **2000/88**.

ERDTMANN, B.-D., WEBER, B., SCHULTZE, H.-P. & EGENHOFF, S. 2000. A possible agnathan plate from the lower Arenig (Lower Ordovician) of south Bolivia. *Journal of Vertebrate Paleontology*, **20**, 394–399.

FOREY, P. L. 1995. Agnathans recent and fossil, and the origin of jawed vertebrates. *Reviews in Fish Biology and Fisheries*, **5**, 267–303.

FOREY, P. L. & JANVIER, P. 1993. Agnathans and the origin of jawed vertebrates. *Nature*, **361**, 129–134.

FOREY, P. L. & JANVIER, P. 1994. Evolution of the early vertebrates. *American Scientist*, **82**, 554–565.

FREEMAN, S. R., BUTLER, R. W. H., CLIFF, R. A. & REX, D. C. 1998. Direct dating of mylonite evolution: a multi-disciplinary geochronological study from the Moine Thrust Zone, NW Scotland. *Journal of the Geological Society, London*, **155**, 745–758.

GABBOTT, S. E., ALDRIDGE, R. J. & THERON, J. N. 1995. A giant conodont with preserved muscle tissue from the Upper Ordovician of South Africa. *Nature*, **374**, 800–803.

GAGNIER, P. Y., BLIECK, A. & RODRIGO, G. 1986. First Ordovician vertebrate from South America. *Geobios*, **19**, 629–634.

GANS, C. & NORTHCUTT, R. G. 1983. Neural crest and the origin of vertebrates: a new head. *Science*, **220**, 268–274.

GAUTHIER, J., KLUGE, A. G. & ROWE, T. 1988. Amniote phylogeny and the importance of fossils. *Cladistics*, **4**, 105–209.

GRAFFIN, G. 1992. A new locality of fossiliferous Harding Sandstone: evidence for freshwater Ordovician vertebrates. *Journal of Vertebrate Paleontology*, **12**, 1–10.

GRIFFITH, R. W. 1994. The life of the first vertebrates. *BioScience*, **44**, 408–417.

GUTMANN, W. F. 1967. Das Dermalskelett der fossilen 'Panzerfische' funktionell und phylogenetisch interpretiert. *Senckenbergiana Lethaea*, **48**, 277–283.

GUTMANN, W. F. & BONIK, K. 1983. Die Entstehung des Dermalskelettes: Biomechanische Grundbedingungen des fossilen Schuppenpanzers. *Senckenbergiana Lethaea*, **64**, 179–197.

HALLAM, A. 1994. *An Outline of Phanerozoic Biogeography*. Oxford University Press, Oxford.

HARTZ, E. B., ANDRESEN, A., HODGES, K. V. & MARTIN, M. W. 2001. Syncontractional extension and exhumation of deep crustal rocks in the east Greenland Caledonides. *Tectonics*, **20**, 58–77.

HIGGINS, A. K., LESLIE, A. G. & SMITH, M. P. 2001. Neoproterozoic–Lower Palaeozoic stratigraphical relationships in the marginal thin-skinned thrust belt of the East Greenland Caledonides: comparisons with the foreland in Scotland. *Geological Magazine*, **138**, 143–160.

HOFFMAN, P. F. 1991. Did the breakout of Laurentia turn Gondwana inside out? *Science*, **252**, 1409–1412.

HOLLAND, N. D. & CHEN, JUN-YUAN. 2001. Origin and early evolution of the vertebrates: new insights from advances in molecular biology, anatomy, and palaeontology. *Bioessays*, **23**, 142–151.

JANVIER, P. 1996*a*. *Early Vertebrates*. Oxford University Press, Oxford.

JANVIER, P. 1996*b*. The dawn of the vertebrates: characters versus common ascent in the rise of current vertebrate phylogenies. *Palaeontology*, **39**, 259–287.

KARATAJŪTĖ-TALIMAA, V. 1996. The Mongolepida: scale structure and systematic position. *Geobios Mémoire Spécial*, **19**, 35–37.

KARATAJŪTĖ-TALIMAA, V. 1997. *Lugalepis* – a new genus of elasmobranchs from Devonian of the western part of the main Devonian field. *Geologija*, **21**, 24–29.

KUMAR, S. & HEDGES, S. B. 1998. A molecular timescale for vertebrate evolution. *Nature*, **392**, 917–920.

MCKERROW, W. S., DEWEY, J. F. & SCOTESE, C. R. 1991. The Ordovician and Silurian development of the Iapetus Ocean. *In:* BASSETT, M.G. *et al.* (eds) *The Murchison Symposium. Proceedings of an International Conference on the Silurian System*. Special Papers in Palaeontology, **44**, 165–178.

MCKERROW, W. S., MAC NIOCAILL, C. & DEWEY, J. F. 2000. The Caledonian Orogeny redefined. *Journal of the Geological Society, London*, **157**, 1149–1154.

MAC NIOCAILL, C. 2000. A new Silurian palaeolatitude for eastern Avalonia and evidence for crustal rotations in the Avalonian margin of southwestern Ireland. *Geophysical Journal International*, **141**, 661–671.

MILLER, J. F. 1984. Cambrian and earliest Ordovician conodont evolution, biofacies, and provincialism. *In:* CLARK, D.L. (ed.) *Conodont Biofacies and Provincialism*. Geological Society of America, Special Papers, **196**, 43–68.

MOORES, E. M. 1991. Southwest U.S.–East Antarctica (SWEAT) connection: A hypothesis. *Geology*, **24**, 917–920.

NORELL, M. A. 1992. Taxic origin and temporal diversity: the effect of Phylogeny. *In:* NOVACEK, M. J. & WHEELER, Q. D. (eds) *Extinction and Phylogeny*. Columbia University Press, New York, 89–118.

P'AN, JIANG, LU, LIWU. & JI, SHU'AN 1996. A brief review of studies on palaeogeography of Middle Paleozoic vertebrates in China. *Journal of Southeast Asian Earth Sciences*, **13**, 185–190.

PURNELL, M. A. 2001. Scenarios, selection, and the ecology of early vertebrates. *In:* AHLBERG, P. E. (ed.) *Major Events in Early Vertebrate Evolution: Palaeontology, Phylogeny, Genetics and Development*. Taylor & Francis, London, 187–208.

RASMUSSEN, J. A. 1998. A reinterpretation of the conodont Atlantic Realm in the late Early Ordovician (Early Llanvirn). Proceedings of the Sixth European Conodont Symposium (ECOS VI). *Palaeontologica Polonica*, **58**, 67–77.

REPETSKI, J. E. 1978. A fish from the Upper Cambrian of North America. *Science*, **200**, 529–531.

RITCHIE, A. & GILBERT-TOMLINSON, J. 1977. First Ordovician vertebrates from the Southern Hemisphere. *Alcheringa*, **1**, 351–368.

SANSOM, I. J. 1996. *Pseudooneotodus*: a histological study of an Ordovician to Devonian vertebrate lineage. *Zoological Journal of the Linnean Society*, **118**, 47–57.

SANSOM, I. J., SMITH, M. P., ARMSTRONG, H. A. & SMITH, M. M. 1992. Presence of the earliest vertebrate hard tissues in conodonts. *Science*, **256**, 1308–1311.

SANSOM, I. J., SMITH, M. P. & SMITH, M. M. 1994. Dentine in conodonts. *Nature*, **368**, 591.

SANSOM, I. J., SMITH, M. P., SMITH, M. M. & TURNER, P. 1995. The Harding Sandstone revisited – a new look at some old bones. *Geobios*, **19**, 57–59.

SANSOM, I. J., SMITH, M. P. & SMITH, M. M. 1996. Scales of thelodont and shark-like fishes from the Ordovician. *Nature*, **379**, 628–630.

SANSOM, I. J., ALDRIDGE, R. J. & SMITH, M. M. 2000. A microvertebrate fauna from the Llandovery of South China. *Transactions of the Royal Society of Edinburgh: Earth Sciences*, **90**, 255–272.

SANSOM, I. J., SMITH, M. M. & SMITH, M. P. 2001. The Ordovician radiation of vertebrates. *In:* AHLBERG, P. E. (ed.) *Major Events in Early Vertebrate Evolution: Palaeontology, Phylogeny, Genetics and Development.* Taylor & Francis, London, 156–171.

SCOTESE, C. R. 1986. Phanerozoic reconstructions: a new look at the assembly of Asia. *University of Texas Institute for Geophysics Technical Report*, **66**, 1–54.

SCOTESE, C. R. & MCKERROW, W. S. 1990. Revised World maps and introduction. *In:* MCKERROW, W. S. & SCOTESE, C. R. (eds) *Palaeozoic Palaeogeography and Biogeography*. Geological Society, London, Memoirs, **12**, 1–21.

SHU, D.-G., LUO, H.-L., CONWAY MORRIS, S., ET AL. 1999*a*. Lower Cambrian vertebrates from south China. *Nature*, **402**, 42–46.

SHU, D.-G., CONWAY MORRIS, S., ZHANG, X.-L., CHEN, L., LI, Y. & HAN, J. 1999*b*. A pipiscid-like fossil from the Lower Cambrian of south China. *Nature*, **400**, 746–749.

SIMONETTA, A. M. & INSOM, E. 1993. New animals from the Burgess Shale (Middle Cambrian) and their possible significance in understanding the Bilateria. *Bollettino Zoologica*, **60**, 97–107.

SMETHURST, M. A., KHRAMOV, A. N. & TORSVIK, T. H. 1998. The Neoproterozoic and Palaeozoic palaeomagnetic data for the Siberian Platform: From Rodinia to Pangea. *Earth Science Reviews*, **43**, 1–24.

SMITH, M. M. 1991. Putative skeletal neural crest cells in early Late Ordovician vertebrates from Colorado. *Science*, **251**, 301–303.

SMITH, M. M. & SANSOM, I. J. 1997. Exoskeletal microremains of an Ordovician fish from the Harding Sandstone of Colorado. *Palaeontology*, **40**, 645–658.

SMITH, M. M., SANSOM, I. J. & SMITH, M. P. 1995. Diversity of the dermal skeleton in Ordovician to Silurian vertebrate taxa from North America: histology, skeletogenesis and relationships. *Geobios*, **19**, 65–70.

SMITH, M. P., BRIGGS, D. E. G. & ALDRIDGE, R. J. 1987. A conodont animal from the lower Silurian of Wisconsin, USA, and the apparatus architecture of panderodontid conodonts. *In:* ALDRIDGE, R. J. (ed.) *Palaeobiology of Conodonts*. Ellis Horwood, Chichester, 91–104.

SMITH, M. P., SANSOM, I. J. & REPETSKI, J. E. 1996. Histology of the first fish. *Nature*, **380**, 702–704.

SMITH, M. P., SANSOM, I. J. & COCHRANE, K. D. 2001. The Cambrian origin of vertebrates. *In:* AHLBERG, P.E. (ed.) *Major Events in Early Vertebrate Evolution: Palaeontology, Phylogeny, Genetics and Development.* Taylor & Francis, London, 67–84.

SOEHN, K. L. & WILSON, M. V. H. 1990. A complete, articulated heterostracan from Wenlockian (Silurian) beds of the Delorme Group, Mackenzie Mountains, Northwest Territories, Canada. *Journal of Vertebrate Paleontology*, **10**, 405–419.

SWEET, W. C. 1984. Graphic correlation of upper Middle and Upper Ordovician rocks of the United States Midcontinent. *In:* Bruton, D.L. (ed.) *Aspects of the Ordovician System*. Palaeontological Contributions of the University of Oslo, **295**, 23–35.

SWEET, W. C. 1988. *The Conodonta: Morphology, Taxonomy, Paleoecology, and Evolutionary History of a Long-extinct Animal Phylum*. Clarendon Press, Oxford.

THANH, TONG-DZUY, PHUONG, TA HOA, BOUCOT, A. J., GOUJET, D. & JANVIER, P. 1997. Vertébrés siluriens du Viêt-nam central. *Comptes rendus de l'Académie des sciences, Série II a*, **324**, 1023–1030.

THOMAS, W. A. & ASTINI, R. A. 1996. The Argentine Precordillera: a traveler from the Ouachita Embayment of North American Laurentia. *Science*, **283**, 752–757.

TORSVIK, T. H., TAIT, J., MORALEV, V. M., MCKERROW, W. S., STURT, B. A. & ROBERTS, D. 1995. Ordovician palaeogeography of Siberia and adjacent continents. *Journal of the Geological Society, London*, **152**, 279–287.

TORSVIK, T. H., SMETHURST, M. A., MEERT, J. G., ET AL. 1996. Continental break-up and collision in the Neoproterozoic and Palaeozoic – A tale of Baltica and Laurentia. *Earth Science Reviews*, **40**, 229–258.

VAN STAAL, C. R., DEWEY, J. F., MACNIOCAILL, C. & MCKERROW, W. S. 1998. The Cambrian–Silurian tectonic evolution of the northern Appalachians and British Caledonides: history of complex west and southwest Pacific-type segment of Iapetus. *In:* BLUNDELL, D. J. & SCOTT, A. C. (eds) *Lyell: the Past is the Key to the Present*. Geological Society, London, Special Publications, **143**, 199–242.

WAGNER, P. J. 1998. Phylogenetic analysis and the quality of the fossil record. *In:* DONOVAN, S. K. and PAUL, C. R. C. (eds) *The Adequacy of the Fossil Record*. John Wiley & Sons, Chichester, 165–187.

WALCOTT, C. D. 1892. Preliminary notes on the discovery of a vertebrate fauna in Silurian (Ordovician) strata. *Bulletin of the Geological Society of America*, **3**, 153–171.

WANG, N. Z. 1995. Silurian and Devonian jawless craniates (Galeaspida, Thelodonti) and their habitats in China. *Bulletin du Muséum national d'Histoire naturelle, Paris*, **17**, 57–84.

WILSON, J. T. 1966. Did the Atlantic close and then re-open? *Nature*, **211**, 676–681.

YOUNG, G. C. 1991. Devonian vertebrate distribution patterns and cladistic analysis of palaeogeographic hypotheses. *In:* MCKERROW, W. S. & SCOTESE, C. R. (eds) *Palaeozoic Palaeogeography and Biogeography*, Geological Society, London, Memoirs, **12**, 243–255.

YOUNG, G. C. 1993. Vertebrate faunal provinces in the Middle Palaeozoic. *In:* LONG, J. A. (ed.) *Palaeozoic Vertebrate Biostratigraphy and Biogeography*. Bellhaven Press, London, 293–323.

YOUNG, G. C. 1997. Ordovician microvertebrate remains from the Amadeus Basin, central Australia. *Journal of Vertebrate Paleontology*, **17**, 1–25.

Euconodont diversity changes in a cooling and closing Iapetus Ocean

H. A. ARMSTRONG[1] & A. W. OWEN[2]

[1]*Department of Geological Sciences, University of Durham, Durham DH1 3LE, UK (e-mail:* H.A.Armstrong@durham.ac.uk*)*

[2]*Division of Earth Sciences, University of Glasgow, Gregory Building, Glasgow G12 8QQ, UK*

Abstract: Constrained seriation of euconodont generic presence–absence matrices for four time slices between the late Llanvirn and late Llandovery provides a qualitative method for defining shelf and oceanic biofacies, reconstructing biofacies architectures and analysing biodiversity within a regional context.

We propose many North Atlantic Province taxa had a pelagic mode of life and ranged widely across the Iapetus Ocean. Oceanic biofacies are considered to reflect water mass structure. Changes in vertical distribution of one such biofacies (including *Amorphognathus* and *Spinodus*) suggest adaptation to cold, nutrient-rich, oxygen-poor upwelling water. Biofacies distributions suggest that upwelling occurred along the Avalonian margin throughout the Ashgill, but was only initiated along the Laurentian margin immediately prior to the Hirnantian glacial maximum.

Clade diversities and trajectories differ between biofacies and latitudes, reflecting different causal mechanisms. In Laurentia, diversity fell in the early Ashgill, coincident with the onset of ocean cooling. Diversity declined in Avalonia when the microcontinent drifted into tropical latitudes. The stability of euconodont biofacies architecture during the Late Ordovician indicates that global cooling and plate reorganization had a low palaeoecological impact despite decreases in alpha and beta diversity.

Over the past 20 years the differential success of clades has been attributed to either intrinsic properties of the clades themselves (e.g. competitive ability or origination rates) or to changes in their biogeographical, environmental and palaeoecological context (Erwin 1998; Jablonski 1998). If intrinsic properties of the clade are the primary control on diversity then clade diversity is considered independent from geological setting. Attempting to correlate changes in clade diversity with independently established changes in environmental conditions would provide a test of these competing hypotheses.

Two recent studies on Palaeozoic clades support the hypothesis that diversity reflects changing palaeoenvironmental conditions at a regional scale. Miller (1997) compared the early Ordovician radiation in six palaeocontinental regions and found regional differences in the evolutionary history of trilobites, brachiopods and molluscs. Miller & Mao (1995) showed that in a generic dataset, corrected for included species number, there was a correlation between Ordovician diversity trends and the extent of siliciclastic sedimentation, considered by them to be a proxy for mountain-building activity. However, aggregate map areas of marine tectonic provinces through the Ordovician indicate a slight increase in the areal extent of the rock record during the period of maximum diversity decline (Miller & Mao 1995). There is therefore no correlation between decrease in diversity and total preserved rock record during the Late Ordovician.

The changing patterns of Phanerozoic biotic diversity have been recognized almost entirely from synoptic global datasets that, by their nature, average very different palaeoenvironmental signals (see Miller 2000 and references therein). Detailed studies of local successions can be ecologically well constrained but raise questions about the quality of the record, particularly with respect to preservation, restricted sampling and the response of the biota to rapidly changing substrates. Regional comparative studies provide the best compromise of taxonomic scope and palaeoenvironmental acuity (Sepkoski 1993; Miller & Mao 1995; Miller 2000).

The late Ordovician was a time of major global environmental change with the late Ordovician glaciation punctuating a period of prolonged global greenhouse climate (Hambrey 1985; Berner 1990, 1992; Crowley &

From: CRAME, J. A. & OWEN, A. W. (eds) 2002. *Palaeobiogeography and Biodiversity Change: the Ordovician and Mesozoic–Cenozoic Radiations*. Geological Society, London, Special Publications, **194**, 85–98.
0305-8719/02/$15.00

Baum 1991). Sedimentological evidence from Dob's Linn, located on the Laurentian slope, indicates that at least in low latitudes ocean cooling persisted from the early Ashgill, with the glacial maximum ranging from the late Rawtheyan to mid-Hirnantian (Armstrong & Coe 1997). This glaciation has been considered a major causal factor in the late Ordovician mass extinction (see reviews in Brenchley 1988; Brenchley *et al.* 1995*a*,*b*; Armstrong 1995; Owen & Robertson 1995). At the same time plate configurations were changing dramatically associated with the closure of the Iapetus Ocean (Scotese & McKerrow 1990; van Staal *et al.* 1998).

We have conducted an analysis of Late Ordovician euconodont generic diversity in the circum-Iapetus region. Different diversity trajectories are found in different biofacies in each of low-latitude Laurentia and mid-temperate Avalonia and Baltica, and can be attributed to different causal mechanisms. Laurentian diversity appears to have been moderated by changing climate, associated with the onset of the latest-Ordovician glaciation, whereas Avalonian diversity reflects the change from temperate to tropical conditions as the plate drifted northwards. This analysis thus supports the hypothesis that changing palaeoenvironmental setting is the primary influence on clade diversity within the euconodonts.

The method we have applied allows regional-scale diversity patterns to be elucidated from existing presence–absence data and is thus widely applicable. The generic associations established by the method also provide an opportunity to analyse biofacies architecture and stability in a number of different regional settings and hence examine the palaeoecological impact of global-scale events (see also Droser *et al.* 1997, 2000).

Method

Presence–absence matrices for genera have been compiled from sections in Laurentia and Avalonia for the upper *serra* and *anserinus* biozones (Llandeilian–Aurelucian), *velicuspis* Chron (equivalent to the upper *superbus* Biozone in the British biostratigraphical scheme), upper *ordovicicus* Biozone and the *celloni* Biozone. Genera are considered taxonomically stable and are more likely to be identified correctly. Each genus comprises less than five species; several are monospecific. The results of a cladistic analysis of euconodont genera are not currently available but many are likely to be paraphyletic (P. C. J. Donoghue, pers. comm., 2001). This does not detract from our analysis as paraphyletic taxa are more likely to have had narrower palaeoecological ranges than monophyletic taxa, in the same way that many paraphyletic fish genera have highly restricted ecological distributions (Sepkoski & Kendrick 1993; Kemp 1999; Miller 2000).

Two major Ordovician conodont faunal provinces have previously been identified in the Iapetus region: the Midcontinent Province, including Laurentia, Siberia and parts of China; and the North Atlantic Province, including Baltoscandia and eastern Laurentia (Barnes *et al.* 1973; Sweet & Bergström 1974, 1984). Cluster analysis of Jaccard and Dice coefficients for euconodont faunas of the lower Llanvirn indicate that the faunal similarities between Baltoscandia and eastern Laurentia are due entirely to the presence of a cosmopolitan deep-water fauna, the *Protopanderodus–Periodon* Biofacies (comparable to OB2 herein) (Rasmussen 1998). Once this biofacies is removed from the analysis then the North Atlantic Province is restricted to Baltoscandia and neighbouring areas. Genera considered representative of the North Atlantic Province (including coniform taxa) and the Midcontinent Province have therefore been separated in the presence – absence matrix for each region.

The original matrices were compiled with the localities arranged with an onshore–offshore distribution (Figs 1–5). The matrices were subjected to manual constrained seriation with the onshore to offshore distribution of localities maintained (Brower & Kile 1988). The result therefore reflects an offshore deepening palaeoecological gradient. Generic biofacies, identified as blocks of taxa within the matrix, were named, and the included genera were coded for biofacies for each time slice. The assumption was made that a genus was more likely to stay within the same biofacies in the subsequent time slice. Anomalous occurrences in the initial coding were corrected by *a posteriori* revision of the seriated matrix. The generic coding is illustrated in Figure 6; biofacies diversity counts were made from this matrix for each time slice and region. The juxtaposition of biofacies at each locality enables the construction of a biofacies architecture diagram (Figs 4b, 5b, 7) for each time slice and region. Biofacies architecture can then be compared through time. There is no correlation of the shelf biofacies with the described shelly invertebrate biofacies such as the substrate-controlled palaeocommunities of the late Llanvirn – early Caradoc of the Welsh basin summarized by Lockley (1983).

Laurentia (LL-C)	Oklahoma #1	Oklahoma #2	Alabama	Girvan	Culzean	Tweed	
Ansella	1	1	1	1	1	0	
Walliserodus	1	1	1	1	1	1	
Panderodus	1	1	1	1	1	1	
Drepanoistodus	1	1	1	1	1	1	
Belodina	1	1	1	1	1	1	
Eoplacognathus	1	0	1	0	1	1	1b
Baltoniodus	0	1	1	0	1	1	1b
Periodon	0	0	1	1	1	1	
Protopanderodus	0	0	1	1	1	1	
Pygodus	0	0	1	1	1	1	
Dapsilodus	0	0	1	0	1	1	
Strachanognathus	0	0	0	1	1	1	
Amorphognathus	0	0	0	0	1	1	
Spinodus	0	0	0	0	1	1	
Cornuodus	0	0	0	0	1	1	
Pteracontiodus	1	1	0	0	0	0	
Curtognathus	1	0	0	0	0	0	
Appalachignathus	1	0	0	0	0	0	
Erismodus	1	0	0	0	1	0	
Oulodus	1	0	0	0	0	0	
Oneotodus?	1	0	0	0	0	0	
Plectodina	1	0	0	0	0	0	
Stauferella	1	0	0	0	0	0	
Cahabagnathus	1	0	1	1	1	0	
Phragmodus	1	1	1	0	0	0	
Coelocerodontus	0	0	1	1	0	0	
Acontiodus	0	0	1	1	0	0	
Erraticodon	0	0	1	1	0	0	
Prattognathus	0	0	1	0	0	0	
Ozarkodina	0	0	1	0	0	0	
Leptochirognathus	0	0	1	0	0	0	

velicuspisChron	2s FH	2s BI	5s WW	8s F	10s KM	11s KC	12s KE	12s KG	12s M	13s KD	15s MA	16s VH	16s VA	16s W	16s BU	18s CO	18s C	18s BV	19s CI	20s MB	20s T	
Belodina	1	1	1	1	1	1	1	1	1	1	1	1	1	1	0	1	1	1	1	1	1	
Culumbodina	1	1	1	1	1	1	0	1	1	1	0	1	1	1	0	0	0	1	0	0	0	
Drepanoistodus	1	1	1	1	1	1	1	1	1	1	1	1	1	1	1	1	1	1	1	1	1	
Panderodus	0	1	1	1	1	1	1	1	1	1	1	0	0	1	1	1	0	0	0	0	0	
Dapsilodus	1	1	0	0	0	0	0	0	0	0	1	1	1	0	1	1	1	0	1	0	1	
Protopanderodus	1	0	1	0	0	0	0	0	1	0	1	1	1	0	1	1	1	1	1	0	1	
Periodon	0	0	1	0	1	1	0	1	1	1	1	1	1	1	0	1	1	0	0	0	1	
Amorphognathus	0	0	0	0	0	0	1	0	1	1	1	1	1	1	1	1	1	1	1	0	1	
Plectodina	1	1	0	1	0	0	0	0	1	0	1	1	0	0	0	0	0	0	0	0	0	
Plegagnathus	1	1	0	0	0	0	0	0	0	0	0	0	0	0	0	0	0	0	0	0	0	
Oulodus																						
Pseudobelodina	1	1	1	1	0	1	1	1	1	1	1	1	1	0	0	1	0	0	0	0	1	
Stauferella	1	0	1	1	0	0	0	0	1	0	0	1	1	1	1	0	1	1	1	1	1	
Aphelognathus	1	1	0	1	0	0	0	0	1	0	0	0	0	0	1	0	0	1	1	1	0	
Phragmodus	0	1	0	0	0	1	1	1	1	1	1	1	1	1	1	1	1	1	1	1	1	
Icriodella	0	1	0	0	0	0	0	0	0	0	0	1	1	0	0	1	1	1	1	0	1	
Coelocerodontus	0	1	0	0	0	0	0	0	0	0	0	0	1	0	1	1	0	0	0	0	1	
Rhodesgnathus	0	0	0	0	0	0	0	0	0	0	0	0	1	0	0	0	1	0	1	0	1	?
Bryantodina?	0	0	0	0	0	0	0	0	0	0	0	0	0	0	0	0	0	0	1	1	0	SB1
Rhipidognathus	0	0	0	0	0	0	0	0	0	0	0	0	0	0	0	0	0	0	1	0	0	

ordovicicus	Anticosti	
Belodina	1	
Panderodus	1	
Drepanoistodus	1	
Walliserodus	1	
Amorphognathus	1	
Dapsilodus	1	
Birksfeldia	1	
Aphelognathus	1	
Pseudooneotodus	1	?
Stauferella	1	
Oulodus	1	
Plectodina	1	
Plegagnathus	1	
Ozarkodina?	1	
Phragmodus	1	?

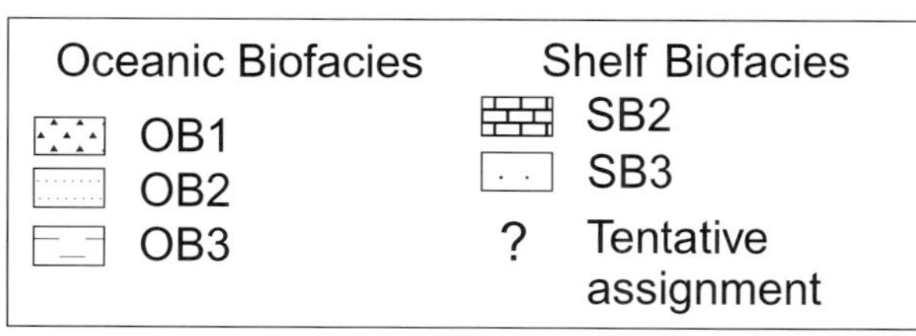

Fig. 1. Constrained seriation matrix for the Laurentian shelf and marginal terranes for each of the Ordovician time slices described in the text. Source of locality data: Oklahoma from Bauer (1994); Alabama from Bergström (1990); Girvan from Bergström (1990); Culzean from Armstrong (2000); Tweed from Armstrong (1997); *velicuspis* Chron data from Sweet & Bergström (1984); *ordovicicus* Biozone data from Nowlan & Barnes (1981) and Radcliffe (1998). Localities have been located onshore (to the left) to offshore with reference to palaeogeographical maps in Scotese & McKerrow (1990) and Bevins *et al.* (1992). Shaded boxes indicate a presence; biofacies for generic association in Llanvin–Caradoc and *velicuspis* Chron have been defined by blocks of genera in the matrix. Biofacies were allocated for *ordovicicus* Biozone genera based on generic occurrences in previous time slices (see Fig. 6 and text for explanation).

Wales (Ll-C)	Bullhill Gutter Stream S. Shropshire (1)	Ffairfach Railway (2)	Keepers Lodge (3)	Lampeter Valley (4)	Dynevor Castle (5)	Bryn-banc Qy Narbeth (6)	Abereiddi Bay (7)	Biofacies
Drepanoistodus	1	1	1	1	1	1	1	
Panderodus	1	1	1	1	1	1	0	
Eoplacognathus	0	1	0	1	0	1	0	OB1b
Baltoniodus	0	1	1	1	0	1	0	OB1b
Amorphognathus	0	1	1	1	1	1	1	
Icriodella	0	0	0	1	1	1	1	
Plectodina	1	1	1	1	1	1	1	?
Prioniodus	1	0	0	0	0	0	0	
'Erismodus'	0	0	1	1	1	1	0	
Chirognathus	0	0	1	0	0	0	0	
Complexodus	0	0	0	1	0	1	0	
Phragmodus	0	0	0	0	0	0	1	

Wales (velicuspis)	Robeston Wathen (Haverfordwest) (8)	Birdshill Qy (Llandeilo) (9)	Crug Farm (Crug Lst Llandeilo) (10)	Gelli Farm (Dolhir Lst Berwyn Hills) (11)
Walliserodus	?	0	0	0
Drepanoistodus	1	1	1	?
Panderodus	1	1	1	1
Amorphognathus	1	1	1	1
Icriodella	1	1	1	1
Protopanderodus	0	1	0	0
Birksfeldia	1	0	0	0
Eocarniodus	1	1	0	0
Aphelognathus	1	0	1	0
Pseudooneotodus	1	1	0	0
?Ozarkodina	0	1	1	1
Rhodesgnathus	0	1	1	0
Prioniodus	0	1	0	0

Wales (ordovicicus)	Pistyll Gwyn Waterfalls (Bala) (12)	Ty-draw Hill Qy (Berwyn Hills) (13)	Llangower (Rhiwlas Lst Bala) (14)
Panderodus	1	1	1
Walliserodus	1	0	0
Amorphognathus	1	1	1
Icriodella	1	1	1
Hamarodus	1	0	1
Scabbardella	1	0	0
Protopanderodus	0	1	0
Birksfeldia	1	0	1
Eocarniodus	1	0	0
Ozarkodina?	1	0	0
Phragmodus	0	1	0

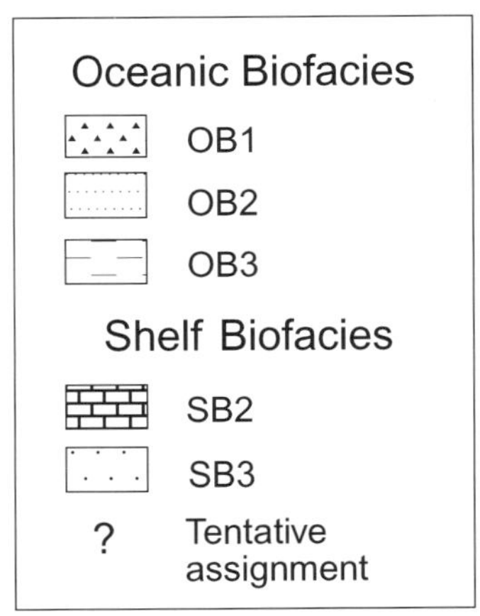

Fig. 2. Constrained seriation matrix for the Avalonian shelf (Wales) for each of the time slices described in the text. Source of locality data: (1) Savage & Bassett (1985); (2–7) Bergström & Orchard (1985), Bergström *et al.* (1987); (8) Savage & Bassett (1985); (9) Savage & Bassett (1985); (10) Orchard (1980), Bergström & Orchard (1985), Savage & Bassett (1985); (11) Bergström & Orchard (1985), Savage & Bassett (1985); (12–14) Orchard (1980), Bergström & Orchard (1985). Localities have been located onshore (to the left) to offshore with reference to Bevins *et al.* (1992). Shaded boxes indicate a presence; biofacies for generic associations have been defined by blocks of generic associations and occurrences in previous time slices (see Fig. 6 and text for explanation).

Results

Biofacies architecture

Six biofacies have been identified. Oceanic Biofacies (OB1, OB2 and OB3) contain pelagic taxa, which range onshore to offshore or in the data matrix have ranges that terminate onshore. These biofacies include taxa considered to represent the North Atlantic Province (NAP) plus many coniform genera. NAP taxa are known in both low and mid-temperate latitude sections

Lakesman(velicuspis)	Murthwaite Cautley (15)	Rawthey (16)	Taythes Cautley (17)	Lake District (18)	Keisley (19)
Panderodus	1	1	1	1	1
Scabbardella	1	0	1	1	1
Amorphognathus	1	1	1	1	1
Icriodella	1	0	0	1	1
Belodina	0	0	0	1	0
Belodella	0	0	0	1	0
Drepanoistodus	0	0	0	0	1
Walliserodus	0	0	0	0	1
Hamarodus	0	0	0	0	1
Dapsilodus	0	0	0	0	1
Protopanderodus	0	0	0	0	1
Strachanognathus	0	0	0	0	1
Birksfeldia	1	1	1	1	1
Aphelognathus	1	1	1	1	1
Plectodina	1	0	1	0	1
Ozarkodina	1	0	1	0	0
Rhodesgnathus	1	0	0	1	0
Pseudooneotodus	0	1	1	1	0
Staufferella	0	0	0	?	0

Lakesman (Raw.)	R. Rawthey (20)	Lake Disrict (21)	High Sprint Gill (22)	BirksfieldBeck (23)	Backside Beck (24)
Panderodus	1	1	1	1	1
Scabbardella	1	1	1	1	1
Amorphognathus	1	1	1	1	1
Walliserodus	1	1	0	1	0
Hamarodus	1	1	0	0	0
Drepanoistodus	0	1	1	1	1
Protopanderodus	0	1	1	1	0
Strachanognathus	0	1	1	1	0
Birksfeldia	1	1	0	1	1
Eocarniodus	0	1	1	1	1

Oceanic Biofacies	Shelf Biofacies
OB1	SB2
OB2	SB3
OB3	? Tentative assignment

Fig. 3. Constrained seriation matrix for the Avalonian shelf (Lakesman Terrane) for each of the time slices described in the text. Locality data from Orchard (1980). Localities have been located onshore (to the left) to offshore with reference to Bevins *et al.* (1992). Shaded boxes indicate a presence; biofacies for generic associations have been defined by blocks of generic associations and occurrences in previous time slices (see Fig. 6 and text for explanation).

and 25% of genera recorded in Laurentia during the *velicuspis* Chron belong to this province. Many of these genera are monospecific or contained few species and are found to range from inner to outer shelf locations indicating their distribution was independent of benthic facies.

Shelf Biofacies (SB1, SB2 and SB3) contain nektobenthic taxa that have a distribution in the matrices that terminate offshore and form discrete blocks within each matrix. SB1 occurs only in the *velicuspis* Chron in Laurentia and includes ?*Bryantodina* and *Rhipidognathus*, genera commonly found in supratidal, or even hypersaline environments (Armstrong 1990) and this biofacies is thought to be representative of local hypersaline conditions. Increasing biofacies number indicates greater distance offshore and hence deeper-water setting.

The relative positions of OB1, OB2 and SB2, SB3 remain stable across the Iapetus Ocean through the time slices, but differences do occur in the relative position of OB3 and the juxtaposition of oceanic and shelf biofacies. On the Avalonian and Baltic margins (Fig. 7), OB3 consistently lies in an intermediate position between (or within) OB1 and OB2 from the *velicuspis* Chron onwards. OB3 moved into this intermediate position in Laurentia during the Rawtheyan, just prior to the Hirnantian global eustatic sea-level fall (Brenchley *et al.* 1995*b*; Crowley & Baum 1991). At the same time the OB1–OB2 boundary across the Iapetus Ocean moved from a position coincident with SB3 to within SB2.

Constrained seriation enables biofacies to be defined from presence–absence data and based upon the overlapping ecological ranges of taxa. This allows a generalized biofacies architecture to be reconstructed but lacks the detailed lithological and palaeoecological data required for the precise definition of the environmental criteria constraining each biofacies. Such data are invariably lacking in euconodont monographs and historical descriptions of collections. Our analysis does however include Laurentian sections utilized by Sweet & Bergström (1984) in their classic study of Late Ordovician

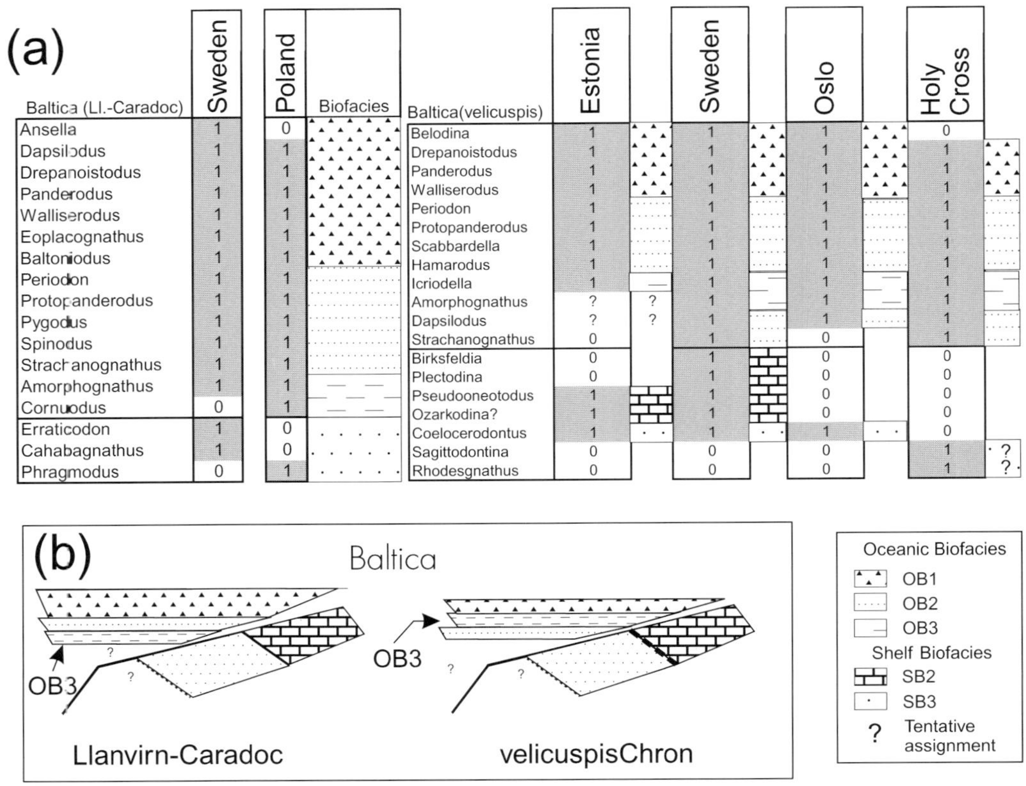

Fig. 4. (**a**)Constrained seriation matrices for the Baltica shelf for each of the time slices described in the text. Source of locality data for Estonia from Sweet & Bergström (1984); Sweden from Bergström (1990); Poland from Dzik (1994); Oslo from Smith (1999). Balto-Scandian locations are constrained onshore to offshore from the confacies belts proposed by Jaanusson (1976). (**b**) Biofacies architecture diagrams; note the intermediate position of OB3 in the *velicuspis* Chron data.

euconodont biofacies. They defined biofacies based upon cluster analysis assuming *a priori* that all euconodonts were pelagic. Their approach clusters localities based on included euconodonts, grouping both pelagic and nektobenthic taxa. Constrained seriation enables the separation of pelagic and nektobenthic taxa and our reinterpretation of only North Atlantic Province and coniform taxa as pelagic makes direct comparison with this work difficult. For example SB2 includes genera common with the *Aphelognathus–Oulodus* and *Plectodina* Biofacies of Sweet & Bergström (1984), SB3 has elements in common with the *Phragmodus–Icriodella–Plectodina* Biofacies though is not a direct correlative. OB2 has genera in common with the *Hamarodus-Dapsilodus-Scabbardella* (HDS) Biofacies and OB3 has taxa in common with the *Amorphognathus* Biofacies, but neither is directly comparable.

Jaanusson & Bergström (1980) recognized three confacies belts in the Middle Ordovician of the Appalachians, based upon shelly faunas and euconodonts. Their inshore Lee and Tazewell belts contain almost identical euconodont faunas including *Belodella*, *Belodina*, *Appalachignathus*, *Curtognathus*, *Erismodus*, *Phragmodus* and *Plectodina*. This fauna contains a mixture of OB1 (coniform) and SB2 (non-coniform) taxa. The offshore Blount Belt contains the euconodonts of OB2, plus *Protopanderodus* (previously *Cordylodus)* and '*Polyplacognathus*'. They also noted the similarity at species level of their Blount Belt fauna with Baltoscandian faunas of the same age and the horizontal uniformity of this fauna across various confacies belts. Both of these observations support our contention that these taxa were pelagic.

Stouge & Rasmussen (1996) identified the HDS Biofacies in the lower Ashgill of Bornholm and concluded that widespread geographical distribution indicated that this faunal assemblage

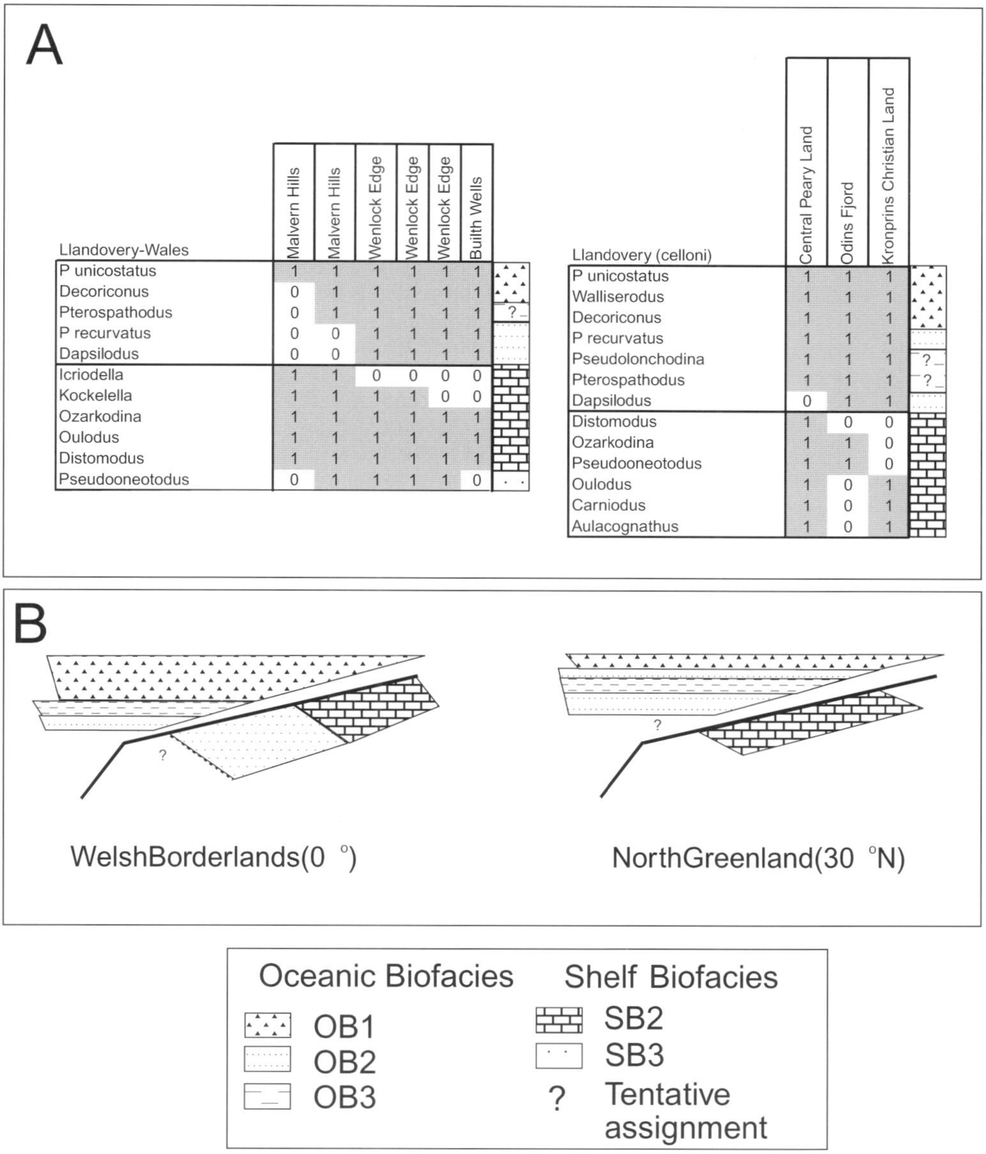

Fig. 5. (**A**) Constrained seriation matrices and (**B**) biofacies architecture diagram for *celloni* Biozone euconodont faunas of the Welsh Borderland (Avalonia) and North Greenland (northern Laurentia) shelves. Welsh data from Aldridge & Mabillard (1981); North Greenland data from Armstrong (1990). *Pterospathodus* and *Pseudolonchodina* are only tentatively assigned to OB3. *Ozarkodina*, *Oulodus* and *Distomodus* though ranging across the Welsh shelf have their acmes in nearshore settings (Aldridge & Mabillard 1981, fig. 1.4).

was indifferent to bottom conditions and occupied a deep-shelf to marginal setting.

Diversity patterns: Laurentia

Generic diversity patterns are illustrated in Figure 8a, b. Higher diversity is found in SB2 and OB1, in the inner shelf and upper water column. By the early Ashgill diversity had declined in all biofacies and low diversity continued into the Silurian. Inner shelf SB2 biofacies shows a slight increase in diversity in the upper *ordovicicus* Biozone with the return of *Oulodus* and *Plegagnathus* but was further

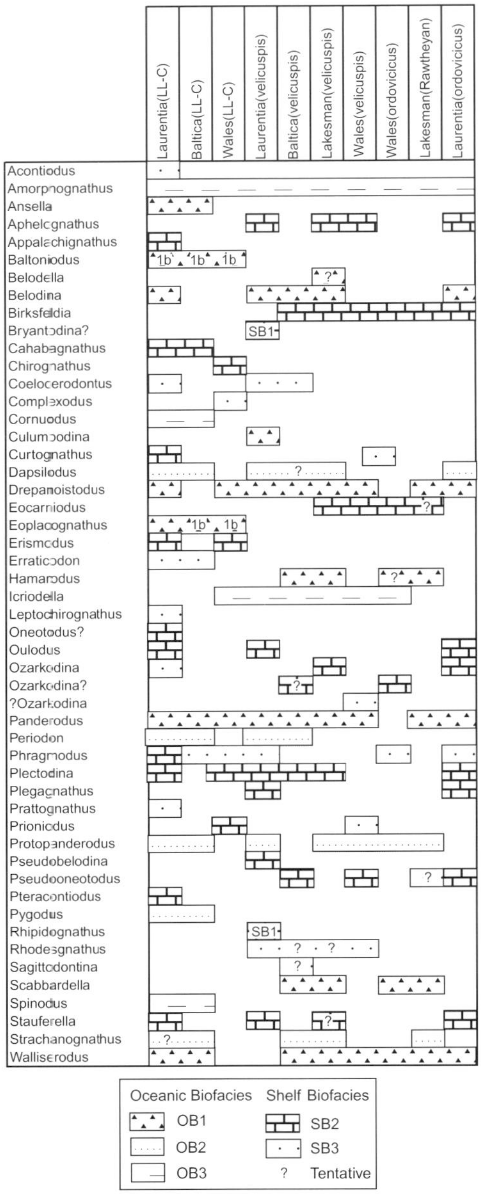

Fig. 6. Generic biofacies for each Ordovician time slice. The assumption was made that a genus was more likely stay within the same biofacies in the subsequent time slice. Anomalous occurrences in the initial coding were corrected by *a posteriori* revision of the seriated matrix, ? is retained for occurrences where this could not be achieved.

reduced into the Early Silurian. This increase may be a sampling artefact. The Ordovician decline in diversity continued for longer in the deeper-water SB3 and OB2 biofacies.

Diversity patterns: Avalonia

Generic diversity patterns are illustrated in Figure 8c, d. Higher diversity is found in SB2 and OB1, in the inner shelf and upper water column, as found in the Laurentia data. Diversity declined in all biofacies during the mid-Ashgill, coincident with the northward drift of Avalonia in tropical latitudes (Fig. 9). Low diversity continued into the early Silurian in oceanic biofacies whilst diversity increased in shelf faunas. Shelf faunas in the Early Silurian contained a mixture of incumbents (*Icriodella* and *Ozarkodina*), recruits (including *Oulodus*), which had an evolutionary origin in the Laurentian inner shelf, plus *Kockelella* and *Distomodus*, new genera with cryptic ancestry. Diversity trajectories in shelf and oceanic biofacies are parallel and the decline in diversity continued for longer in the oceanic biofacies, implying the causal mechanism persisted for longer in deeper water and affected taxa in all habitats.

Diversity patterns: Baltica

Generic diversity (Fig. 8e, f) is low in the Baltic succession, comparable to that in Avalonia, and the paucity of data makes interpretation difficult. SB2 and SB3 biofacies show slight rises in diversity into the early Ashgill, OB1 remains static and OB2 shows a slight decline. Llandovery diversity is higher than in the early Ashgill in shelf biofacies and is lower in oceanic biofacies. Diversity patterns in Baltica most closely compare with those from Avalonia.

Diversity in the early Silurian

Our analysis is restricted to the *celloni* Biozone of Avalonia (Welsh Borderland data from Aldridge & Mabillard 1981) and a comparative section from northern Laurentia (North Greenland data from Armstrong 1990; though not part of the Iapetus margin at this time these data are taken as typical for the Laurentian shelf). A full review of Silurian successions is beyond the scope of this paper and the Iapetus Ocean was virtually closed by the Early Silurian (Armstrong & Owen 2001). Oceanic and shelf biofacies identical to those in the Ordovician can be identified in this limited dataset (Fig. 5). Oceanic biofacies comprise predominantly coniform taxa. *Pterospathodus* and *Pseudolonchodina* are tentatively assigned to OB3 as both range across the Greenland shelf. *Pseudolonchodina* is not present in the Welsh dataset. *Ozarkodina*, *Oulodus* and *Distomodus* also

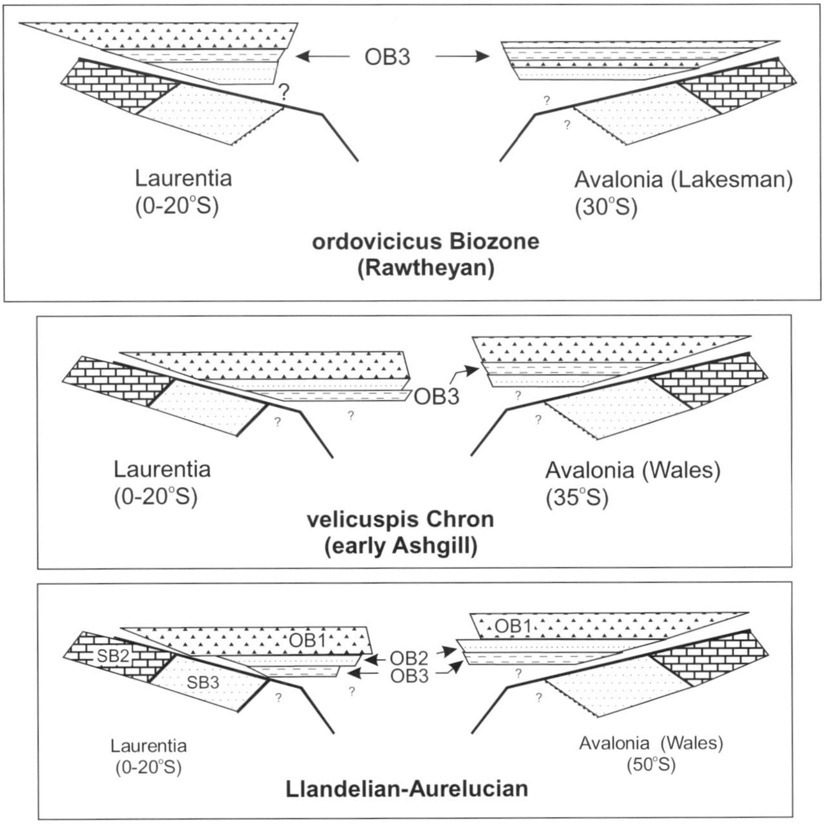

Fig. 7. Biofacies architecture diagrams for Laurentia and Avalonia at each of the Ordovician time slices. See Figure 1 for key.

range across the Welsh shelf but Aldridge & Mabillard (1981) recognized these taxa had their acmes in nearshore settings and declined rapidly in abundance offshore. This pattern suggests that these genera included wide-ranging, nektobenthic, ecological generalists that preferred nearshore habitats. *Icriodella* is restricted to SB2 in the Silurian and OB3 in the Ordovician. This suggests either a change from a pelagic mode of life, or that Silurian and Ordovician *Icriodella* are not congeneric. Silurian nektobenthic genera appear to have much broader ecological ranges than their Ordovician counterparts.

Interpretation

Biofacies architecture

In the modern ocean planktonic and pelagic species inhabit depth-related water masses defined by oxygen content, salinity, temperature and nutrient availability, comparable to biofacies in continental shelf settings, though often of much larger geographical extent. Water mass character is partly set by the shape of the basin, the direction of surface winds and patterns of evaporation and precipitation over the oceans (Emery & Meincke 1986). In general terms temperature and salinity fall with depth whilst nutrients increase (see review in Norris 2000). Specific water masses and their included fauna can move both vertically and horizontally depending upon oceanography and climatic factors, and are divided by physiochemical surfaces along which species or prey congregate for growth or reproduction (Norris 2000). Vertical movements of water masses occur as a result of changes in temperature and density and are most marked in thermally stratified oceans in continental margin upwelling zones. The structure of the upwelling zone off the SW

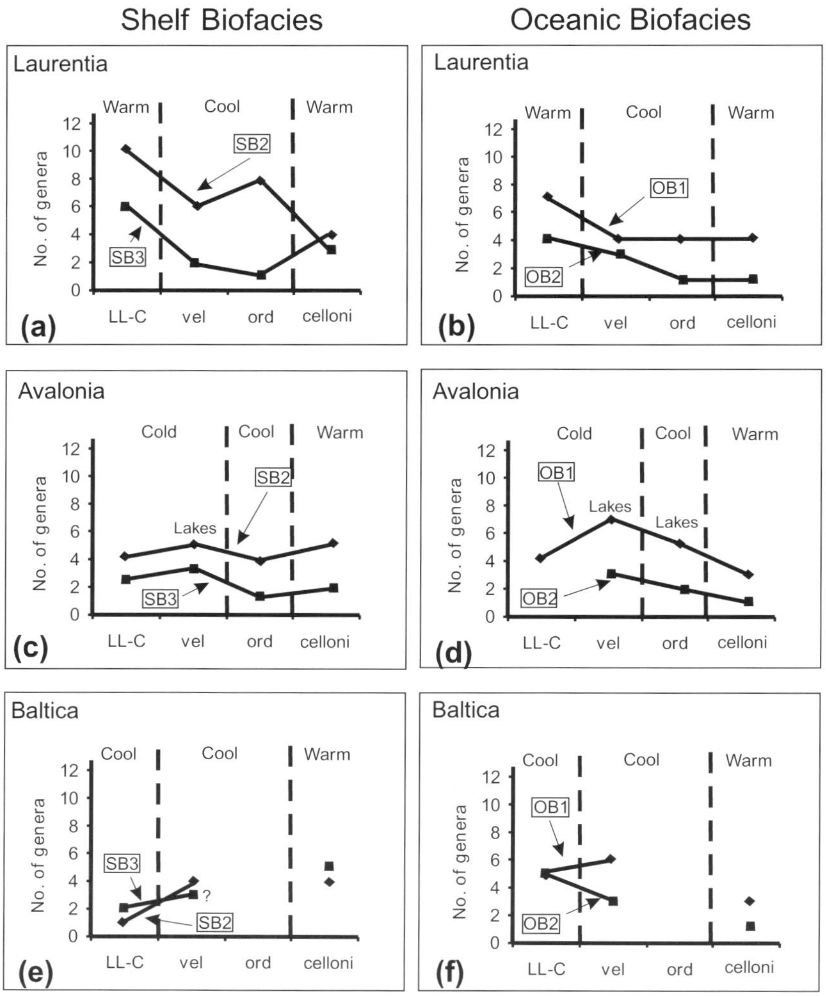

Fig. 8. Plots of maximum recorded generic diversity for Laurentia (**a**, **b**), Avalonia (**c**, **d**) and Baltica (**e**, **f**) for biofacies through time. Ocean temperature for Laurentia is known directly from stable oxygen isotope data (Brenchley *et al.* 1995*b*). Global cooling was initiated in the early Ashgill (Armstrong & Coe 1997) with the glacial maximum in the late Ashgill (Hirnantian). Temperatures for Avalonia and Baltica are inferred from palaeogeographical reconstruction and the drift northwards of Avalonia into the cooling tropics in the mid-Ashgill (Fig. 9). Abbreviations: LL-C, Llanvirn–Caradoc; vel, *velicuspis* Biozone; ord, *ordovicicus* Biozone; celloni, *celloni* Biozone.

African shelf, associated with the cold Benguela Current, is illustrated in Figure 10. Here, upwelling is driven by persistent offshore winds that skim off warmer surface water allowing cold, nutrient-rich, oxygen-poor subsurface water to ascend from intermediate depth (Demaison & Moore 1980; Fig. 10). In comparison with this modern analogue we propose that the oceanic biofacies identified in our analysis represent water-mass-restricted faunas and the distribution of OB3 biofacies reflects an area of cold, nutrient-rich, oxygen-poor water (Fig. 10). The differential vertical movement of OB3 relative to OB1 and OB2 (Fig. 7) therefore probably reflects upwelling adjacent to the Avalonian and Baltic margins of the Iapetus Ocean.

The presence of vigorous upwelling would have a profound effect on the sedimentology of outer shelf areas, with the deposition of black shales and potentially phosphate enrichment of the sediment. The high percentage of phosphate in limestone of the late Caradoc Nod Glas

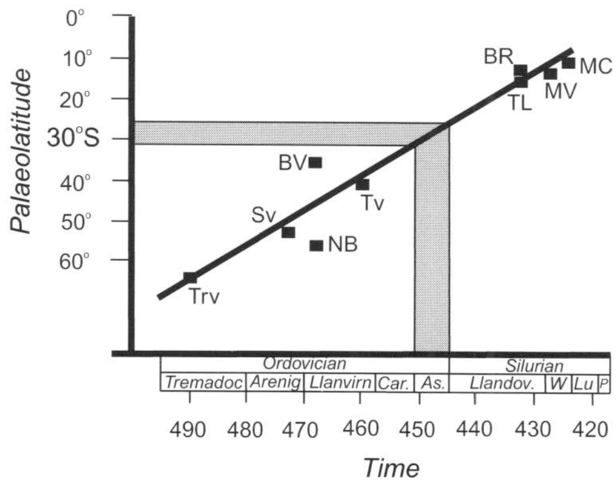

Fig. 9. Palaeolatitudes for East Avalonia (based on MacNiocail 2000). These data indicate a closure rate of *c*.3 cm a^{-1}. Timescale from Tucker & McKerrow (1995). Abbreviations: BR, Browgill Redbeds; BV, Builth Volcanics; MC, Mill Cove Redbeds; MV, Mendips Volcanics; NB, North Builth; Sv, Stapley Volcanics; TL, Tortworth Lavas; Trv, Treffgarne Volcanics; Tv, Tramore Volcanics.

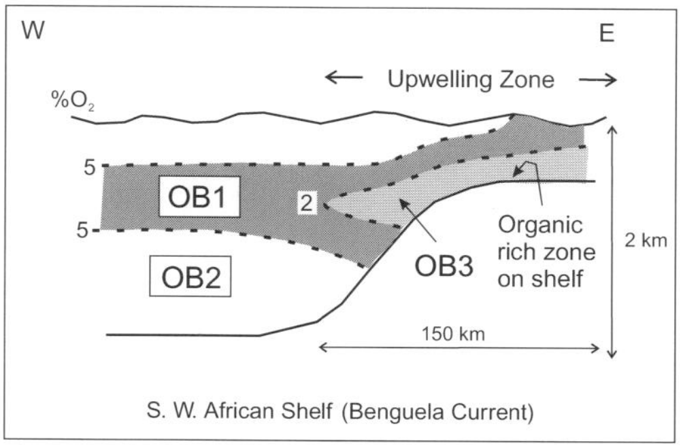

Fig. 10. Water mass structure off the SW African shelf based upon percentage oxygen content of the water (based on Demaison & Moore 1980). Upwelling water is generated by onshore to offshore winds skimming off warm surface water and thus allowing cold, nutrient-rich water to well up from intermediate depths. A similar structure can be found off the Peruvian margin associated with the Humboldt Current (Demaison & Moore 1980). We hypothesize that oceanic euconodont oceanic biofacies reflect water masses and have superimposed these biofacies onto the water mass structure of the SW African shelf. The pattern generated is similar to that found in the biofacies architecture diagram for Avalonia (Llanvirn–Ashgill) and Laurentia (*ordovicicus* Biozone) in Figure 7.

Formation of mid-Wales (Cave 1965; Smith 1999) and thin phosphorite conglomerate at the base of the Venstøp Formation in the Oslo–Asker district (Williams & Bruton 1983; Owen *et al.* 1990) indicates that the water along the southern margin of the Iapetus Ocean was rich in nutrients and supports the upwelling hypothesis.

Upwelling persisted along the southern margin of the Iapetus Ocean at least from the late Caradoc–early Ashgill and perhaps as early as the uppermost Tremadoc or early Arenig (Lindström & Vortisch 1983) Upwelling appears not to have been initiated at the low-latitude Laurentian margin until the Rawtheyan, just predating the glacial maximum and coincident with the general upward movement of oceanic biofacies across the Iapetus Ocean.

Regional diversity trajectories

A major reduction in diversity in Laurentian biofacies correlates with the early Ashgill onset of ocean cooling along the Laurentian margin (Armstrong & Coe 1997) and represents the extinction of taxa adapted to warm, tropical conditions. A second decline in diversity in SB2 correlates with a return to greenhouse conditions, and extinctions in taxa interpreted as being adapted to cooler 'glacial' conditions (see also Brenchley 1988; Brenchley *et al.* 1995*a*).

The major decline in diversity in Avalonian biofacies was coincident with the drift of Avalonia into warmer (but cooling) tropical water (Figs 8, 9). This suggests extinction of taxa adapted to cold water conditions, introduced into cooling but none-the-less warmer water of the tropics.

Baltica moved from intermediate latitudes to the tropics from the Arenig to the Ashgill and sutured to Avalonia during the late Ordovican (see reviews in Cocks & Fortey 1998; Cocks 2001). The Late Ordovician euconodont diversity trajectories (Figs 8e-f) are more closely similar to those of Avalonia, suggesting similar underlying controls. The onset of upwelling and close proximity of Avalonia and Baltica in the late Caradoc–early Ashgill corresponds with a rise in diversity largely the consequence of migrant shelf taxa from Avalonia.

Conclusions

Constrained seriation of presence–absence matrices provides a method of qualitatively defining generic associations or biofacies and hence reconstructing biofacies architectures without recourse to extensive recollecting. Upper Ordovician euconodonts can be divided into shelf and oceanic biofacies by this method. Euconodonts have traditionally been considered to be markedly provincial during the Late Ordovician with warm, tropical-water taxa restricted to the American Midcontinent Province and cooler-water taxa representative of the North Atlantic Province. Our analysis indicates that many genera previously considered to be representative of the North Atlantic Province had a pelagic mode of life and ranged widely across the Iapetus Ocean. This interpretation challenges the accepted view of euconodont provinciality and supports the province biofacies model introduced by Rasmussen (1998) for early Llanvirn euconodonts of the Iapetus Ocean.

Oceanic biofacies are considered to have reflected water mass structure. The changing vertical distribution of OB3 biofacies taxa suggests that they were adapted to cold, nutrient-rich, oxygen-poor upwelling water. If this is confirmed then the southern margin of the Iapetus Ocean was characterized by upwelling currents during the Ashgill. Upwelling was initiated along the Laurentian margin during the late Ashgill, coincided with an upward movement of oceanic biofacies and just predates the glacial maximum. Biofacies architecture patterns established in the Late Ordovician appear to have remained into the Early Silurian.

Clade diversities and trajectories in separate biofacies and latitudes can be attributed to different causal mechanisms. In Laurentia, the maximum diversity decline in all biofacies occurred in the early Ashgill and was coincident with the onset of ocean cooling. Maximum diversity decline in Avalonia coincided with its northward drift into tropical latitudes. The stability of euconodont biofacies architecture during Late Ordovician global cooling and plate reorganization indicates that these events had a low palaeoecological impact, causing little disruption of marine habitats despite a substantial decrease in both alpha and beta diversity (see also Droser *et al.* 1997, 2000).

Late Ordovician euconodont diversity reflects the unique environmental conditions specific to a region and not intrinsic properties of the clade. The post-extinction recovery of euconodonts was largely restricted to nektobenthic shelf taxa. The failure of many long-ranging pelagic clades to recover from the latest-Ordovician mass extinction suggests intrinsic properties of these clades may have suppressed their rediversification.

Our observations have implications for the alpha taxonomy of euconodonts. Genetic evidence for numerous cryptic species in morphologically identical pelagic organisms is consistent with the idea that they are much more specialized than would be inferred from their geographical distributions (Norris 2000). We may now have to fundamentally revise our morphological species concepts for pelagic euconodont clades that are likely to have been much more diverse than has been generally accepted in traditional taxonomies.

The authors acknowledge funding from the NERC (GR3/11834 to A.W.O.; GR9/02834 to H.A.A. and A.W.O.). The comments provided by the referees M. P. Smith and J. A. Rasmussen have improved this paper which is a contribution to IGCP 410, 'The Great Ordovician Biodiversification Event'.

References

ALDRIDGE, R. J. & MABILLARD, J. E. 1981. Local variations in the distribution of Silurian conodonts: an example from the *amorphognathoides* interval of the Welsh Basin. *In*: NEALE, J. W. & BRASIER, M. D. (eds) *Microfossils from Recent and Fossil Shelf Seas* Ellis Horwood, Chichester, 10–18.

ARMSTRONG, H. A. 1990. Conodonts from the Lower Silurian of the North Greenland carbonate platform. *Bulletin Grønlands Geologiske Undersøgelse*, **159**.

ARMSTRONG, H. A. 1995. High-resolution biostratigraphy (conodonts and graptolites) of the Upper Ordovician and Lower Silurian – a chronometric framework in which to evaluate the causes of the mass extinction. *Modern Geology,* **20**, 1–28.

ARMSTRONG, H. A. 1997. Conodonts from the Shinnel Formation, Tweeddale Member (middle Ordovician), Southern Uplands, Scotland. *Palaeontology,* **40**, 763–799.

ARMSTRONG, H. A. 2000. Conodont micropalaeontology of mid-Ordovician aged limestone clasts from LORS conglomerates, Lanark and Strathmore basins, Midland Valley, Scotland. *Journal of Micropalaeontology*, **19**, 45–59.

ARMSTRONG, H. A. & COE, A. L. 1997. Deep sea sediments record the geophysiology of the end Ordovician glaciation. *Journal of the Geological Society, London,* **154**, 929–934.

ARMSTRONG, H. A. & OWEN, A. W. 2001. Terrane evolution of the paratectonic Caledonides of northern Britain. *Journal of the Geological Society, London*, **158**, 475–486.

BARNES, C. R., REXROAD, C. B. & MILLER, J. F. 1973. Lower Paleozoic provincialism. *In*: RHODES, F. H. T. (ed.) *Conodont Zoology*. Geological Society of America, Special Papers, **141**, 156–190.

BAUER, J. A. 1994. Conodonts from the Bromide Formation (Middle Ordovician, south-central Oklahoma. *Journal of Paleontology,* **68**, 358–376.

BERGSTRÖM, S. M. 1990. Biostratigraphic and biogeographic significance of Middle and Upper Ordovician conodonts in the Girvan succession, South-West Scotland. *Courier Forschungs Institute Senckenberg*, **118**, 1–43.

BERGSTRÖM, S. M. & ORCHARD, M. 1985. Conodonts of the Cambrian and Ordovician Systems from the British Isles. *In*: HIGGINS, A. C. & AUSTIN, R. L. (eds) *A Stratigraphical Index of Conodonts.* Ellis Horwood, Chichester, 32–68.

BERGSTRÖM, S. M., RHODES, F. H. T. & LINDSTRÖM, M. 1987. Conodont biostratigraphy of the Llanvirn–Llandeilo and Llandeilo–Caradoc Series boundaries in the Ordovician System of Wales and the Welsh Borderland. *In*: AUSTIN, R. L. (ed.) *Conodonts: Investigative Techniques and Applications*. Ellis Horwood, Chichester, 294–316.

BERNER, R. A. 1990. Atmospheric carbon dioxide levels over Phanerozoic time. *Science,* **249**, 1382–1386.

BERNER, R. A. 1992. Palaeo-CO_2 and climate. *Nature*, **358**, 114.

BEVINS, R. E., BLUCK, B. J., BRENCHLEY, P. J., FORTEY, R. A., HUGHES, C. P., INGHAM, J. K. & RUSHTON, A. W. A. 1992. Ordovician. *In*: COPE, J. W. C., INGHAM, A. J. K. & RAWSON, P. F (eds) *Atlas of Palaeogeography and Lithofacies*. Geological Society, London, Memoirs, **13**, 19–36.

BRENCHLEY, P. J. 1988. Environmental changes close to the Ordovician–Silurian boundary. *Bulletin of the British Museum of Natural History (Geology),* **43**, 377–385.

BRENCHLEY, P. J., CARDEN, G. A. F. & MARSHALL, J. D. 1995*a*. Environmental changes associated with the 'first strike' of the late Ordovician mass extinction. *Modern Geology*, **20**, 69–82.

BRENCHLEY, P. J., MARSHALL, J. D., CARDEN, G. A. F., *ET AL.* 1995*b*. Bathymetric and isotopic evidence for a short-lived late Ordovician glaciation in a greenhouse period. *Geology,* **22**, 295–298.

BROWER, J. C. & KILE, K. M. 1988. Seriation of an original data matrix as applied to paleoecology. *Lethaia,* **21**, 79–92.

CAVE, R. 1965. The Nod Glas sediments of Caradoc age in North Wales. *Geological Journal*, **4**, 278–298.

COCKS, L. R. M. 2001. Ordovician and Silurian global geology. *Journal of the Geological Society, London*, **158**, 197–210.

COCKS, L. R. M. & FORTEY, R. A. 1998. The Lower Palaeozoic margins of Baltica. *Geologiska Föreningens i Stockholm Förhandlingar*, **120**, 173–179.

CROWLEY, T. J. & BAUM, S. K. 1991. Towards reconciliation of Late Ordovician (~440Ma) glaciation with very high CO_2 levels. *Journal of Geophysical Research*, **96**, 22597–22610.

DEMAISON, G. J. & MOORE, G. T. 1980. Anoxic environments and oil source bed genesis. *AAPG Bulletin*, **64**, 1179–1209.

DROSER, M. L., BOTTJER, D. J. & SHEEHAN, P. M. 1997. Evaluating the ecological architecture of major events in the Phanerozoic history of marine invertebrate life. *Geology,* **25**, 167–170.

DROSER, M. L., BOTTJER, D. J., SHEEHAN, P. M. & MCGEE, G. R. JR. 2000. Decoupling of taxonomic and ecological severity of Phanerozoic marine mass extinctions. *Geology*, **28**, 675–678.

DZIK, J. 1994. Conodonts of the Mójcza Limestone. *Palaeontologica Polonica*, **53**, 43–128.

EMERY, W. J. & MEINCKE, J. 1986. Global water masses: summary and review. *Oceanologica Acta,* **9**, 383–391.

ERWIN, D. H. 1998. After the end: Recovery from extinction. *Science*, **279**, 1324–1325.

HAMBREY, M. J. 1985. The Late Ordovician–Silurian glacial period. *Palaeogeography, Palaeoclimatology, Palaeoecology*, **51**, 273–289.

JAANUSSON, V. 1976. Faunal dynamics in the Middle Ordovician (Viruan) of Balto-Scandia. *In*: BASSETT, M. G. (ed.) *The Ordovician System, Proceedings of a Palaeontological Association Symposium. Birmingham, September 1974.* University of Wales and National Museum Wales, Cardiff, 301–326.

JAANUSSON, V. & BERGSTRÖM, S.M. 1980. Middle Ordovician faunal spatial differentiation in Baltoscandia and the Appalachians. *Alcheringa*, **4**, 89–110.

JABLONSKI, D. 1998. Geographic variation in the molluscan recovery from the end-Cretaceous Extinction. *Science,* **279**, 1327–1329.

KEMP, T. S. 1999. *Fossils and Evolution*. Oxford University Press, Oxford.

LINDSTRÖM, M. & VORTISCH W. 1983. Indications of upwelling in the Lower Ordovician of Scandinavia. *In:* THIEDE, J. & SUESS, E. (eds) *Coastal Upwelling*. Plenum, New York, 535–551.

LOCKLEY, M. G. 1983. A review of brachiopod dominated palaeocommunities from the type Ordovician. *Palaeontology*, **26**, 111–145.

MACNIOCAILL, C. 2000. A new Silurian palaeolatitude for eastern Avalonia and evidence for crustal rotations in the Avalonian margin of southwestern Ireland. *Geophysical Journal International*, **141**, 661–671.

MILLER, A. I. 1997. Comparative diversification dynamics among palaeocontinents during the Ordovician Radiation. *Geobios Mémoire Spécial*, **20**, 397–406.

MILLER, A. I. 2000. Conversations about Phanerozoic global diversity. *In*: ERWIN, D. H. & WING, S. L. (eds) *Deep Time: Paleobiology's Perspective. Paleobiology* (Supplement), **26**, 53–74.

MILLER, A. I. & MAO, SHUGUANG. 1995. Association of orogenic activity with the Ordovician radiation of marine life. *Geology,* **23**, 305–308.

NORRIS, R. D. 2000. Pelagic species diversity, biogeography and evolution. *In*: ERWIN, D. H. & WING, S. L. (eds) *Deep Time: Paleobiology's Perspective. Paleobiology* (Supplement), **26**, 236–259.

NOWLAN, G. S. & BARNES, C. R. 1981. Late Ordovician conodonts from the Vauréal Formation, Anticosti Island, Québec. *Geological Survey of Canada Bulletin*, **329**, 1–49.

ORCHARD, M. 1980. Upper Ordovician conodonts from England and Wales. *Geologica et Palaeontologica*, **14**, 9–44.

OWEN, A. W., BRUTON, D. L., BOCKELIE, J. F. & BOCKELIE, T. G. 1990. *The Ordovician successions of the Oslo Region, Norway*. Norges Geologiske Undersøkelese, Special Publications, **4**.

OWEN, A. W. & ROBERTSON, D. B. R. 1995. Ecological changes during the end-Ordovician extinction. *Modern Geology*, **20**, 21–39.

RADCLIFFE, G. 1998. *Biotic Recovery of Conodonts Following the end-Ordovician Mass Extinction*. PhD thesis, University of Durham.

RASMUSSEN, J. A. 1998. A reinterpretation of the conodont Atlantic Realm in the late Early Ordovician (early Llanvirn). *In*: SZANIAWSKI, H. (ed.) *Proceedings of the Sixth European Conodont Symposium (ECOS VI). Palaeontologica Polonica*, **58**, 67–77.

SAVAGE, N. M. & BASSETT, M. G. 1985. Caradoc–Ashgill conodont faunas from Wales and the Welsh Borderland. *Palaeontology*, **28**, 679–715.

SCOTESE, C. R. & MCKERROW, W. S. 1990. Revised world maps and introduction. *In*: MCKERROW, W. S. S. & SCOTESE, C. R. (eds) *Palaeozoic Palaeogeography and Biogeography*. Geological Society, London, Memoirs, **No. 12**, 1–25.

SEPKOSKI J. J. JR 1993. Ten years in the library: New data confirm paleontological patterns. *Paleobiology*, **19**, 43–51.

SEPKOSKI J. J. JR & KENDRICK, D. C. 1993. Numerical experiments with model monophyletic and paraphyletic taxa. *Paleobiology*, **19**, 168–184.

SMITH, C. J. 1999. *Evolutionary Palaeobiology of Deep-water Conodonts*. PhD thesis, University of Durham.

STOUGE, S. & RASMUSSEN, J. A. 1996. Upper Ordovician conodonts from Bornholm and possible migration routes into the Palaeotethys Ocean. *Bulletin of the Geological Society of Denmark*, **43**, 54–67.

SWEET, W. C. & BERGSTRÖM, S. M. 1974. Provincialism exhibited by Ordovician conodont faunas. Society of Economic Paleontologists and Mineralogists, Special Publications, **21**, 189–202.

SWEET, W. C. & BERGSTRÖM, S. M. 1984. Conodont provinces and biofacies of the Late Ordovician. Geological Society of America, Special Papers, **196**, 69–87.

TUCKER, R. D. & MCKERROW, W. S. 1995. Early Paleozoic chronology: a review in light of new U-Pb zircon ages from Newfoundland and Britain. *Canadian Journal of Earth Sciences*, **32**, 368–379.

VAN STAAL, C. R., DEWEY, J. F., MAC NIOCIALL, C. & MCKERROW, W. S. 1998. The Cambrian–Silurian tectonic evolution of the northern Appalachians and British Caledonides: history of complex, west and southwest Pacific-type segment of Iapetus. *In*: BLUNDELL, D. J. & SCOTT, A. C. (eds) *Lyell: the Past is the Key to the Present.* Geological Society, London, Special Publications, **143**, 199–242.

WILLIAMS, S. H. & BRUTON, D. L. 1983. The Caradoc–Ashgill boundary in the central Oslo Region and associated graptolite faunas. *Norsk Geologisk Tidsskrift*, **23**, 147–191.

The role of pyroclastic volcanism in Ordovician diversification

JOSEPH P. BOTTING

Department of Earth Sciences, University of Cambridge, Downing Street, Cambridge CB2 3EQ, UK (e-mail: joseph00@esc.cam.ac.uk*)*

Abstract: Pyroclastic volcanism has been tentatively correlated with Ordovician faunal diversification, but it is unclear whether the volcanism itself or associated tectonic effects were more influential. Elevated nutrient flux from uplifted areas has been widely discussed, but the link between enhanced productivity and diversification is unclear. Ecosystem compartmentalization due to irregular topography was probably significant, but it is debatable whether this was sufficient to explain the entire diversification. This paper introduces a further volcanic process, which may have been more significant than these established hypotheses, although it is emphasized that all relevant factors acted in combination. Recent studies of the local ecological effects of ash-fall have revealed dramatic post-depositional plankton and benthic blooms associated with overturning of a stratified water mass. The data are consistent through all studied sections from the Builth Inlier, Central Wales, while deeper-water sediments showed plankton blooms only. Repeated bloom events over localized dysaerobic shelf areas may have produced temporary benthic 'islands', enhanced genetic heterogeneity within affected regions, and increased speciation rates. In order to maximize bloom intensity (and possible effects on evolution), this process requires widespread pyroclastic volcanism combined with extensive shelf areas with well-developed stratification. The hypothesis potentially allows explanation of the differential rates of evolution between different groups, particularly the decoupled planktic–benthic patterns. A preliminary test using Welsh Basin ostracodes is presented, with predictions for the global database results.

The Ordovician Period contained the most dramatic, sustained increases in global diversity during the Phanerozoic, but its cause (or causes) remains elusive. Unlike most other episodes of diversification, it did not follow a major extinction event. There are marked differences regarding the onset and extent of diversity changes between palaeocontinents, and a strong ecological influence. The majority of the diversity increase is accommodated by sessile (and relatively immobile) benthos, and manifested partly as alpha and sometimes gamma diversity (Miller 1997*a*; Patzkowsky 1995). Beta (inter-community) diversity may even have decreased during the Late Ordovician (Miller 1997*b*). Gamma (inter-regional) diversity increases are generally considered to have been limited (Sepkoski 1988), despite the hypothesized habitat-fragmentation influence of numerous separate continents and volcanic islands (but see Patzkowski 1995). Sepkoski (1988) suggested that much of the Ordovician diversity increase was incorporated through the origination of novel, biogenic subenvironments such as reefs, bioherms and crinoid thickets. Data presented by McCormick & Owen (2001), describing Ordovician trilobite diversity in the Welsh Basin, show a significant increase in the species: genus ratio during diversity peaks. Genus-level data are more usually employed in diversity analyses, but the species approach should be encouraged where possible, since the diversification operated primarily at low taxonomic levels. However, the known fossil record of several classes (particularly sessile epifauna) begins in the Ordovician, suggesting some high-level taxonomic origination (Conway Morris 1999). This is unlikely to be due entirely to the origination of preservable skeletons in previously soft-bodied lineages. The relationships between the different diversity measures are poorly understood (e.g. Adrain *et al.* 2000), and the type of diversification observed (alpha, beta or gamma) may also depend on the taxonomic level examined. A similar pattern is apparent in the post-Permian recovery, in which high-level taxonomy showed few changes, with diversification concentrated at family level and below (Erwin *et al.* 1987).

The primary difficulty in interpreting the Ordovician Radiation (i.e. extreme complexity) may also allow its ultimate solution. Differential diversity patterns occur between different regions and taxonomic groups, and there are putative correlations with changes in climate, sea level, palaeogeography and volcanism. All these factors probably affected diversity, in addition to internal, ecological interactions: to

From: CRAME, J. A. & OWEN, A. W. (eds) 2002. *Palaeobiogeography and Biodiversity Change: the Ordovician and Mesozoic–Cenozoic Radiations*. Geological Society, London, Special Publications, **194,** 99–113.
0305-8719/02/$15.00

some extent, origination of novel life-habits provides additional subsidiary niches, most obviously in the development of epifaunal tiering. Only through understanding detailed diversity patterns can these influences possibly be separated, a procedure that has been advocated strongly by Miller (1997*a*, 1998, 2000).

The approach of Sepkoski (e.g. 1993) and most others to global diversity changes has been that of large-scale datasets, at intermediate taxonomic levels (usually family or genus), but also primarily in undivided continental or global format. Unless specifically utilized in subsets, global databases will generally provide only broad patterns, where the influence of each factor is disguised by its combination with others. Complex patterns of local and regional migration also confuse data on precise geographic origins; it has not yet been shown whether high-diversity, stable late Ordovician communities occupied the same environments as the centres of diversification, or whether the development of long-term global diversity is independent of areas where rapid speciation and overturn take place. Swain (1996), for example, described exceptionally rapid ostracode turnover in a bentonite-rich succession, where diversity at the top of the section did not greatly exceed the initial, but standing diversity at intermediate levels was much higher. Thus, in order to recognize meaningful diversification patterns reliably, overall diversity changes must be examined on regional or even intra-basinal scales (Miller 2000). Then, to identify the significance of any given process, a precise ecological–evolutionary signature must be established, where possible, and then tested. This is best examined by testing hypotheses in limited regions, under the specific conditions required for the testing. If a meaningful effect is observed, locally or regionally, then large-scale databases (e.g. Owen & McCormick 1999) can allow assessment of the wider significance of established local effects. Although many factors may have acted simultaneously in promoting the Ordovician Radiation, specific predictions of major regional processes should be recognizable on larger scales. Indeed, the ultimate challenge in understanding the Ordovician Radiation will be the extrapolation of known small-scale patterns into a global situation, where interactions between regions and processes will require complex palaeogeography-based analyses.

Among the many possible influences on Ordovician diversity, the apparent correlation with global pyroclastic volcanism has retained a perennial but uncertain significance. Miller (1997*a*) provided a correlation with global tectonic/volcanic intensity, in which Palaeozoic standing diversity appeared approximately to match levels of orogenic activity. However, correlation of this nature may represent a relationship between standing diversity and instantaneous tectonism, or between rapid speciation and tectonism plus another factor, which was restricted to the Ordovician. Extreme fluctuations in tectonism during the Permian, in a relatively stable part of the diversity curve, and the lack of correlation in the Mesozoic and Tertiary, lend some support to the latter; the correlation is much more precise for the Ordovician than for subsequent periods. Similarly, Miller & Mao (1995) presented a geographical correlation of diversification centres with forearc basins. Assuming some causal relationship, it is unclear whether the volcanism itself or associated tectonic activity was the more significant influence on diversity. Miller (1997*a*) suggested habitat partitioning as a primary cause of diversification, in addition to considering substrate transitions and nutrients, while Swain (1996) preferred an increase in nutrient supply to explain ostracode diversification following major ash-falls. Increased nutrient supply from run-off in uplifted areas is perhaps inevitable, but the possible link to diversification is unclear; in modern oceans, abundant nutrients generally encourage low diversity and very high dominance (e.g. Boyd *et al.* 2000). In particular, a regional increase in nutrient availability would tend to homogenize subfacies variations for which scarcity of food supply was critical for their definition. Parapatric or sympatric speciation based largely on behavioural food-acquisition procedures is, intuitively, more likely to arise in situations where food sources are relatively rare and competition most intense.

This chapter introduces a mechanism by which pyroclastic volcanism could have directly influenced the speciation rate on regional scales, with predictions for the evolutionary patterns to be expected. The available global data subdivided by taxonomy (e.g. Sepkoski 1995) appear to be consistent with these predictions regarding the differences in diversity changes between ecological groups. The hypothesis potentially allows explanation of the differential evolutionary rates between ecological/taxonomic groups, particularly the decoupled planktic–mobile benthic-sessile benthic diversity curves. A preliminary test of the hypothesis is included, using Welsh Basin ostracodes, and similar methods are encouraged for those investigating other hypotheses.

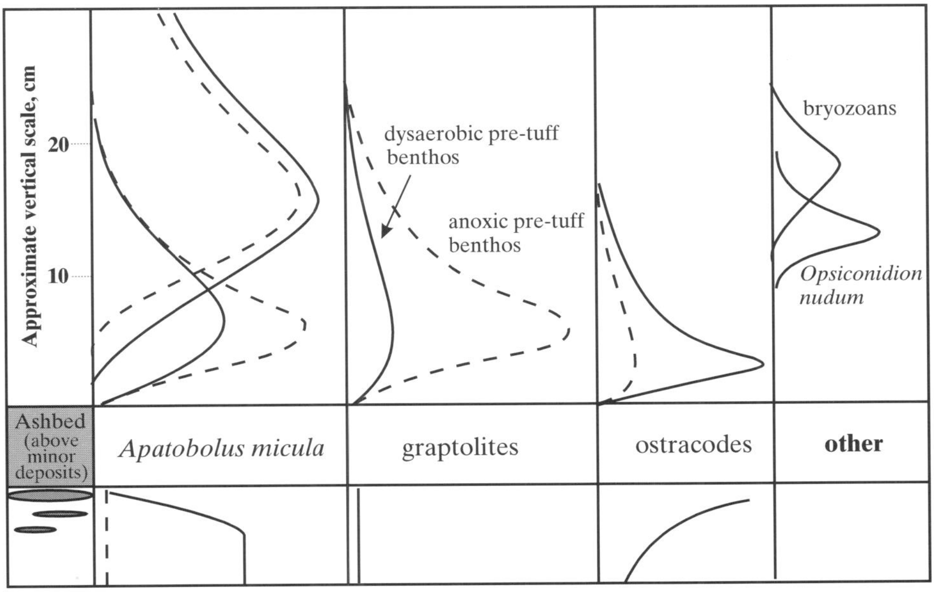

Fig. 1. Summary of ecological responses of plankton and benthos to volcanic ash deposition in Ordovician fine sediments of Llandrindod, central Wales. Vertical scale (with samples 2 cm thick) is dependent on sedimentation rates, and should be treated as an example rather than absolute. Originally anoxic (dashed) and dysaerobic (solid) situations are distinguished; benthic populations of *Opsiconidion nudum* and bryozoans only occur under relatively high oxygenation and food supply.

Ecological effects of local ash deposition

To establish the effects of volcanic eruptions on regional or global diversity, a detailed understanding of the ecological results of normal volcanic activity is required. Eruptions sufficiently large to affect climate significantly are believed to have dominantly destructive effects on biota (e.g. Javoy & Courtillot 1989), but Huff *et al.* (1992) found no evidence for this. Although massive eruptions undoubtedly occurred during the Ordovician (e.g. Huff *et al.* 1992), they represent a rare influence beyond the normal range of variability. The great majority of sedimentary ash layers are less than 10 cm thick outside the immediate vicinity of the source, suggesting eruptions of similar size to the present day. However, in many Ordovician successions, ash beds are extremely abundant, implying a generally higher eruption frequency resulting from widespread continental convergence (Stillman 1984).

Detailed, small-scale logs showing the ecological effects of discrete ash beds in Ordovician sediments of the Welsh Basin (Botting 2000) have allowed the reconstruction of local population dynamics. Dysaerobic Llanvirn and basal Caradoc siltstones of the Builth-Llandrindod Inlier, Central Wales, have yielded consistent patterns of faunal abundance. Palaeoenvironmental analyses are dominated by the local development of the Builth volcanic cone in the upper part of the Llanvirn, with the spectacular but controversial shoreline unconformity of Jones & Pugh (1949) underlying *teretiusculus* Biozone siltstones near Builth. Water depth during the late Llanvirn at Llandrindod is estimated at equivalent to mid to outer shelf (50–150 m), based on the dominance of the Raphiophorid Community of Fortey & Owens (1978), proximity to the Builth cone and palaeocontinental margin (from Cope *et al.* 1992), and sedimentology (Botting 2000); Cope (1999) inferred intertidal or shallow sublittoral conditions for beds underlying the volcanic succession. Many of the sections are organic-rich, with faunas indicative of oxygen limitation, including persistent benthic anoxia in parts of the *murchisoni* and *teretiusculus* Biozones. The faunal logs are mostly dominated by the partly pseudoplanktic inarticulate brachiopod *Apatobolus micula* (Botting & Thomas 1999), with abundant graptolites, ostracodes and chitinozoans in some sections. The occurrence of all

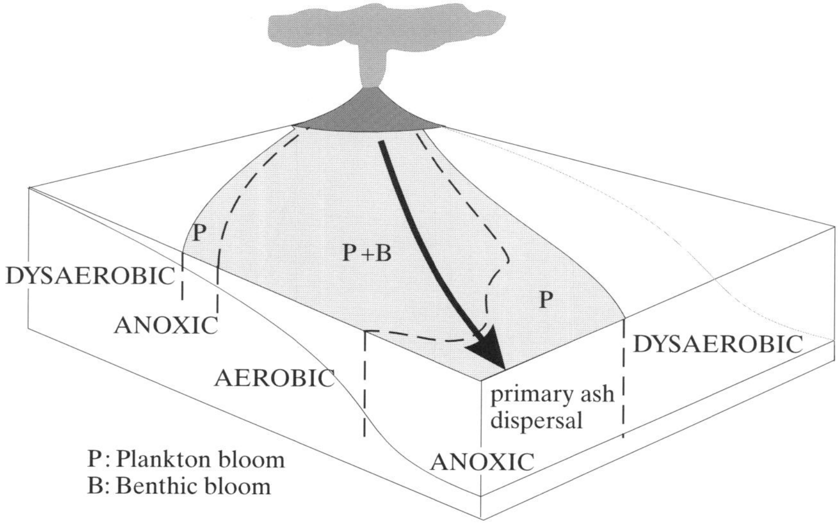

Fig. 2. Idealized distribution of oxygenation effects around ash-fall deposits in a shelf region with immobile water mass: central area of benthic oxygenation by strong overturning, and deoxygenation in lateral regions due to partial overturning and plankton bloom. Anoxic regions begin to disperse following decline of plankton bloom (one year), allowing immigration of benthos from exterior.

other faunal elements was noted, although data were usually insufficient for meaningful abundance comparisons. The complex overall patterns are consistent throughout all studied sections, minimizing the possibility of an artificial distribution. A generalization of the results is shown in Figure 1 and explained below. Full details of these data, including deeper-water facies from which benthic blooms were absent, will be published elsewhere, with full interpretation, and are also available in Botting (2000).

The ash bed is usually barren, with only occasional chitinozoans recorded. A brief, usually minor bloom of small mobile benthos (ostracodes and simple burrows) immediately followed ash deposition, overlapping with a subsequent dramatic plankton bloom of *Apatobolus micula* and graptolites. The plankton bloom showed a pseudo-logarithmic decline, while *A. micula* also formed a partly overlapping second bloom, of longer duration. Thereafter, abundance gradually declined to, or below, initial levels, unless benthic oxygenation became established. The ratio of ostracode to graptolite abundance varied according to initial conditions; in originally anoxic facies, the plankton (*Apatobolus micula* plus graptolites) bloom was by far the greater, and *vice versa* for originally dysaerobic substrates (Fig. 1). A horizontal sequence, sampling the 2 cm immediately above the ash upper surface, revealed steep faunal abundance gradients (>300% over 10 m), corresponding with up to a 20% change in ash thickness. A strong maximum in the immediately post-ash plankton abundance overlay the thickest part of the bentonite.

Multiple lines of evidence were used to constrain the processes operating, and a coherent interpretation developed, based on vertical circulation of a stratified water column. Fine ash deposited onto a standing body of water descended initially as turbid flows rather than discrete particles (Carey 1997). Provided a critical water depth (strongly dependent on several parameters) was not exceeded, these flows replaced deep, dysaerobic water with aerobic surface water, enabling a bloom of small mobile benthos, restricted by low food supplies. Corresponding upwelling induced a large-scale plankton bloom, continuing until inertial circulation ceased. This circulation could have been maintained directly by ash deposition for several weeks, through continuous fragmentation of floating pumice and slow sinking of fine particles. Following ash deposition, circulation would be prolonged by the temperature inversion of ash-free fluid, and perhaps enhanced by large-scale particulate phytoplankton production at the surface. Upwelling is expected to have continued weakly for a few months, sufficiently long to establish a substantial plankton bloom involving several trophic levels. The organic rain allowed a benthic bloom of sessile suspension-feeders to develop until oxygenation was depleted below tolerable limits.

The total duration of these bloom events is reconstructed, based on a comparison with modern bloom events (e.g. Gallardo *et al.* 1977) and post-seismic sedimentation rates (Goff 1997), as five to ten years, with up to one year for the plankton bloom (Botting 2000). The area affected depends on the extent of ash dispersal. Downwelling flows would decay rapidly through lateral entrainment of fluid, unless concentrated by continued ash input from above. Distal deposits in deep shelf settings would thus have been unable to host benthic blooms, although the plankton bloom would be unaffected. This would have provided elevated organic rain and encouraged benthic anoxia at the ash dispersal margins, where the high organic input was not offset by oxygenated downwelling (Fig. 2). While large eruptions affect wide areas, a typical pyroclastic event may induce benthic blooms in areas of a few hundred to a few thousand square kilometres. Coarse ash and crystal-tuff beds are less likely to initiate vertical circulation, since the critical particle density for mass flow is less likely to be exceeded, and individual particle settling is much more rapid and more varied.

Population genetics

The presence of repeated regional bloom events, following the removal of small or sessile benthos, potentially influenced population genetics in several ways. For general discussion of relevant concepts, refer to Harwood & Amos (1999), Amos & Harwood (1998), and references therein. Primarily, bloom events encourage the retention of novel characters, via an increased surviving proportion within each generation. Although the survival probabilities of such mutations are not elevated after the bloom maximum, the reduced juvenile mortality during the bloom growth phase increases the chance of persistent variations. For example, beneficial genotypic changes that require two mutations, one neutral or detrimental, have very low probability of arising under normal population conditions. During a bloom increase phase, carriers of each mutation are much more abundant, allowing carriers of both to arise, and thereafter become preferentially selected. Ford & Ford (1930) described abundance and variation fluctuations in an isolated butterfly population (of duration 10–20 years); variation was maximized during a population increase, with the subsequent maximum and declining populations being more homogenous. The final population was morphologically distinct from the initial. While a single bloom event is probably insufficient for speciation to occur from a homogeneous population, the result is to produce genetic heterogeneity on a subregional scale. This may allow distinct species to appear rapidly under subsequent blooms, as disparate parts of an already heterogeneous population are brought into immediate contact during recolonization (discussed below). Although some intraspecific homogenization would be expected to occur during intervening quiescent periods, migration of genetic characteristics through a stable population is much slower than physical migration of a population into a barren region.

Following the eradication of small mobile and sessile benthos by ash deposition, recolonization occurred by some combination of exponential population increase of rare survivors, and immigration from the surrounding area. The delayed onset of the sessile benthic bloom relative to the mobile bloom, interpreted above as resulting from low food supplies inhibiting suspension-feeders, may also suggest a gradual immigration. In reality, the bloom population was almost certainly derived from geographically separated lateral immigrants, combined with a remnant endemic population; lateral separation was probably tens to hundreds of kilometres, depending on the eruption size. The genetic diversity of the benthic bloom biota should thus be much greater than where the population is derived from a single endemic community. This disparity is further exaggerated by the process discussed above, whereby previous bloom events induced strong regional genetic variability. The populations at opposite edges of the affected area would be expected to differ more than in a stable environment, with these populations becoming immediately adjacent during recolonization. Within the resultant population, hybridization and segregation between end-members would result in a higher rate of speciation than in more homogeneous communities. Hybridization would further promote the appearance of individuals with distinctive characteristics by providing novel gene combinations, while the initiation of non-interbreeding subpopulations is a necessary first stage in sympatric speciation. In a gradually varying population, there is little incentive for segregation, but behavioural divisions can exist immediately when disparate parts of that population are brought into contact.

The general situation of colonization of an ecologically depauperate region promotes genetic variation under many circumstances

(Whitlock & McCauley 1990; McCauley 1991). Among the most critical parameters controlling genetic variability is whether recolonization occurs through population fusion (differentiation decreased) or permanent fission (differentiation increased). In many cases, however, it appears that subspecific diversification results from fission of a population, followed by refusion, and further factors are presumed to be operating in order to account for observed diversification. One possibly is unidirectional 'meiotic drive,' discussed by Amos & Harwood (1998), but present understanding is very limited. The critical point when considering the increases in genetic diversity and potential speciation rate is that the bloom events were repeated. Several thousand events, each separated by the order of decades to centuries, have the potential to exaggerate slight probabilistic changes, perhaps producing noticeable effects on global diversity curves.

A further element that strongly encourages speciation in modern faunas (e.g. cichlids: Barlow *et al.* 1984; Russock 1986; Knight *et al.* 1998; Seehausen & van Alphen 1999) is sexual selection based on parental imprinting, which can potentially lead to sympatric speciation in heterogeneous populations. However, interpreting the reproductive habits of extinct organisms is notoriously difficult, despite ambiguous evidence that trilobites, for example, provided some degree of guardianship over their offspring (Davis *et al.* 2001). Uniformitarian assumptions suggest that many arthropods provided the potential for imprinting (e.g. Barnes 1987), but direct evidence is lacking. Overall, the potential for parental imprinting and pure sympatric speciation by this mechanism in invertebrates was probably limited.

The effects of subsequent mixing under stable environmental conditions must also be considered. Taxa showing highly specific behavioural traits may not fall into direct competition, and the increased diversity may be conserved. This is also true of inhabitants of facies with a diverse range of ecological niches, a feature enhanced by the development of tiering during the Ordovician (e.g. Droser *et al.* 1995). Soft-substrate shelf facies with low faunal abundance and diversity may provide the focus for diversity increases, but some diversity is likely to have been lost by merging or competition during long-term quiescence, in the absence of complex ecosystems that could maintain niche separation. Although the centres of diversification would have been in these environments, sustained long-term diversity increases may have initially occurred by range expansion into shallower-water facies, where diverse physical conditions and complex ecosystems provided permanent habitats. (Trends of this nature should be recognizable through comparison of temporal ranges and morphological variability between closely related species occurring in different environments.) Abundant specific niches could then have allowed preservation of lineages that would rapidly have become extinct in their original environment. The observed onshore–offshore trend during the Palaeozoic (e.g. Bottjer *et al.* 1988) would reflect the gradual expansion of this complex, stable ecosystem into deeper water, while most of the explosive diversity increase at speciation centres was subsequently lost from the original environment.

Since the geographic mixing of distinctive populations is critical to this process, planktic organisms should not be significantly affected. The point relating to preservation of novel characters within the blooms would still stand, but constant mixing of populations would reduce the significance of these effects; diversity increases should be expected, but to a much lesser extent than in the benthos.

Implications for Ordovician diversity

The Ordovician Period was characterized by widespread explosive volcanism in many parts of the world, as discussed above. The results of ash deposition in shelf regions represent a conceptual argument whereby an increased speciation rate could have operated in volcanically active areas, and which provides objective predictions for several aspects of the diversity patterns expected. The temporal, geographical and environmental distributions should show three basic correlations; maximum rates of diversification should correspond to:

(a) areas and times of extensive pyroclastic volcanism, emphasizing numerous small eruptions rather than few large ones;
(b) benthic oxygen limitation and well-developed stratification, leading to high-amplitude bloom events;
(c) high eustatic sea level, increasing the shelf area over which small eruptions could produce significant effects.

Volcanism and sea level are usually related, on a global scale, through increased tectonic activity and ocean ridge volume, while their relationship to benthic anoxia is presently uncertain. However, all three factors particularly characterize the Ordovician in general, and the coincidence of all three is predicted to correlate with diversification centres on a regional to basin scale. If the proposed process was significant, this should be supported by the global database results; for example, areas of diversification should be dominated by volcanically active regions showing stratification. The diversity increases in other areas should then be shown to occur largely through immigration of previously existing taxa.

A further set of predictions relates to the differential rates of evolution between different ecological and taxonomic groups. In these cases, the evolutionary response of each group can be predicted from the established ecological response, in that those groups showing the highest abundance fluctuations and most limited individual geographic ranges should respond most strongly to the variations in population genetics outlined above. Temporary local eradication followed by high-amplitude blooms from rapid recolonization should have maximized the evolutionary response. These ecological responses were presumably dependent on several factors, including larval and adult life habits, trophic group, size, reproductive strategy and developmental time-scales, most of which can only be estimated. Observations of palaeoecological response remove the complications of reasoning about factors that may be difficult to assess in extinct taxa. The predicted evolutionary responses are described according to broad ecological divisions.

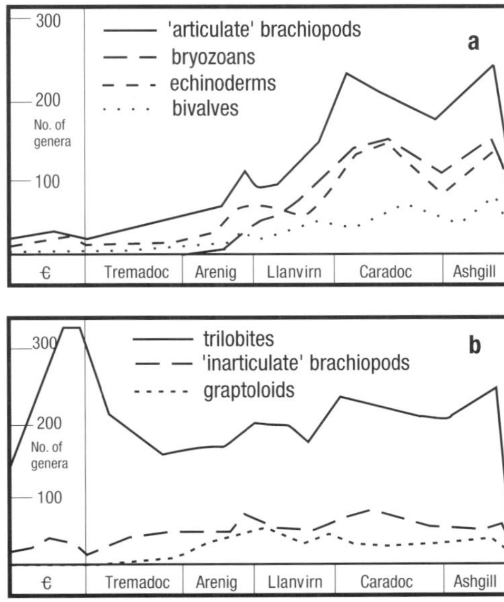

Fig. 3. Summary of differing diversification patterns by phylum based on Sepkoski (1995), reproduced by kind permission of the Pacific Section – SEPM. (**a**) Taxa in groups 1, 2 and 3 (part) (see text), exhibiting strong ecological response to ash-fall; (**b**) taxa in groups 3 (part), 4 and 5 (see text), exhibiting lesser ecological response to ash-fall, or existing as plankton only.

(1) *Small sessile benthos and essentially immobile epibenthos*: extensive responses, apparently completely eradicated by substantial ash deposition, but with rapid recolonization in a high-amplitude bloom. Includes bryozoans, small brachiopods ('articulate' and 'inarticulate'), ostracodes, sphenothallids. (Ostracodes are included here because their small size allows very restricted movement. However, they were apparently able to escape minor ash deposition, while small sessile taxa were not, and recolonized a region more rapidly.)

(2) *Intermediate to large sessile benthos*: significant responses, with local eradication and high-magnitude blooms. May escape slight ash deposition. Includes crinoids, sponges, bryozoans, larger brachiopods, conulariids. Although generally rare, some examples of this group (sponges, brachiopods, bryozoans) gave rise to brief, intense blooms (Botting 2000).

(3) *Weakly mobile epi- and endobenthos (excluding ostracodes)*: often able to escape deposition of small ash volumes, but bloom events not obvious due to small population size. Includes bivalves, gastropods, palaeoscolecids, larger lingulids. Their response may be underestimated here because of lower preservation potential of aragonitic and multi-element skeletons.

(4) *Mobile epibenthos/nekton*: limited response; little noticeable effect on abundance across ash beds. Includes trilobites, nautiloids, although aragonitic nautiloids may be underestimated because of reduced preservation.

(5) *Plankton and pelagic nekton*: affected by extensive blooms, but little or no prior destruction; rapid migration prevents segregation into genetically distinct communities, and isolated populations implausible. Includes graptolites, pelagic trilobites, acritarchs (and some chitinozoans?).

Further complications include the tolerance of some taxa to low oxygenation, and ambiguous interpretations of life habits; 'inarticulate' brachiopods often appear in dysaerobic settings, and some were almost certainly pseudoplanktic

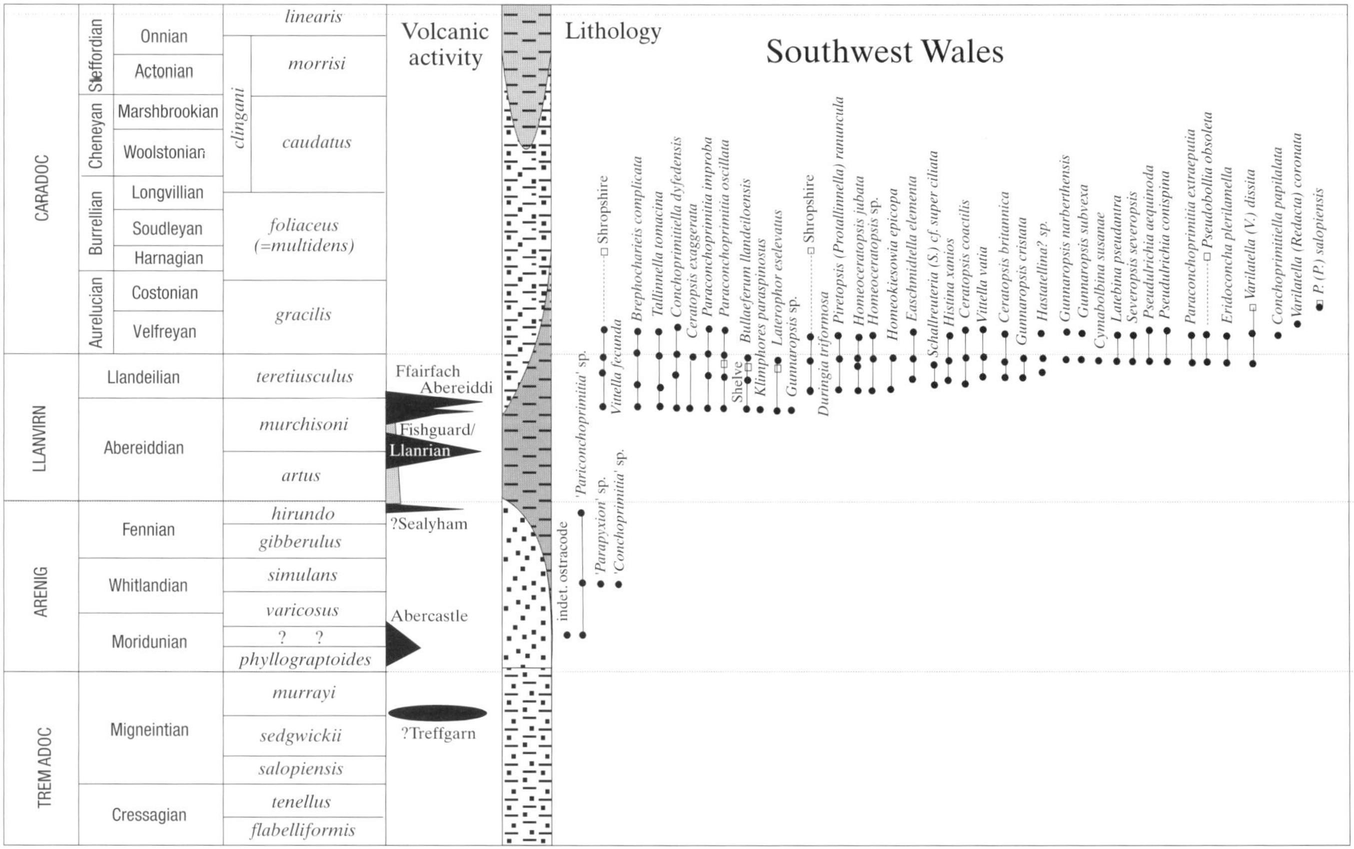

Fig. 4. Beyrichiocope ostracode range chart for the pre-Ashgill of SW Wales, correlated with volcanic activity (local volcanic centres = black fill; abundant ash beds = grey background) and generalized dominant lithology (dots, sandstone; dots + dashes, siltstone; dashes, shale; stratification/benthic anoxia indicated by shading). Ostracodes: solid circles, occurrence of species in this region; open squares, occurrence of species in other region of Welsh Basin; open circles, presence of species as immigrant to this region. No species occur outside the Welsh Basin, although some genera are shared with Baltica, Europe and North America. Data from Jones (1986–1987), Siveter (in press) and personal observation.

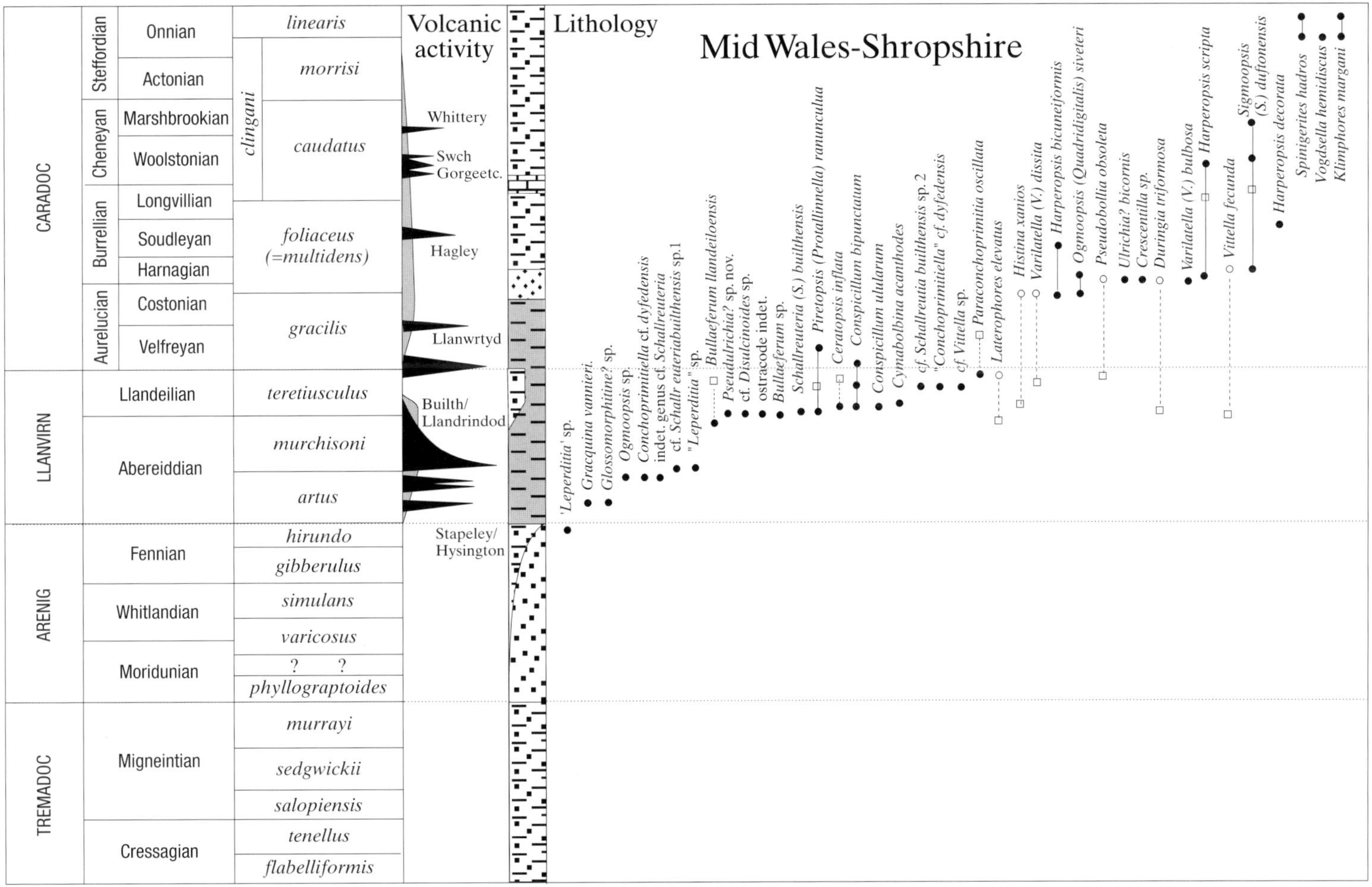

Fig. 5. Ostracode range chart for the pre-Ashgill of Mid-Wales (Powys) and Shropshire. See Figure 4 for explanation.

(Lockley & Antia 1980; Gabbott 1999; Botting & Thomas 1999). For example, *A. micula* benthic populations would have been less strongly affected than predicted from the benthic bloom, because of population mixing under pseudoplanktic conditions. In contrast, 'articulate' brachiopods are generally facies-restricted, and appear among predictable faunal assemblages. It is significant that Sepkoski (1995) found much greater net diversification among 'articulates' than 'inarticulates,' since many linguloids were capable of burrowing, while 'articulates' were almost exclusively sessile epibenthos.

Published global diversity curves differentiated into taxonomic groups (Sepkoski 1995) appear to record patterns that are consistent with these predictions (Fig. 3). In general, sessile taxa (brachiopods, bryozoans, corals) show the greatest diversification, with a highly consistent pattern between phyla. Bivalves and gastropods first achieved significant diversity during the Ordovician, but then show very rapid diversification, while trilobites fluctuated irregularly (confirmed by Adrain & Westrop 2000). Graptolites show little overall diversity increase, although with significant fluctuations (Cooper 1999); a similar pattern is observed among acritarchs and chitinozoans (e.g. Kaljo *et al.* 1995). Among the groups predicted to show greatest diversification are the ostracodes, based on their ecological response. Swain (1996) reported extremely elevated ostracode turnover, with substantial overall diversification, in a bentonite-rich Ordovician sequence of North America. He tentatively attributed this to nutrient input but, as discussed above, a plausible precise mechanism is not obvious.

Welsh Basin beyrichiocope ostracode diversity

A summary of the Ordovician diversity of pre-Ashgill beyrichiocope ostracodes in the Welsh Basin follows; other 'ostracode' groups are not included, because of uncertainty over mutual relationships. The area is subdivided into three subregions (SW Wales, Mid-Wales and Shropshire, North Wales and northern England) and correlated with major volcanic episodes within each area. An estimate of bentonite frequency is included, based on limited descriptions in stratigraphic literature and personal observation, although in many cases thin beds are not obvious in outcrop, and have not been mentioned in publications. Where bentonites appear frequently within an interval lacking a local volcanic source, they can usually be correlated with episodes of major volcanism in one of the other subregions. In addition, a schematic log of primary lithology is included for each region, although extreme local variation may occur; apparent stratification (benthic anoxia) is denoted by shading (Figs 4, 5, 6).

Ostracode data were obtained primarily from Jones (1986–1987), with additional Arenig information from Siveter (in press), and personal observations. The rarity of pre-Llanvirn taxa is apparently real, although the simple morphology of described examples may disguise greater taxonomic variation than is practically identifiable. However, the same underestimate should operate in later unornamented taxa, suggesting that the observed morphological diversification accurately reflects a real pattern. Data regarding the stratigraphic position of volcanic episodes, and dominant lithologies were obtained from Smith & George (1961), George (1970), Fortey *et al.* (2000) and Rushton *et al.* (2000) as primary sources, supplemented by additional local references (e.g. Stamp & Wooldridge 1923; Davies 1933; Jones & Pugh 1941, 1949; Diggens & Romano 1968; Earp & Haines 1971; Lynas 1983; Cave & Rushton 1996; Davies *et al.* 1997).

In SW Wales (Fig. 4), massive diversification over a brief interval (Llanvirn) is correlated with black shales and extensive volcanism of the Fishguard, St David's and Strumble Head regions. Arenig volcanism in aerobic sand facies produced little response, although a few morphologically conservative taxa appeared at this time. The low diversity in the upper Caradoc may be partly an artefact; diverse Ashgill faunas are known to exist, but are yet to be studied intensively (Jones 1987). However, although Caradoc ostracodes above the Aurelucian are almost certainly present (some taxa, such as *Duringia triformosa*, reappear in Shropshire during the Burrellian), they are apparently very rare; no identifiable specimens from the area have been described.

In Mid-Wales and Shropshire (Fig. 5), speciation occurred over the Llanvirn and Caradoc, corresponding to penecontemporaneous volcanic centres and restricted oxygenation. A few species were immigrants from SW Wales, but most were endemic.

There are limited data for North Wales and northern England (Fig. 6) but the majority of species correspond to the Borrowdale volcanic episode (Caradoc); a high proportion was immigrant. Additional faunas are predicted to be discovered from the Llanvirn and lower Caradoc.

Overall, the patterns appear to show

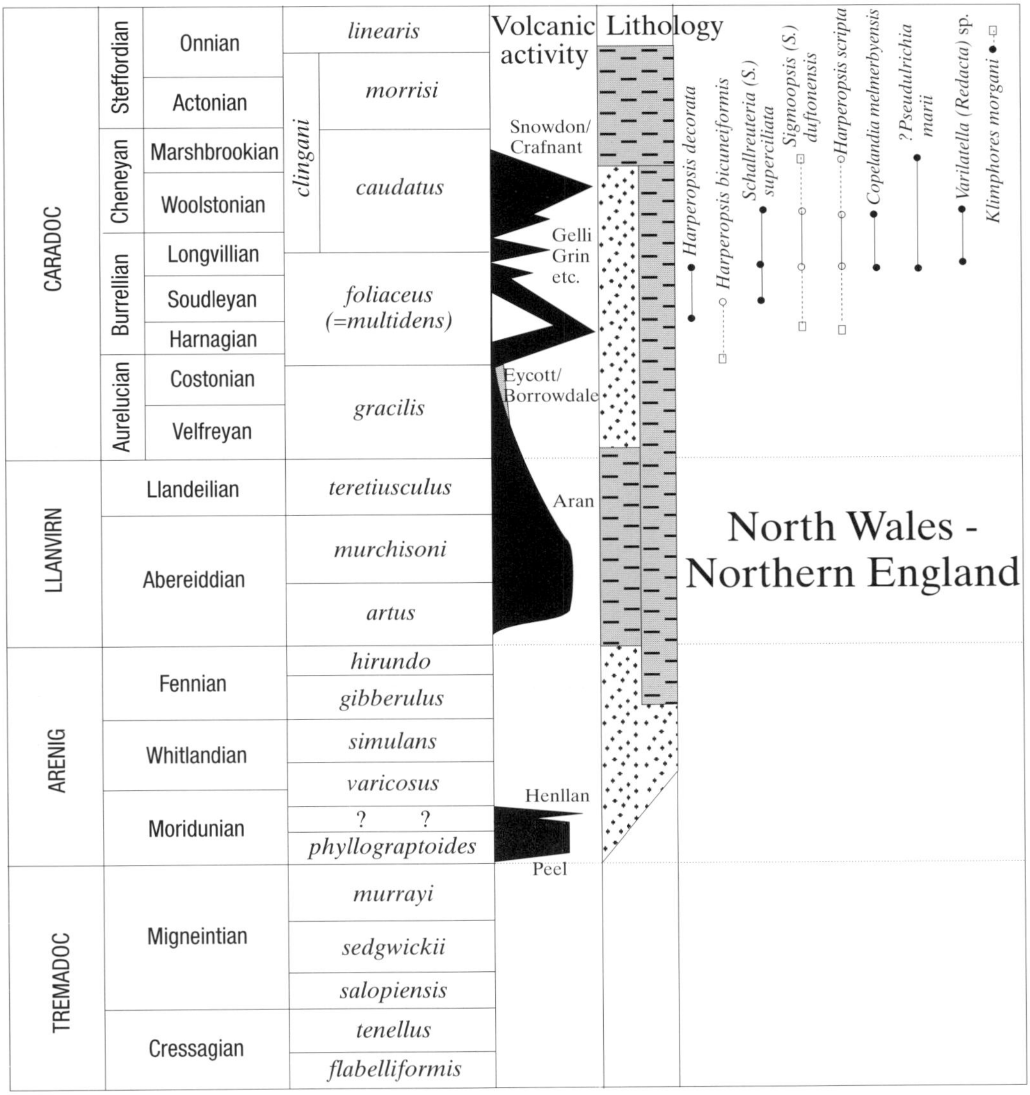

Fig. 6. Ostracode range chart for North Wales and northern England. See Figure 4 for explanation.

enhanced speciation over the intervals containing frequent ash beds in each area, and low diversity outside these episodes. Although no similar comparison has been made with areas lacking ash beds, the correlation of diversity with volcanism appears to be precise in the current data; this view is reinforced by the marked differences between adjacent regions.

The question of sampling bias must be addressed in this situation, since volcanic areas tend to preserve more resistant lithologies, which may aid the preservation and exposure of fossils. However, if this is true, identical patterns should emerge among taxonomic groups that shared similar environments and are extracted by similar processes. The ostracode data derive largely from silicified specimens extracted by acid digestion of limestones, and from mouldic material in siliciclastic sedimentary rocks. Sutton *et al.* (1999–2000) monographed Welsh Basin lingulid brachiopods, using very similar material from a similarly comprehensive

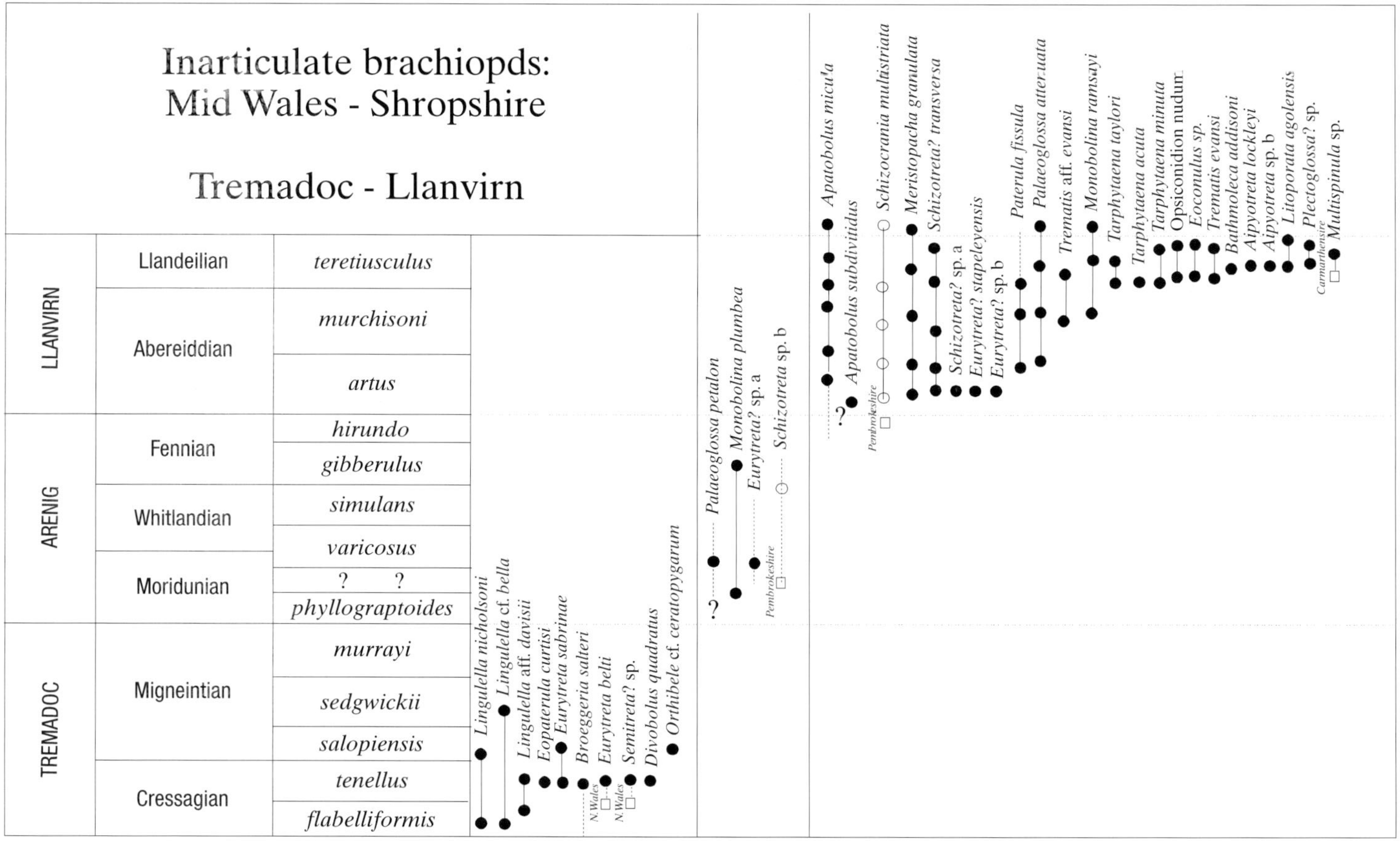

Fig. 7. Pre-Caradoc range chart for Mid-Wales (Powys) and Shropshire 'inarticulate' brachiopods, based on data from Sutton *et al.* (1999–2000). Compare with ostracode data in Figure 5, including lithology and volcanism.

number of sites. If the patterns observed for ostracodes reflect sampling bias resulting from areal exposure and preservation, the same patterns should be seen among 'inarticulate' brachiopods. Figure 7 shows the pre-Caradoc 'inarticulate' data for Mid-Wales and Shropshire. This reveals a more gradual increase in diversity, with a significant number of Tremadoc and Arenig species, although the rate of species appearance also increased at the time of Llanvirn volcanism, including similar absolute origination. However, it is likely that many more ostracode species remain undiscovered than brachiopods, due to their inconspicuousness. A significant proportion of the ostracodes encountered during fieldwork were undescribed, whereas brachiopods were almost always identifiable. Although at an early stage, this initial analysis suggests a local influence of volcanism on diversification, which affected ecological or taxonomic groups in different ways, these differences being consistent with ecological studies of the local effects of volcanic ash deposition.

Conclusions

The local effects of volcanic ash deposition on planktic and benthic communities in dysaerobic shelf/basin settings have been summarized. The resultant elimination of benthos, recolonization from lateral areas, and bloom events in both plankton and benthos, have been analysed in terms of population genetics, and shown to comprise a conceptual method for encouraging speciation. Physical conditions necessary for this process were maximized during the Ordovician, leading to comparisons with the Ordovician Radiation. If the process were a significant factor in this diversification event, predictable differences would be expected between taxonomic and ecological groups, which may be assessed by comparison with observed palaeoecological responses, and reasoning from modern ecological models. A series of such predictions is provided, with caveats in the case of ambiguous groups. This is illustrated by a subbasin-scale analysis of ostracode diversity, correlated with volcanism, and compared with the less dramatic diversification of 'inarticulate' brachiopods to reduce concerns of sampling bias. A possible distinction between those facies showing very high speciation and extinction rates, and others able to adopt more modest long-term diversity increases is also suggested. Speciation is viewed as occurring largely in the mid to outer shelf regions, but within an ecosystem insufficiently complex to support this diversity following environmental stabilization. However, onshore expansion of new species may have contributed to the increasing diversity of more complex shallow-water communities that typify the Palaeozoic Evolutionary Fauna, resulting in a gradual offshore expansion of these ecosystems.

This chapter presents a mechanism that potentially explains large parts of the Ordovician diversification, with predictions. Readers are encouraged to present additional data as a test of the hypothesis, in order to establish its validity. Even if the hypothesis is disproved, the use of fine-scale diversity and palaeobiogeography databases is suggested as an essential procedure for understanding processes involved in the diversity changes. Extrapolation and averaging over larger scales will hopefully then allow elucidation of global database results.

Development of the population genetics section was greatly aided by discussion with W. Amos (Zoology, University of Cambridge), while D. J. Siveter (University of Leicester) and A. W. Owen generously provided 'in press' data regarding ostracodes and trilobites respectively. The palaeoecological study was performed in partial fulfilment of a PhD at the University of Birmingham, during which discussions with A. T. Thomas and M. P. Smith are particularly acknowledged. The manuscript was improved by constructive reviews from A. I. Miller and D. J. Siveter. Research is currently funded by Christ's College, Cambridge. This paper is a contribution to IGCP 410 'The Great Ordovician Biodiversification Event'.

References

ADRAIN, J. M. & WESTROP, S. R. 2000. An empirical assessment of taxic paleobiology. *Science*, **289**, 110–112.

ADRAIN, J. M., WESTROP, S. R., CHATTERTON, B. D. E. & RAMSKOLD, L. 2000. Silurian trilobite alpha diversity and the end-Ordovician mass extinction. *Paleobiology*, **26**, 625–646.

AMOS, W. & HARWOOD, J. 1998. Factors affecting levels of genetic diversity in natural populations. *Philosophical Transactions of the Royal Society of London*, **B 353**, 177–186.

BARLOW, G. W., VALERIO, M & HOLDER, J. 1984. Does the polychromatic midas cichlid show sexual imprinting? *American Zoologist*, **24**, A51 (abstract).

BARNES, R. D. 1987. *Invertebrate Zoology*. Saunders College Publishing, USA.

BOTTING, J. P. 2000. *Palaeoecology and Systematics of Ordovician Biotas from Welsh Volcaniclastic Deposits*. PhD Thesis, University of Birmingham.

BOTTING, J. P. & THOMAS, A. T. 1999. A pseudoplanktonic inarticulate brachiopod attached to graptolites and algae. *Acta Universitatis Carolinae: Geologica*, **43**, 333–335.

BOTTJER, D. J., DROSER, M. L. & JABLONSKI, D. 1988. Palaeonenvironmental trends in the history of trace fossils. *Nature*, **33**, 252–255.

BOYD, P. W., WATSON, A. J.. LAW, C. S., *ET AL.* 2000. A mesoscale phytoplankton bloom in the polar Southern Ocean stimulated by iron fertilization. *Nature*, **407**, 695–702.

CAREY, S. N. 1997. Influence of convective sedimentation on the formation of widespread tephra fall layers in the deep sea. *Geology*, **25**, 839–842.

CAVE, R. & RUSHTON., A. W. A. 1996. The Llandeilo (Ordovician) Series in the core of the Tywi Anticline, Llanwrtyd, Powys, UK. *Geological Journal*, **31**, 47–60.

CONWAY MORRIS, S. 1999. The evolution of diversity in ancient ecosystems: a review. *In*: MAGURRAN, A. E. & MAY, R. M. (eds.) *Evolution of Biological Diversity*. Oxford University Press, New York, 282–322.

COOPER, R. A. 1999. Graptolites and the great Ordovician biodiversification event. *Acta Universitatis Carolinae: Geologica*, **43**, 441–442.

COPE, J. C. W. 1999. Middle Ordovician bivalves from Mid-Wales and the Welsh borderland. *Palaeontology*, **42**, 467–499.

COPE, J. C. W. INGHAM, J. K. & RAWSON, P. F. 1992. *Atlas of Palaeogeography and Lithofacies*. Geological Society, London, Memoirs, **13**.

DAVIES, J. R., FLETCHER, C. J. N., WATERS, R. N., WILSON, D., WOODHALL, D. G. & ZALASIEWICZ, J. A. 1997. *Geology of the Country around Llanillar and Rhayader.* British Geological Survey, The Stationery Office, London.

DAVIES. K. A. 1933. The geology of the country between Abergwesyn (Breconshire) and Pumpsaint (Carmarthenshire). *Quarterly Journal of the Geological Society, London*, **89**, 172–201.

DAVIS, R. A., FRAAYE, R. H. B. & HOLLAND, C. H. 2001. Trilobites within nautiloid cephalopods. *Lethaia*, **34**, 37–45.

DIGGENS, J. N. & ROMANO, M. 1968. The Caradoc rocks around Llyn Cowlyd, North Wales. *Geological Journal*, **6**, 31–48.

DROSER, M. L., SHEEHAN, P. M., FORTEY, R. A. & XING LI, 1995. The nature of diversification and palaeoecology of the Ordovician Radiation with evidence from the Great Basin. *In*: COOPER, J. D., DROSER, M. L. & FINNEY, S. C. (eds) *Ordovician Odyssey: Short Papers for the Seventh International Symposium on the Ordovician System*. Pacific Section SEPM, Fullerton, California, 415–418.

EARP, J. R. & HAINES, B. A. 1971. *British Regional Geology: the Welsh Borderland* (third edition). Institute of Geological Sciences, London.

ERWIN, D. H., VALENTINE, J. W. & SEPKOSKI, J. J. JR. 1987. A comparative study of diversification events: the Early Palaeozoic versus the Mesozoic. *Evolution*, **41**, 1177–1186.

FORD, H. D. & FORD, E. B. 1930. Fluctuation in numbers, and its influence on variation, in *Melitaea aurinia*, Rott. (Lepidoptera). *Transactions of the Entomological Society of London*, **78**, 345–351.

FORTEY, R. A. & OWENS, R. M. 1978. Early Ordovician (Arenig) stratigraphy and faunas of the Carmarthen district, south-west Wales. *Bulletin of the British Museum Natural History (Geology)*, **30**, 225–294.

FORTEY, R. A., HARPER, D. A. T., INGHAM, J. K., OWEN, A. W. & RUSHTON, A. W. A. 2000. *A revised correlation of Ordovician rocks in the British Isles*. Geological Society, London, Special Reports, **5**.

GABBOTT, S. E. 1999. Orthoconic cephalopods and associated fauna from the late Ordovician Soom Shale lagerstätte, South Africa. *Palaeontology*, **42**, 123–148.

GALLARDO, V. A., CASTILLO, J. G., RETAMAL, M. A., YANEZ, A., MOYANO, H. I., & HERMOSILLA, J. G. 1977. Quantitative studies on the soft-bottom macrobenthic animal communities of shallow Antarctic bays. *In*: LLANO, G. A. (ed.) *Adaptations within Antarctic Ecosystems*. Proceedings of the 3rd SCAR Symposium on Antarctic Biology, Smithsonian Institution, Washington D.C., 361–387.

GEORGE, T. N. 1970. *British Regional Geology: South Wales*. Institute of Geological Sciences, London.

GOFF, J. R. 1997. A chronology of natural and anthropogenic influences on coastal sedimentation, New Zealand. *Marine Geology*, **138**, 105–117.

HARWOOD, J. & AMOS, W. 1999. Genetic diversity in natural populations. *In*: MAGURRAN, A. E. & MAY, R. M. (eds) *Evolution of Biological Diversity*. Oxford University Press, New York, 56–69.

HUFF, W. D., BERGSTROM, S. M. & KOLATA, D. R. 1992. Gigantic Ordovician volcanic ash fall in North America and Europe: biological, tectonomagmatic and event-stratigraphic significance. Geology, **20**, 875–878.

JAVOY, M. & COURTILLOT, V. 1989. Intense acidic volcanism at the Cretaceous–Tertiary boundary. *Earth and Planetary Science Letters*, **94**, 409–416.

JONES, C. R. 1986–1987. Ordovician (Llandeilo and Caradoc) beyrichiocope Ostracoda from England and Wales. *Monograph of the Palaeontographical Society, London*, 1, 1986: **138**, 1–76; 2, 1987: **139**, 77–114.

JONES, O. T. & PUGH, W. J. 1941. The Ordovician rocks of the Builth district. *Geological Magazine*, **78**, 185–191.

JONES, O. T. & PUGH, W. J. 1949. An Early Ordovician shoreline in Radnorshire, near Builth Wells. *Quarterly Journal of the Geological Society of London*, **104**, 43–70.

KALJO, D., NÕLVAK, J. & UUTELA, A. 1995. Ordovician microfossil diversity in the Rapla section, northern Estonia. *In*: COOPER, J. D., DROSER, M. L. & FINNEY, S. C. (eds) *Ordovician Odyssey: Short Papers for the Seventh International Symposium on the Ordovician System*. Pacific Section SEPM, Fullerton, California, 415–418.

KNIGHT, M. E., TURNER, G. F., RICO, C., VAN OPPEN, M. J. H. & HEWITT, G. M. 1998. Microsatellite paternity analysis on captive Lake Malawi cichlids supports reproductive isolation by direct mate choice. *Molecular Ecology*, **7**, 1605–1610.

LOCKLEY, M. G. & ANTIA, D. D. J. 1980. Anomalous occurrence of the Lower Palaeozoic brachiopod *Schizocrania*. *Palaeontology*, **23**, 707–713.

LYNAS, B. D. T. 1983. Two new Ordovician volcanic centres in the Shelve Inlier, Powys, Wales. *Geological Magazine*, **120**, 535–542.

MCCAULEY, D. E. 1991. Genetic consequences of local population extinction and recolonization. *Trends in Ecology and Evolution*, **6**, 5–8.

MCCORMICK, T. & OWEN, A. W. 2001. Assessing trilobite biodiversity in the Ordovician of the British Isles. *Geological Journal*, **36**, 279–290.

MILLER, A. I. 1997*a*. Dissecting global diversity patterns: examples from the Ordovician Radiation. *Annual Review of Ecology and Systematics*, **28**, 85–104.

MILLER, A. I. 1997*b*. A new look at age and area: the geographic and environmental expansion of genera during the Ordovician Radiation. *Paleobiology*, **23**, 410–419.

MILLER, A. I. 1998. Biotic transitions in global marine diversity. *Science*, **281**, 1157–1160.

MILLER, A. I. 2000. Conversations about Phanerozoic global diversity. *In*: ERWIN, D. G. & WING, S. L. (eds) *Deep Time, Paleobiology (supplement)*, **26**, 53–73.

MILLER, A. I. & MAO, SHUGUANG. 1995. Association of orogenic activity with the Ordovician radiation of marine life. *Geology*, **23**, 305–308.

OWEN, A. W. & MCCORMICK, T. 1999. Ordovician biodiversity change in the British Isles: a database approach. I. Rationale and database structure. *Acta Universitatis Carolinae: Geologica*, **43**, 523–526.

PATZKOWSKY, M. E. 1995. Ecologic aspects of the Ordovician Radiation of articulate brachiopods. *In*: COOPER, J. D., DROSER, M. L. & FINNEY, S. C. (eds) *Ordovician Odyssey: Short Papers for the Seventh International Symposium on the Ordovician System*. Pacific Section SEPM, Fullerton, California, 413–414.

RUSHTON, A. W. A., OWEN, A. W., OWENS, R. M. & PRIGMORE, J. K. 2000. *British Cambrian to Ordovician Stratigraphy.* Joint Nature Conservation Committee, Peterborough.

RUSSOCK, H. I. 1986. Preferential behavior of *Sarotherodon (Oreochromis) mossambicus* (Pisces, Cichlidae) fry to maternal models and its relevance to the concept of imprinting. *Behaviour*, **96**, 304–321.

SEEHAUSEN, O. & VAN ALPHEN, J. M. 1999. Can sympatric speciation by disruptive sexual selection explain rapid evolution of cichlid diversity in Lake Victoria? *Ecology Letters*, **2**, 262–271.

SEPKOSKI, J. J. JR. 1988. Alpha, beta or gamma: where does all the diversity go? *Paleobiology*, **14**, 221–234.

SEPKOSKI, J. J. JR. 1993. Ten years in the library – new data confirm paleontological patterns. *Paleobiology*, **19**, 43–51.

SEPKOSKI, J. J. JR. 1995. The Ordovician radiations: diversification and extinction shown by global genus-level data. *In*: COOPER, J. D., DROSER, M. L. & FINNEY, S. C. (eds) *Ordovician Odyssey: Short Papers for the Seventh International Symposium on the Ordovician System*. Pacific Section SEPM, Fullerton, California, 393–396.

SIVETER, D. J. (in press). The Ordovician. *In*: WILKINSON, I. (ed.) *A Stratigraphical Index of British Ostracoda* (second edition). British Micropalaeontological Society.

SMITH, B., & GEORGE, T. N. 1961. *British Regional Geology: North Wales*. Her Majesty's Stationery Office, London.

STAMP, L. D. & WOOLDRIDGE, S. W. 1923. The igneous and associated rocks of Llanwrtyd (Brecon). *Quarterly Journal of the Geological Society, London*, **74**, 16–46.

STILLMAN, C. J. 1984. Ordovician volcanicity. *In*: BRUTON, D. L. (ed.) *Aspects of the Ordovician System*. Palaeontological Contributions from the University of Oslo, **295**, Universitetsforlaget, 183–194.

SUTTON, M. D., BASSETT, M. G. & CHERNS, L. 1999–2000. Lingulate brachiopods from the Lower Ordovician of the Anglo-Welsh Basin. *Monograph of the Palaeontographical Society, London*, 1, 1999: **153**, 1–60; 2, 2000: **154**, 61–114.

SWAIN, F. M. 1996. Ostracode speciation following Middle Ordovician extinction events, north central United States. *In*: HART, M. B. (ed.) *Biotic Recovery from Mass Extinction Events*. Geological Society, London, Special Publications, **102**, 97–104.

WHITLOCK, M. C. & MCCAULEY, D. E. 1990. Some population genetic consequences of colony formation and extinction: genetic correlations within founding groups. *Evolution*, **44**, 1717–1724.

The early evolution and palaeobiogeography of Mesozoic planktonic foraminifera

MALCOLM B. HART, MELISSA J. OXFORD & WENDY HUDSON
Department of Geological Sciences & Plymouth Environmental Research Centre, University of Plymouth, Drake Circus, Plymouth PL4 8AA, UK
(e-mail: mhart@plymouth.ac.uk*)*

Abstract: The planktonic foraminifera almost certainly evolved from benthonic ancestors in the early Jurassic. The meroplanktonic genus *Conoglobigerina*, known from south-central and eastern Europe, appears in the Bajocian and is probably derived from the even more geographically restricted *Praegubkinella*. This genus was represented by a single taxon in the earliest Toarcian but diversified after the Toarcian anoxic event. At the same level *Oberhauserella quadrilobata* Fuchs, 1967 became more inflated and there is some evidence to suggest that the 'anoxic event' was the environmental perturbation that began the transition to a planktonic mode of life. In the Callovian–Oxfordian interval, the planktonic foraminifera are still restricted to a relatively limited area bounded by the North Atlantic Ocean, NW Europe and Eastern Europe and this remained the case even in the earliest Cretaceous. It was only in the Aptian–Albian that the palaeogeographical distribution changed dramatically, probably as a response to the elevated sea levels caused by the increased rate of ocean crust production which began in the Early Aptian. The principal diversification events in the Jurassic (Toarcian, Bajocian, Callovian–Oxfordian) also appear to be related to sea level highstands.

The planktonic foraminifera are a distinctive and abundant part of the modern oceanic fauna. Together with the other planktonic groups (coccolithophorids, diatoms, radiolarians, dinoflagellates, etc), they assist in the chemical/nutrient cycling of the ocean system. While the oceanic plankton, in general, impact on the levels of atmospheric CO_2 in the Mesozoic and Cenozoic, the contribution of the planktonic foraminifera is probably difficult to quantify although, as we will demonstrate, the development of the group was rather limited until the mid-Cretaceous. During the Mesozoic it is known that the planktonic foraminifera:

- evolved from benthonic ancestors;
- diversified;
- expanded to attain an almost global distribution;
- suffered during several major 'events' (including possible bolide impacts and methane escapes from gas hydrates); and
- were reduced to two or three species by the terminal Cretaceous 'events' or 'event' (bolide impact(s), sea-level change, temperature rise/fall, volcanic fall-out, etc.).

Most micropalaeontologists (e.g. Caron 1983; Hart 1999; Premoli Silva & Sliter 1999) have treated the planktonic foraminifera as though they were a single evolutionary plexus, although those working on DNA sequencing of the modern fauna have questioned this basic assumption (e.g. Darling *et al.* 1996, 1999*a*,*b*) and suggested that the evolution of the planktonic foraminifera may be polyphyletic. It is certainly true that one cannot claim that the Jurassic record of the group (Fig.1) provides anything other than a disjointed set of relationships. While preservation and/or collection failure is probably a contributing factor, there are several problems:

- the *Praegubkinella – Conoglobigerina* link;
- the disjointed distribution of *Conoglobigerina*; and
- the origins of *Globuligerina* and other taxa in the mid- to late Jurassic.

Higher, in the Cretaceous succession, we have little evidence for the origin of the *Praehedbergella* lineage or the heterohelicids (Fig. 2) and it is clear that the new evidence from the work on DNA must be taken into account and a polyphyletic history considered alongside other suggestions.

From: CRAME, J. A. & OWEN, A. W. (eds) 2002. *Palaeobiogeography and Biodiversity Change: the Ordovician and Mesozoic–Cenozoic Radiations*. Geological Society, London, Special Publications, **194,** 115–125.
0305-8719/02/$15.00

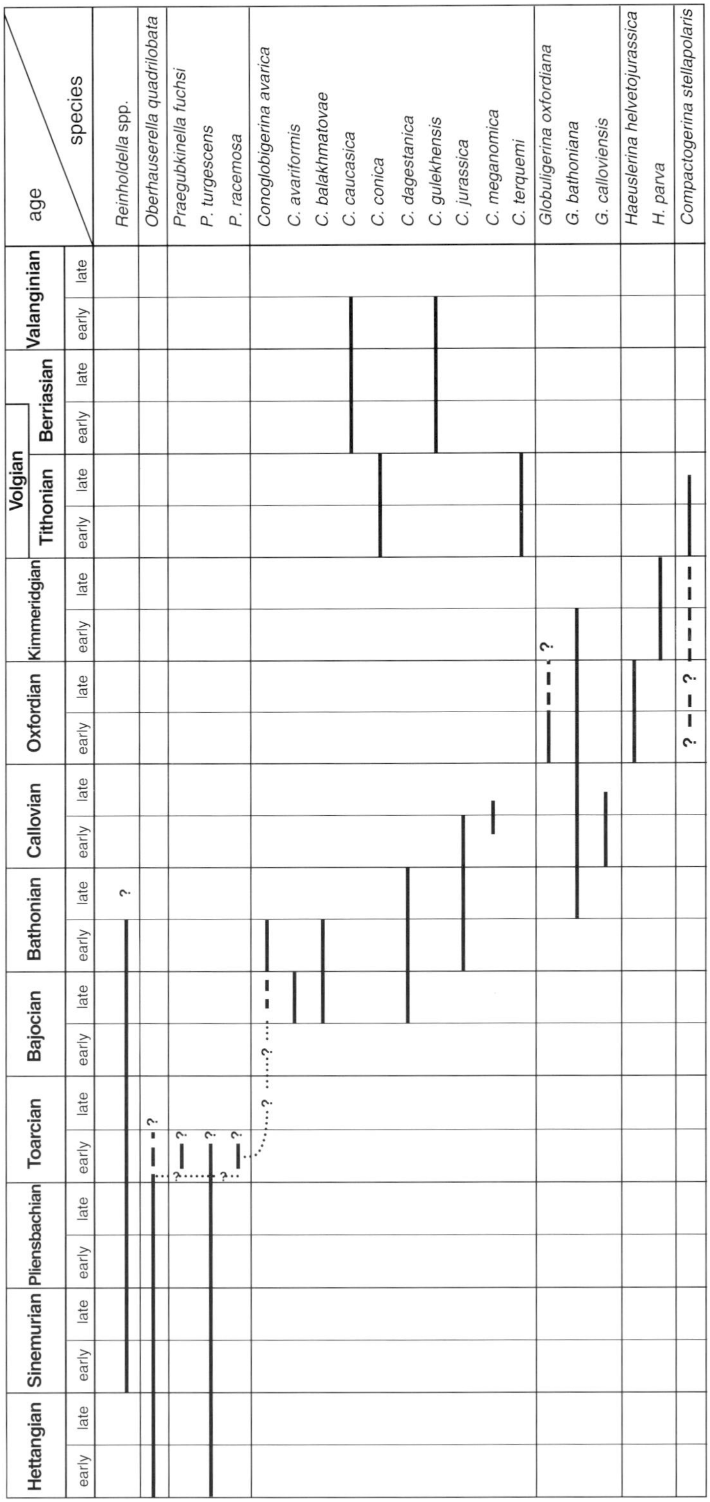

Fig. 1. Stratigraphical ranges of Jurassic and early Cretaceous planktonic foraminifera, partly based on Simmons *et al.* (1997) together with the stratigraphical ranges of related benthonic taxa.

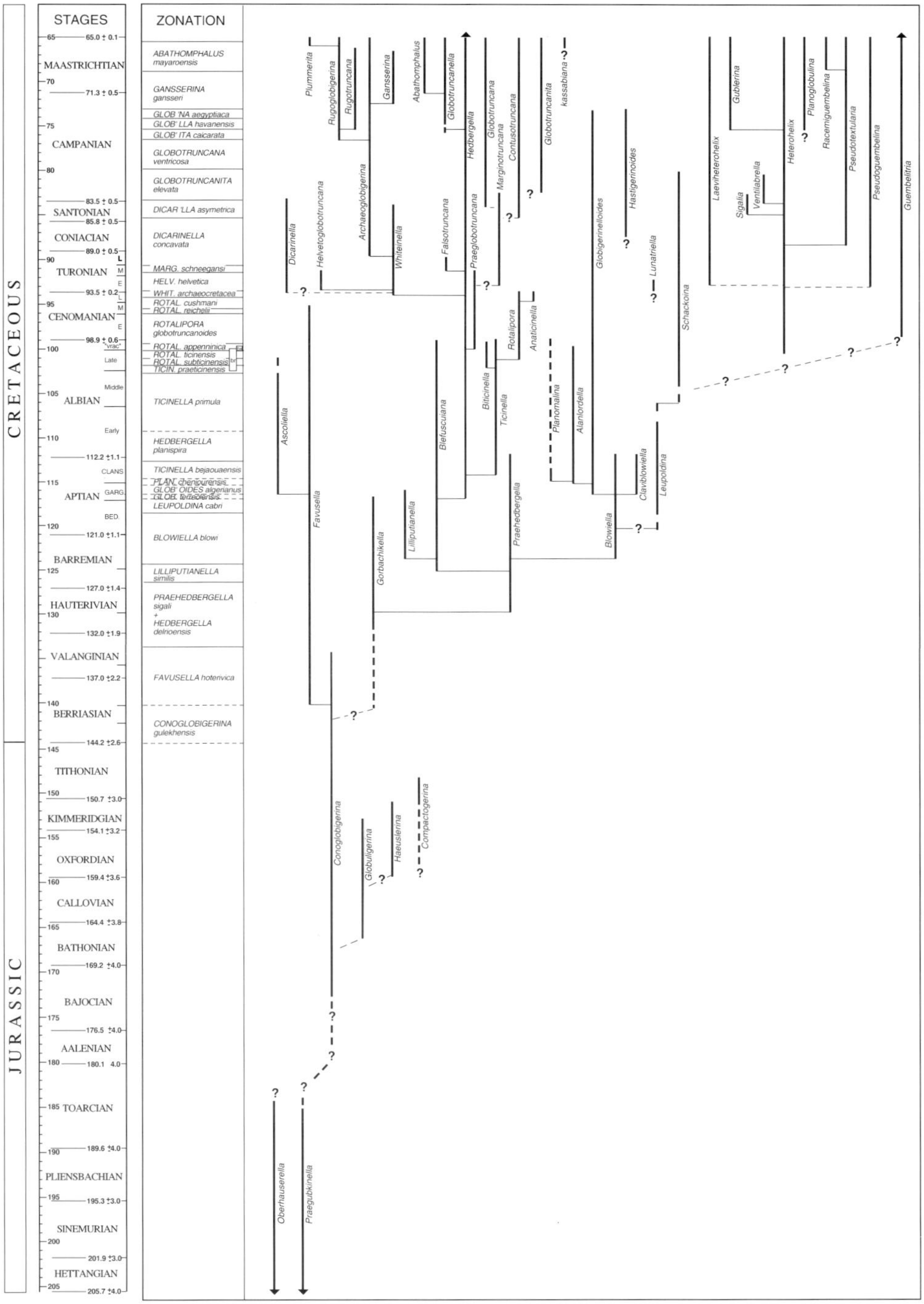

Fig. 2. Generic evolution of the Jurassic and Cretaceous foraminifera (partly after Hart 1999).

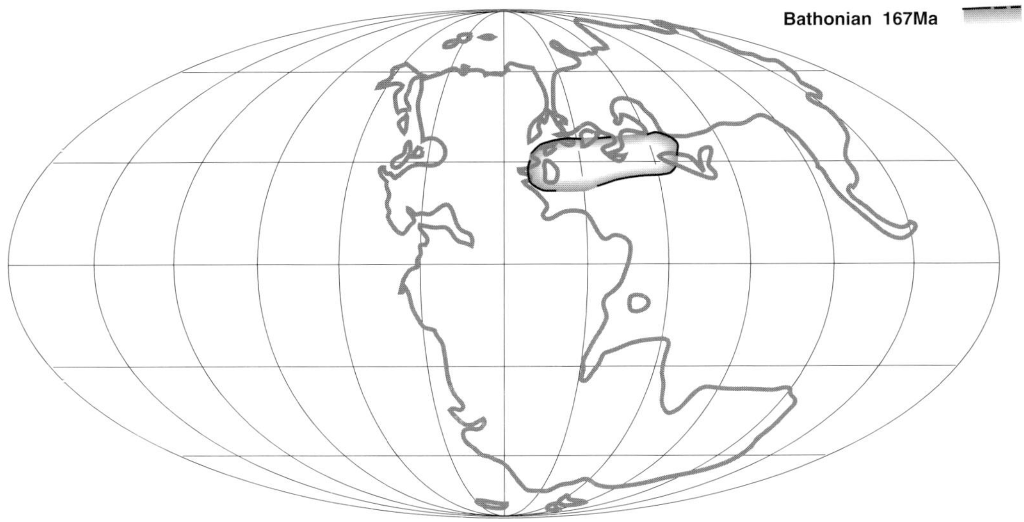

Fig. 3. Distribution of *Conoglobigerina* spp. in the Middle Jurassic.

Origin of the planktonic foraminifera

In their review of this subject, Simmons *et al.* (1997) dismiss the report of Pliensbachian planktonic specimens (Görög 1994) as probable contamination but, more significantly, reject all the taxa described as planktonic foraminifera by Fuchs (1967, 1971, 1973, 1975, 1977). These poorly preserved specimens have been examined by F. Rögl (Natural History Museum, Vienna) and A. Görög (Budapest, Hungary), both of whom indicate that they are not planktonic taxa. Some of this material is currently on loan to The Natural History Museum (London) and it is clear to the senior author that they are benthonic taxa with flattened umbilical sides and a complete lack of the inflated chambers normally associated with a planktonic mode of life.

If the Triassic and early Jurassic records of planktonic foraminifera are rejected then the origins of the group must be pre-Bajocian (Fig. 1) as the species of *Conoglobigerina* recorded from the Bajocian are clearly meroplanktonic (Wernli & Görög 1999, 2000). In the Toarcian sediments of the Creux de l'ours section (Teysachaux, Fribourg, Switzerland) Wernli (1995) has described a fauna of *Praegubkinella* spp. that is associated with *Oberhauserella quadrilobata*, one of the original 'planktonic' taxa of Fuchs (1967). Wernli (1995) demonstrates the possible evolution from *Praegubkinella racemosa* Wernli into *Conoglobigerina* (Fig.1), thereby extending the range of the ancestral forms into the Toarcian. The section of the 'Creux de l'ours' has been described by several authors (e.g. Mettraux *et al.* 1986, 1989) and the recent commentary (Weidmann 1993) on the Swiss Geological Map No.1244, Châtel-St-Denis (Weidmann *et al.* 1993) provides much new information. The famous section (coordinates 565.07/154.30 on Swiss Geological Map No.1244) was improved in 1970 when a new forestry road was constructed. Unfortunately many of the new sections are now degraded but one face still exists on a bend where the road crosses the stream. Previous workers have identified the *elegantulum* and *exaratum* Subzones of the *falciferum* Zone, placing the section within the Toarcian 'anoxic event' (Mettraux *et al.* 1986, 1989, and references therein). The senior author has found specimens of *Harpoceras* and *Dactylioceras* in this part of the section, together with examples of *Posidonia* associated with hard, organic-rich mudstones. M. Mettraux provided the samples for the investigation by Wernli (1995), the foraminifera coming from a sample of the dark mudstones from stratigraphically above the 'anoxic event'. The *Praegubkinella/Oberhauserella* assemblage of Wernli (1995) is, therefore, of early Toarcian age.

Work on the Toarcian foraminifera of the UK and NW Europe (Hylton 2000; Hylton & Hart 2000) has shown that, associated with the Toarcian anoxic event, there is a 'bloom' of small forms of *O. quadrilobata* which possess slightly more inflated chambers than the typical forms

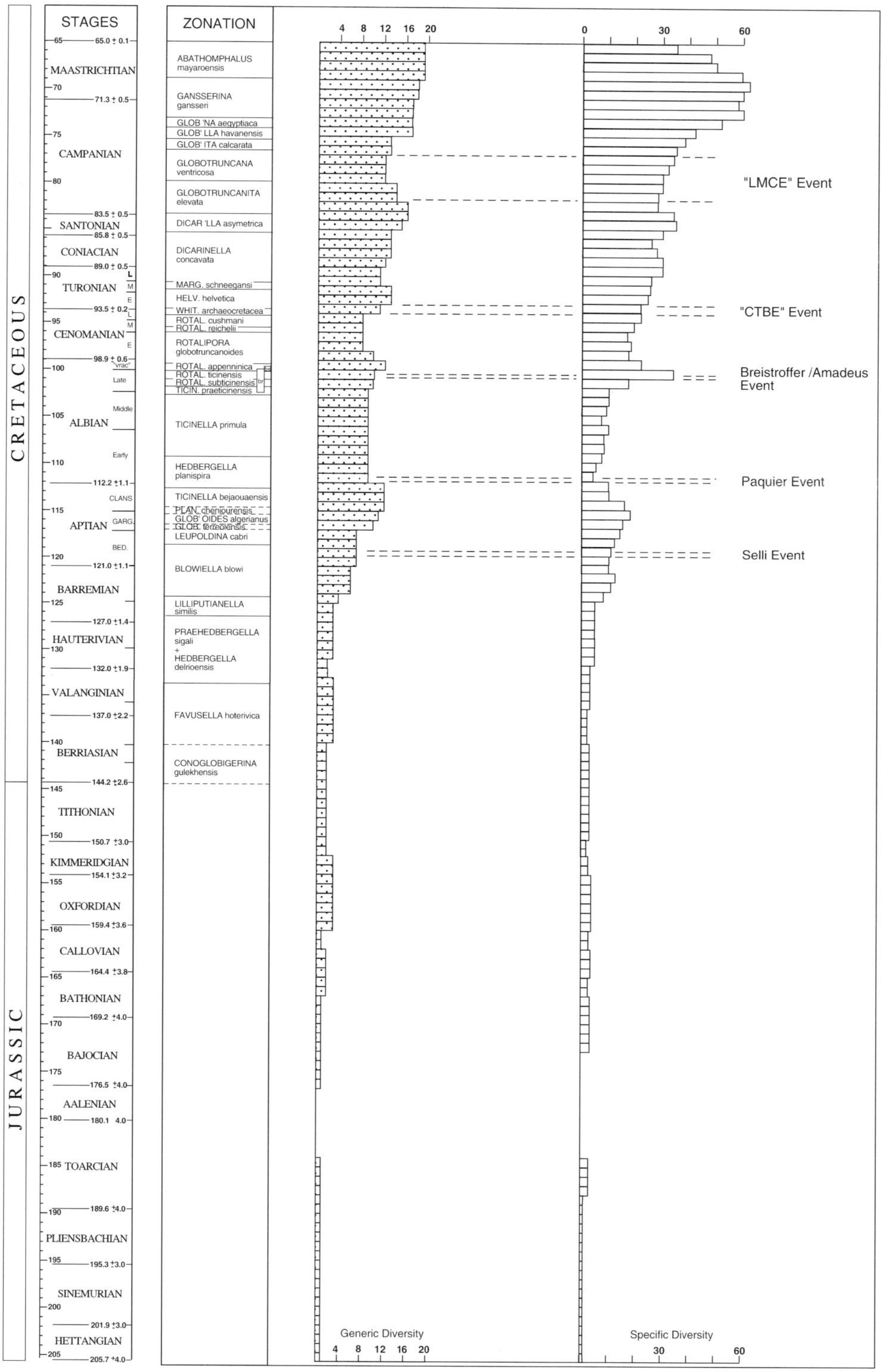

Fig. 4. Generic and specific diversity of planktonic foraminifera throughout the Jurassic and Cretaceous (partly after Hart 1999).

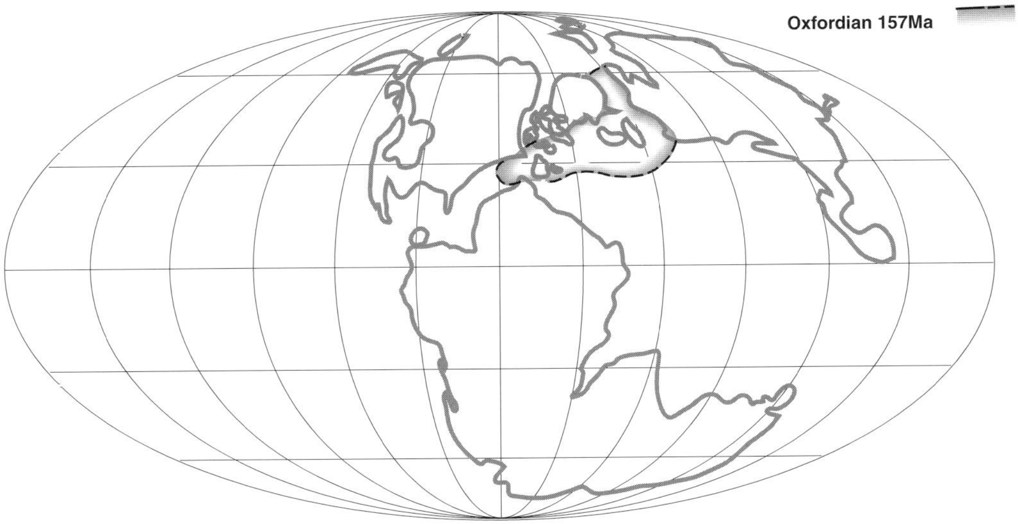

Fig. 5. Distribution of planktonic foraminifera in the Oxfordian.

described by Fuchs (1967) and which have been seen by the senior author in The Natural History Museum. This species appears to be a *disaster* or *opportunist* taxon, according to the definition of Harries *et al.* (1996). These more inflated forms of *O. quadrilobata* are found in the UK, Germany and France (Hylton 2000), while the associated fauna of *Praegubkinella* spp. is only known from Switzerland (Wernli 1995) at the present time.

Our more inflated specimens of *O. quadrilobata* are probably still benthonic forms as the apertural side remains relatively flat. It is clear, however, that chamber inflation has occurred and it is impossible to deny the possibility that, in mature specimens, some degree of 'floating'

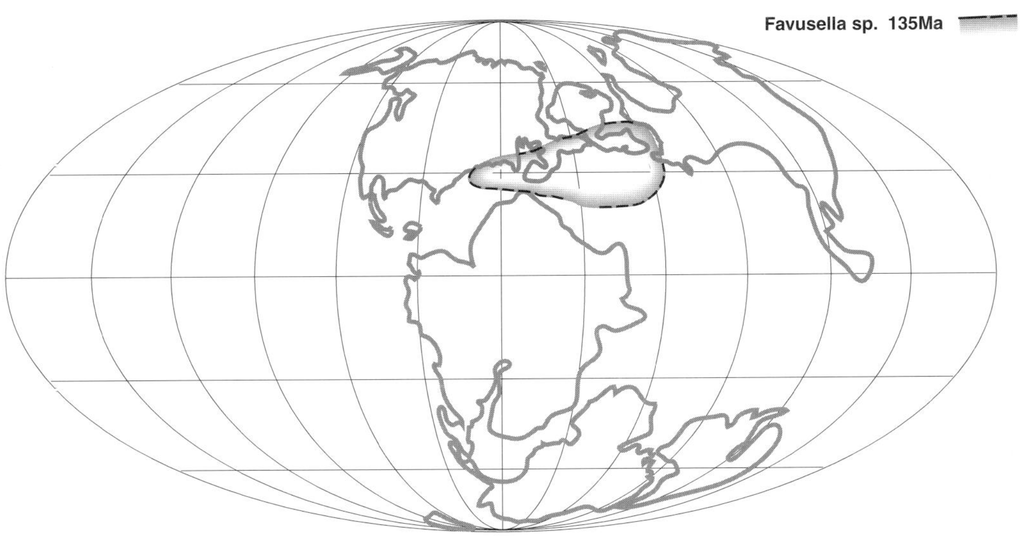

Fig. 6. Distribution of planktonic foraminifera in the Valanginian.

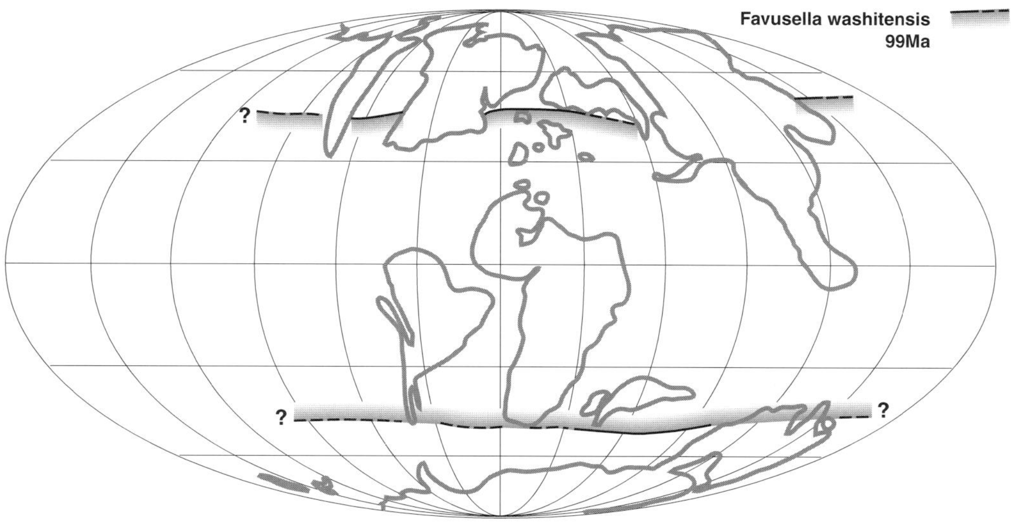

Fig. 7. Distribution of planktonic foraminifera in the mid- to late Albian.

could have taken place. In Wernli's (1995) assemblage of *Praegubkinella* there are some forms with relatively flat apertural faces while *P. racemosa* appears much more inflated. Again there is the possibility that mature individuals may have been 'planktonic'. The time gap between the ranges of *P. racemosa* and the earliest true *Conoglobigerina* is significant and this is one stratigraphic interval that requires urgent investigation.

The *Conoglobigerina* fauna is best described from Hungary (Wernli & Görög 1999, 2000), Morocco (Wernli 1987), the former Soviet Union (Morozova & Moskalenko 1961; Kuznetsova & Gorbatchik 1980; Kasimova & Aliyeva 1984) and Greece (Baumgartner 1985). In the company of A. Görög and J. Svabo, the senior author has visited the Somhegy (Som Hill) section west of Zirc (Hungary). Here 'ammonitico rosso' carbonates are found in fissures within the local succession. Samples collected from the Bajocian/Bathonian sediments have confirmed the presence of abundant *Conoglobigerina*, comprising both thick-walled and thin-walled forms. Whether this is just a specific difference or whether this variation in wall thickness indicates genuine dimorphism requires further investigation. The external morphology of the thick-walled forms (Wernli & Görög 2000) is such that they appear almost identical to the thin-walled forms and it seems unlikely, therefore, that they are simply benthonic/planktonic dimorphic pairs. Species of *Conoglobigerina* are known from many locations in central-southern and eastern Europe and this certainly marks a major expansion of the planktonic fauna within the Middle Jurassic (Fig. 3). If Wernli (1995) was correct in his derivation of *Conoglobigerina* from *Praegubkinella racemosa*, then an origin from the *Praegubkinella* assemblages of the Alpine area would appear to be quite plausible in palaeogeographical terms.

Evolution of the Mesozoic planktonic foraminifera

Figure 1 shows the stratigraphical distribution of the early planktonic foraminifera and while, in many cases, the evolutionary lineages/linkages are unknown it is clear that both specific and generic diversity remained low throughout the Jurassic and early Cretaceous. Some of the lineages are quite disjointed and it is apparent that a considerable amount of research remains to be done on these faunas. One of the principal problems, however, is the question of preservation. The early planktonic foraminifera are all thought to have been aragonitic, very much like the *Reinholdella, Oberhauserella* and *Epistomina* faunas with which Jurassic taxa are often associated in faunal assemblages. Aragonitic foraminifera are best preserved in dense clay lithologies and are rarely well preserved in porous sandstones and siltstones or some carbonate-rich environments. The earliest Globigerinina have been placed in the Superfamily Favusellacea (Simmons *et al.* 1997), a group which continues through until the mid-Cretaceous (Fig. 2). In the Albian and Cenomanian the *Favusella* spp. recorded in the UK (Carter & Hart 1977) do not appear to be aragonitic, being found well-preserved in the chalk facies (in which few other aragonitic fossils are recorded in such a good state of preservation). At the present time there is little evidence as to when the transition from aragonitic to calcitic tests is to be found. The lack of information on the faunas of the Tithonian to Valanginian interval may be critical in this respect.

The major evolutionary changes in the Toarcian and Bajocian are potentially related to sea-level highstands (Haq *et al.* 1988) and appear coincident with intervals of marked ammonite diversification (O'Dogherty *et al.* 2000). If one plots the distribution of the fauna through the Jurassic and Cretaceous (Fig. 2) it is clear that the major change in evolutionary rate occurs in the Aptian (Hart 1999; Premoli Silva & Sliter 1999). Both specific and generic diversity (Fig. 4) also increase at this level and this change appears to be coincident with a major increase in ocean crust production (Larson 1991*a*,*b*; Larson *et al.* 1993)

In the earliest Oxfordian the planktonic foraminifera (mainly *Globuligerina oxfordiana* (Grigelis 1958) and *G. bathoniana* (Pazdrowa 1969)) expanded their distribution (Fig. 5) and are reported from much of Europe, including the UK (Oxford *et al.* in press) and the Grand Banks (Ascoli 1976). The work of Oxford *et al.* (in press) in Dorset has demonstrated the presence of *G. oxfordiana*, *Haeuslerina helvetojurassica* (Haeusler, 1881) and *Compactogerina stellapolaris* (Grigelis in Grigelis *et al.* 1977) in the Furzedown Clays (*Q. mariae* Zone) of the Oxfordian. This is a comparable level to the occurrence of *G. oxfordiana* in the Marnes de Villers of the Vaches Noires cliffs of Normandy and subsurface material from near Le Havre (Bignot & Guyader 1966, 1971; Samson *et al.* 1992). All these occurrences are coincident with the *Q. mariae* Zone highstand of the early Oxfordian (Haq *et al.* 1988; Jacquin *et al.* 1998) and appear to represent a distinctive pulse of migration into the area of NW Europe. Information from the latest Jurassic and earliest Cretaceous is rather limited and the distribution and evolution of the planktonic foraminifera are poorly known. Sea levels during this time interval are, generally, low and many of the shelf areas covered during the Oxfordian and Kimmeridgian are either non-marine or exposed land. Even in the Valanginian the distribution of the planktonic foraminifera (Fig. 6) is essentially that seen in the earliest Oxfordian. In the Barremian stage, however, diversification begins (Figs 2 and 4) and continues through the Aptian and Albian. By the latest Albian (Fig. 7) the near-global distribution of the planktonic foraminifera is established and continues through until the end-Maastrichtian extinction event.

It is during the latest Albian that the distinctive 'keeled' morphotypes appear as, prior to that time, only 'hedbergellid' taxa are known. Within the early Albian a number of lineages appear (*Ticinella, Globigerinelloides,* etc.) that indicate a diversification of the hedbergellid stock, but it is difficult to assess if these taxa were depth-stratified in the same way as the younger faunas (Hart 1999). The single-keeled morphotypes appear in the latest Albian, possibly associated with the sea-level changes in the *S. dispar* Zone. The twin-keeled morphotypes appear in the latest Cenomanian which, again, marks a time of significant sea-level rise. Throughout the mid- to late Cretaceous the succession is punctuated by a number of 'events' beginning with the Faroni Event in the Barremian. All of these events (Selli, Paquier, Amadeus, Bonnarelli, etc.) are associated with black shales and/or mudstones and have often been described as 'anoxic events'. Their effects on the planktonic foraminifera have been documented by a wide range of authors (e.g. Premoli Silva & Sliter 1999) and while the changes at some of these levels are significant, they cannot be described as *controlling* the evolution of the planktonic foraminifera (Hart 1999; Premoli Silva & Sliter 1999). Hart (1999) showed that there were significant changes to the fauna at the level of the Bonnarelli event (latest Cenomanian) with a number of extinctions and first appearances, but other events (e.g. the Selli event in the earliest Aptian) record few, if any, changes.

The palaeolatitudinal distribution of the Cretaceous planktonic foraminifera has recently been documented by Hart (2000) and Hudson (2000). This work continues and a series of maps for the Jurassic and Cretaceous is currently in production. The changes recorded at any of the events during the Cretaceous cannot, however, be compared to that at the end of the Cretaceous (Hart 1999, fig. 5). At the end of the Maastrichtian the fauna was decimated and only two or three survivors provided the foundation for the Cenozoic fauna. Only the tiny, hedgergellid taxa survived and the evolutionary clock is returned almost to Oxfordian levels. The diversification of the Cenozoic fauna following the extinction event is, however, much more rapid than that seen in the Mesozoic.

Summary

The origins of the planktonic foraminifera can be traced back to the Toarcian, although it is only in the Bathonian that meroplanktonic taxa appear. The inflated *O. quadrilobata* and *Praegubkinella* spp. present in Toarcian strata are clearly not fully planktonic and may have been benthonic throughout much of their life cycle. The environmental disruption caused by the early Toarcian sea-level rise and associated 'anoxic event' are potential triggers for this change in lifestyle as may have been the massive release of methane-derived carbon from sub-sea gas hydrates in the *falciferum* Zone of the early Toarcian (Hesselbo *et al.* 2000).

These early faunas, including the *Conoglobigerina* spp. assemblage in the Bathonian, are all limited to the northern side of Tethys (Figs 3, 5 and 6). With such a limited fauna in the restricted areas shown in these maps it is unlikely that many diversification opportunities were available, especially in the latest Jurassic and earliest Cretaceous interval which was characterized by reduced sea levels. Once ocean floor production accelerated during the latest Barremian and earliest Aptian and, as a result, sea levels rose globally, the planktonic foraminifera were able to use the newly formed ocean basins (and expanded lengths of continental margin) to attain a near-global distribution by the mid-Cretaceous (Fig. 7). The continental fragmentation during the Aptian and Albian allowed the hedbergellid faunas to expand rapidly and as the various ocean basins became connected, the stratification of the water column generated the opportunity for the evolution of a depth-stratified planktonic fauna by the mid- to late Albian (see Hart 2000).

The authors acknowledge the helpful discussions of this topic with M. Simmons (Cambridge), M. Caron (Fribourg), A. Görög (Budapest), R. Wernli (Geneva), J. Gregory (St Albans), A. Henderson (London) and J. Whittaker (London). A. Görög and J. Svabo are thanked for their help in the field in Hungary and J.-P. Berger is thanked for help in the field in Switzerland. M.B.H. acknowledges a Research Grant from the Royal Society while M.J.O. acknowledges receipt of a Research Studentship funded by the University of Plymouth and Amerada Hess (London). J. Abraham is thanked for his patient work in preparing the figures.

References

ASCOLI, P. 1976. Foraminiferal and ostracod biostratigraphy of the Mesozoic–Cenozoic Scotian Shelf, Atlantic Canada. *Maritime Sediments, Special Publication,* **1**, 653–771.

BAUMGARTNER, P.O. 1985. Jurassic sedimentary evolution and nappe emplacement in the Argolis Peninsula (Peloponnesus, Greece). *Denkschriften der Schweizerischen Naturforschenden Gesellschaft,* **69**.

BIGNOT, G. & GUYADER, J. 1966. Découverte de Foraminifères planctoniques dans l'Oxfordien du Havre (Seine Maritime). *Revue de Micropaléontologie,* **9**, 104–110.

BIGNOT, G. & GUYADER, J. 1971. Observations nouvelles sur *Globigerina oxfordiana* Grigelis. *In*: FARINACCI, A. (ed.) *Proceedings of the Second Planktonic Conference, Roma 1970,* Volume 1. Edizioni Tecnoscienza, Rome, 79–81.

CARON, M. 1983. Taxonomie et phylogenie de la famille des Globotruncanidae. *Zitteliana,* **10**, 677–681.

CARTER, D. J. & HART, M. B. 1977. Aspects of mid-Cretaceous stratigraphical micropalaeontology. *Bulletin of the British Museum, Natural History (Geology),* **29**, 1–135.

DARLING, K. F., KROON, D., WADE, C. M. & LEIGH BROWN, A. J. 1996. Molecular phylogeny of the planktic Foraminifera. *Journal of Foraminiferal Research,* **26**, 324–330.

DARLING, K. F., WADE, C. M., KROON, D. & LEIGH BROWN, A. J. 1999*a*. Planktic foraminiferal molecular evolution and their polyphyletic origins from benthic taxa. *Marine Micropaleontology,* **30**, 251–266.

DARLING, K. F., WADE, C. M., KROON, D., LEIGH BROWN, A. J. & BIJMA, J. 1999*b*. The diversity and distribution of modern planktic foraminiferal small subunit ribosomal RNA genotypes and their potential as tracers of present and past ocean circulation. *Paleoceanography,* **14**, 3–12.

FUCHS, W. 1967. Űber Ursprung und Phylogenie de Trias- 'Globigerinen' und die Bedeutung Dieses Formenkreises fűr das echte Plankton. *Verhandlungen der Geologischen Bundesanstalt, Wien,* **1967**, 135–176.

FUCHS, W. 1971. Eine alpine Foraminiferenfauna des tieferen Mittel-Barrême aus den Drusbergschichten von Ranzenberg bei Hohenems in Vorarlberg. *Abhandlungen der Geologischen Bundesanstalt, Wien,* **27**, 1–49.

FUCHS, W. 1973. Ein Beitrag zur Kenntnis der Jura – 'Globigerinen' und verwandter Formen an Hand polnischen Materials des Callovien und Oxfordien. *Verhandlungen der Geologischen Bundesanstalt, Wien,* **1973**, 445–487.

FUCHS, W. 1975. Zur Stammesgeschichte der Plankton-foraminiferen und verwandter Formen im Mesozoikum. *Jahrbuch der Geologischen Bundesanstalt, Wien,* **118**, 193–246.

FUCHS, W. 1977. A contribution to the phylogeny of the Mesozoic planktonic foraminifera. *In*: Acte du VIe Colloque Africain de Micropaléontologie.

Annales des Mines et de la Géologie, Tunis, **28**, 71–74.

GRIGELIS, A. A. 1958. *Globigerina oxfordiana* sp. n. – an occurrence of *Globigerina* in the Upper Jurassic strata of Lithuania. *Nauchnye Doklady Vysshei Shkoly, Geologo-Geograficheskie Nauki,* **1958(3)**, 109–111 (in Russian).

GRIGELIS, A. A., Mesezhinikov, M. S., Yakovleva, S. P. & Kozlova, G. E. 1977. The first finds of planktonic foraminifera in the Upper Jurassic of the Pechora River Basin. *Doklady Akademii Nauk SSSR,* **233**, 926–927 (in Russian).

GÖRÖG, A. 1994. Early Jurassic planktonic foraminifera from Hungary. *Micropaleontology*, **40**, 225–260.

HAEUSLER R. 1881. Note sur une zone à globigerines dans le terrain jurassique de la Suisse. *Annales de la Société Royale Malacologique de Belgique* (Serie 3, **1**), **16**, 188–190.

HAQ, B. U., HARDENBOL, J. & VAIL, P. R. 1988. Mesozoic and Cenozoic chronostratigraphy and cycles of sea-level change. *In*: WILGUS, C. K., HASTINGS, B. S., KENDALL, C. G. ST., POSAMENTIER, H. W., ROSS, C. A. & VAN WAGONER, J. C. (eds) *Sea-level Changes: An Integrated Approach, Society of Economic Paleontologists and Mineralologists*, Special Publications, **42**, 71–108.

HARRIES, P. J., KAUFFMAN, E. G. & HANSEN, T. A. 1996. Models for biotic survival following mass extinction. *In*: HART, M. B. (ed.) *Biotic Recovery from Mass Extinction Events.* Geological Society, London, Special Publications, **102**, 41–60.

HART, M. B. 1999. The evolution and biodiversity of Cretaceous planktonic Foraminiferida. *Geobios*, **32**, 247–255.

HART, M. B. 2000. Climatic modelling in the Cretaceous using the distribution of planktonic Foraminiferida. *In*: HART, M. B. (Ed.) *Climates: Past and Present.* Geological Society, London, Special Publications, **181**, 33–41.

HESSELBO, S. P., GRÖCKE, D. R., JENKYNS, H. C., BJERRUM, C. J., FARRIMOND, P. MORGANS BELL, H. S. & GREEN, O. R. 2000. Massive dissociation of gas hydrate during a Jurassic anoxic event. *Nature*, **406**, 392–395.

HUDSON, W. 2000. *A comparison of the Cretaceous and Recent faunas together with their spatial and temporal distributions.* MSc Thesis, University of Plymouth.

HYLTON, M. D. 2000. *Microfaunal investigation of the Early Toarcian (Lower Jurassic) extinction event in N.W.Europe.* PhD Thesis, University of Plymouth.

HYLTON, M. D. & HART, M. B. 2000. Benthic foraminiferal response to Pliensbachian–Toarcian (Lower Jurassic) sea-level change and oceanic anoxia in NW Europe. *GeoResearch Forum*, **6**, 455–462.

JACQUIN, T., DARDEAU, G., DURLET, C., DE GRACIANSKY, P.-C. & HANTZPERGUE, P. 1998. The North Sea Cycle: an overview of 2nd-Order transgressive/regressive facies cycles in Western Europe. *In*: De GRACIANSKY, P.-C., HARDENBOL, J., JACQUIN, T. & VAIL, P.R. (eds) *Mesozoic and Cenozoic Sequence Stratigraphy of European Basins.* Society of Economic Paleontologists and Mineralogists, Special Publications, **60**, 445–466.

KASIMOVA, G. K. & ALIYEVA, D. G. 1984. Planktonic foraminifera of the Middle Jurassic beds of Azerbaijan. *Voprosy Paleontologii i Stratigrafi Azerbaijana,* **479**, 8–19 (in Russian).

KUZNETSOVA, K. I. & GORBATCHIK, T. N. 1980. Novienakhodki planktonnikh foraminifer, v verkhyurskikh otlzhenii Krime. *Doklady Academii Nauk, SSSR,* **254**, 748–751 (in Russian).

LARSON, R. L. 1991*a*. Latest pulse of the Earth: evidence for a mid-Cretaceous superplume. *Geology*, **19**, 547–550.

LARSON, R. L. 1991*b*. Geological consequences of superplumes. *Geology*, **19**, 963–966.

LARSON, R. L., FISCHER, A. G., ERBA, E. & PREMOLI SILVA, I. 1993. Summary of workshop results. *In*: LARSON, R. L. *et al.* (eds) *Apticore-Albicore: A workshop on global events and rhythms of the mid-Cretaceous*, Perugia, October 1992.

METTRAUX, M., DUPASQUIER, C. & HOMEWOOD, P. 1986. Conditions de dépôt et diagenèse précoce du Toarcien inférieur des Préalpes médianes romandes (Suisse). *Documents BRGM,* **110**, 231–237.

METTRAUX, M., WEISSERT, H. & HOMEWOOD, P. 1989. An oxygen-minimum palaeoceanographic Signal from the Early Toarcian cavity fills. *Journal of the Geological Society, London,* **146**, 333–344.

MOROZOVA, V. G. & MOSKALENKO, T. A. 1961. Planktonnie foraminiferi progranichnikh otlozhenii Bayosskogo i Batskogo yarusov tsentral'nogo Dagestana (severo-vostochnii Kavkaz) *Voprosy Mikropaleontologii,* **5**, 3–30 (in Russian).

O'DOGHERTY, L., SANDOVAL, J. & VERA, J. A. 2000. Ammonite faunal turnover tracing sea-level changes During the Jurassic (Betic Cordillera, southern Spain). *Journal of the Geological Society, London,* **157**, 723–736.

OXFORD, M. J., GREGORY, F. J., HART, M. B., HENDERSON, A. S., SIMMONS, M. D. & WATKINSON, M. P. (in press). Jurassic planktonic foraminifera from the United Kingdom. *Terra Nova.*

PAZDROWA, O. 1969. Bathonian *Globigerina* of Poland. *Rocznik Polskiego Towarzystwa Geologicznego,* **39**, 41–56.

PREMOLI SILVA, I. & SLITER, W. V. 1999. Cretaceous paleoceanography: evidence from planktonic foraminiferal evolution. *Geological Society of America, Special Paper,* **332**, 301–328.

SAMSON, Y., JANIN, M.-C. & BIGNOT, G. 1992. Les globuligerines (Foraminifères Planctoniques) de l'Oxfordien Inférieur de Villers-sur-Mer (Calvados, France) dans leur gisement. *Revue de Paléobiologie*, **11**, 409–431.

SIMMONS, M. D., BOUDAGHER-FADEL, M. K., BANNER, F. T. & WHITTAKER, J. E. 1997. The Jurassic Favusellacea, the earliest Globigerinina. *In*: BOUDAGHER-FADEL, M. K., BANNER, F. T. & WHITTAKER, J. E. (eds) *The Early Evolutionary History of Planktonic Foraminifera.* Chapman & Hall, London, 17–51.

WEIDMANN, M. 1993. *Atlas géologique de la Suisse, 1: 25000.* Notice explicative, Feuille 1244, Châtel-St-Denis.

WEIDMANN, M., HOMEWOOD, P., MOREL, R. *ET AL.* 1993. *Atlas Géologique de la Suisse, Feuille 1244, Châtel-St-Denis (Feuille 92 de L'Atlas).* Publié par le Service hydrologique et géologique national.

WERNLI, R. 1987. Les protoglobigérines (Foraminifères) du Bajocien inférieur des Sofs (Rif, Maroc). *Eclogae Geologicae Helvetiae*, **80**, 817–829

WERNLI, R. 1995. Les Foraminifères globigériniformes (Oberhauserellidae) du Toarcien Inférieur de Teysachaux (Préalpes médianes, Fribourg, Suisse). *Revue de Paléobiologie,* **14**, 257–269.

WERNLI, R. & GÖRÖG, A. 1999. Protoglobigerinids (Foraminifera) acid extracted from Bajocian limestones (Hungary). *Revista Española de Micropaleontologia,* **31**, 419–426.

WERNLI, R. & GÖRÖG, A. 2000. Determination of Bajocian protoglobigerinids (Foraminifera) in thin sections. *Revue de Paléobiologie,* **19**, 399–407.

Opening of the Hispanic Corridor and Early Jurassic bivalve biodiversity

MARTIN ABERHAN

Museum für Naturkunde, Institut für Paläontologie, Invalidenstr. 43, D-10115 Berlin, Germany, (e-mail: martin.aberhan@rz.hu-berlin.de*)*

Abstract: The Hispanic Corridor is a postulated marine seaway linking the eastern Pacific and western Tethyan oceans as early as Early Jurassic times. Two existing hypotheses relate the Pliensbachian–Toarcian bivalve extinction and recovery to immigration of bivalve species through the Hispanic Corridor. The extinction hypothesis implies that, in South America, the Pliensbachian–Toarcian extinction can be partly explained by the immigration of bivalves through the Hispanic Corridor and subsequent competitive replacement. The recovery hypothesis states that, in NW Europe, the renewed rise in diversity in the late Toarcian/Aalenian was largely a consequence of immigration of taxa from Andean South America via the Hispanic Corridor.

To test these hypotheses, I calculated immigration and origination rates of bivalves per million years. In both regions, early Pliensbachian to Aalenian immigration rates remained at low levels, thus disproving both hypotheses. By comparison, the origination of new species generally played a much more important role than immigration in controlling overall diversity of both regions. Future research should investigate if this is a more general pattern in the recovery of post-extinction biotas.

The apparently global Pliensbachian–Toarcian diversity crisis may be best explained by a combination of physicochemical factors, invoking intense volcanism, sea-level highstand and widespread anoxia, as well as biological factors. Recovery from this mass extinction commenced when origination rates increased again, which, in the Andean basins, was in the Aalenian and in NW Europe, the late Toarcian.

Investigations of the history of diversity and of evolutionary processes that generate biodiversity not only improve our knowledge of the fossil past, but may contribute significantly to our understanding of the course and consequences of the modern biodiversity crisis. A promising palaeontological approach is the analysis of the geographic variation of patterns observed during extinction episodes and their subsequent recoveries. In the present study, I analyse the biogeography of Early Jurassic bivalves with the aim of testing two hypotheses existing in the literature. These hypotheses relate the Pliensbachian–Toarcian mass extinction to the opening of a marine connection (the so-called Hispanic Corridor) between the eastern Pacific and western Tethyan oceans. The first hypothesis states that, in South America, the preferential disappearance of endemic bivalves across the Pliensbachian–Toarcian boundary can be partly explained by immigration of cosmopolitan species via the Hispanic Corridor and subsequent competitive replacement (Aberhan & Fürsich 1997). The second hypothesis states that, in NW Europe, the renewed rise in bivalve diversity in the late Toarcian/Aalenian was largely a consequence of immigration of taxa from Andean South America through the Hispanic Corridor, presumably filling the ecospace vacated during the extinction event (Hallam 1983, 1996; Hallam & Wignall 1997). I demonstrate that neither of the two hypotheses can be sustained. Whilst the extinction may be best explained by a combination of global palaeoenvironmental changes and biological processes, the recovery appears to be largely controlled by increasing within-region origination rates rather than immigration of already existing species.

Early Jurassic bivalve biodiversity and the Hispanic Corridor

Their high abundance and taxonomic diversity since early Mesozoic times, along with an excellent fossil record, make bivalves an ideal group for palaeodiversity analysis in the marine realm. At the species level, Early Jurassic bivalve diversity is known in detail from NW European epicontinental seas (Hallam 1986, 1987) and from western South America (Damborenea 1996; Aberhan & Fürsich 1997, 2000), where bivalves occur very abundantly in a series of backarc basins, the so-called Andean basins. In both regions, bivalves experienced a

From: CRAME, J. A. & OWEN, A. W. (eds) 2002. *Palaeobiogeography and Biodiversity Change: the Ordovician and Mesozoic–Cenozoic Radiations*. Geological Society, London, Special Publications, **194**, 127–139.
0305-8719/02/$15.00

considerable drop in species diversity from late Pliensbachian to early Toarcian times. This Pliensbachian–Toarcian extinction apparently was of global extent, clearly exceeds background levels of extinction and also affected several other groups of organisms at various taxonomic ranks (e.g. Raup & Sepkoski 1984; Hallam 1987; Little & Benton 1995; Aberhan & Fürsich 1997). For these reasons the Pliensbachian–Toarcian extinction can be termed a second-order mass extinction.

The Hispanic Corridor is a postulated, shallow marine connection between the eastern Pacific and the western Tethys, which preceded the birth of the Atlantic Ocean by many millions of years (Smith 1983). Although there is little direct geological evidence for its existence, palaeontological data suggest that a narrow seaway was established by Early Jurassic times (e.g. Damborenea & Manceñido 1979; Hallam 1983; Smith & Tipper 1986). Biogeographic analysis of Early Jurassic pectinoid bivalves, combined with a review of the relevant literature, supports the view that the Corridor presumably was open from Pliensbachian times onwards (Aberhan 2001). The apparently bidirectional faunal exchange through the Corridor is consistent with the establishment of a megamonsoonal circulation for Pangaea, which may have caused seasonal alternation of flow directions within the Corridor (Aberhan 2001). Thus, the Hispanic Corridor was in operation during the Early Jurassic extinction and recovery intervals, and could have promoted changes in regional diversity at opposite ends of the Corridor.

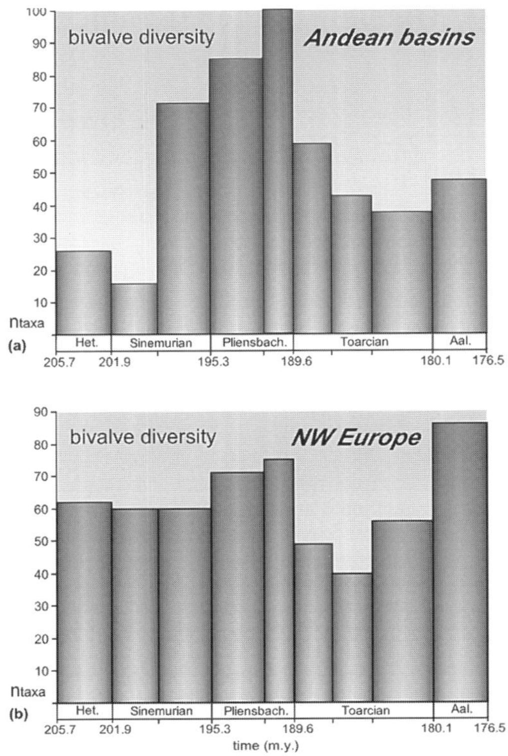

Fig. 1. Bivalve diversity (expressed as number of species) for various Jurassic stages and substages. (**a**) Andean basins, (**b**) NW Europe. Het., Hettangian; Aal., Aalenian.

Database and methods

This study is based on species-level data for bivalves from two regions, western South America (Colombia, Peru, Chile, Argentina) and NW Europe (Great Britain, Sweden, Denmark, Belgium, Luxemburg, Germany, Switzerland, northern France). The South American data (see Appendix 1) are based on extensive field work and many years of detailed taxonomic work (e.g. Aberhan 1994) and also include a wealth of literature citations. The NW European data (see Appendix 2) mainly stem from the work of Hallam (1976, Appendix; 1987, Appendix). This database was modified to include recent work by Johnson (1984), Hodges (1991, 2000), Muster (1995), Röhl (1998) and Harries & Little (1999), and also takes into account my examination of museum collections in The Natural History Museum, London; Goldfuß-Museum, Bonn; Museum für Naturkunde, Berlin; Sedgwick Museum, Cambridge; and Staatliches Museum für Naturkunde, Stuttgart. Whilst the NW European data are resolved to zonal level, the South American data were pooled in substages due to the lack of more precise biostratigraphic information. To permit a comparison, the data were processed at the substage level for both regions.

Species diversity is expressed as the number of species per substage. Total extinction rates for the various substages are calculated as the number of species making their last appearance divided by the duration of the substage (substage durations after Gradstein *et al.* 1994). Per-species extinction rates, analogous to Raup & Sepkoski's (1984) per-family extinction rate, are determined as the total extinction rate divided by the number of species extant during the substage. For each region, extinction metrics are calculated separately for the extinction of endemic bivalves, the regional disappearance

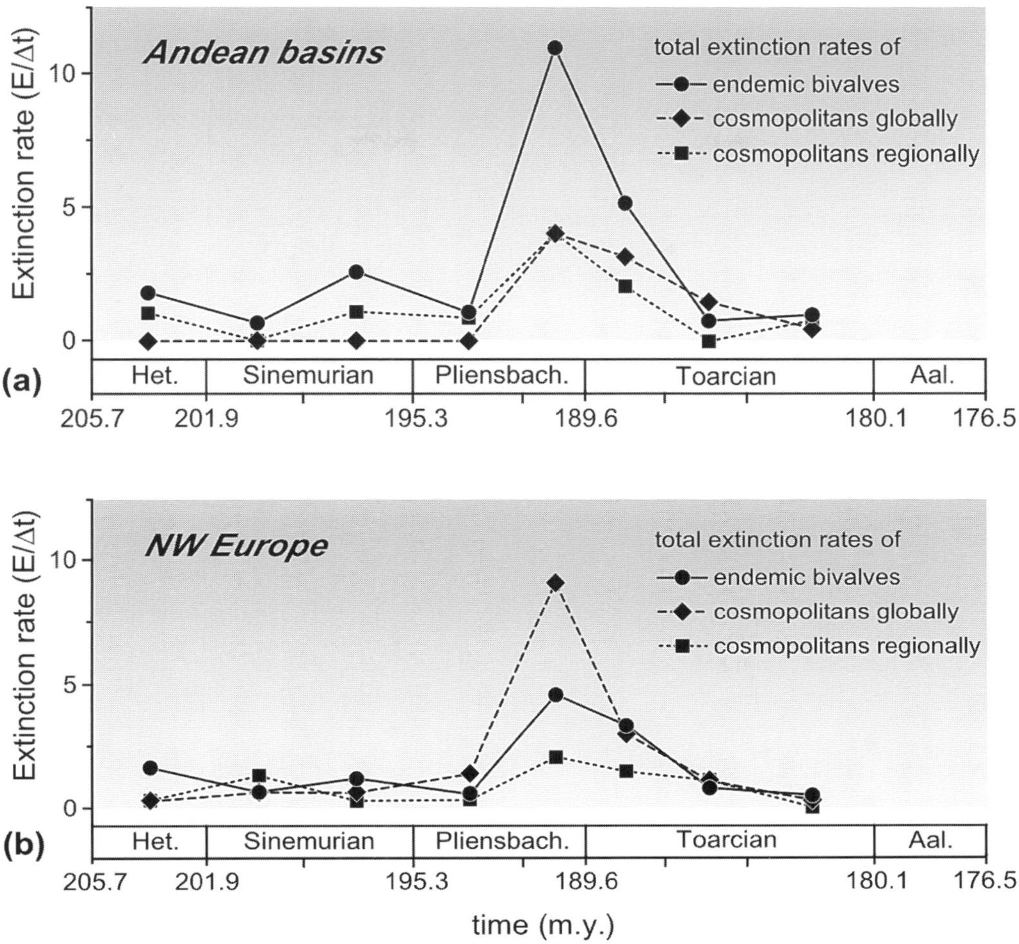

Fig. 2. Total extinction rates of bivalve species per million years through Early Jurassic time. (**a**) Andean basins, (**b**) NW Europe. Key for time units as in Figure 1.

of cosmopolitans and the global extinction of cosmopolitans. Taxa are classified as cosmopolitans if they occurred in western South America and/or NW Europe and in at least one more region such as southern Europe, North America, Siberia etc. Origination rates are expressed as the number of species originating in a substage divided by the duration of the substage.

Finally, I calculated the rate at which species immigrated into NW Europe and the Andean basins, respectively. Immigration rates are defined as the number of immigrating species divided by the duration of the substage. To be classified as an immigrant that utilized the Hispanic Corridor for dispersal a species had to be: (1) present at opposite sides of the Corridor; (2) simultaneously absent in the western Pacific/eastern Tethys; and (3) confined to relatively low palaeolatitudes. Fulfilment of all three criteria renders alternative dispersal routes, such as migration around the periphery of Pangaea or long-range dispersal of larvae across the palaeo-Pacific, unlikely (see also Aberhan 2001). For each bivalve species of the two regions the geographic distribution was determined by a survey of the relevant literature. This was greatly facilitated by the Jurassic bivalve catalogue housed in Würzburg. It consists of copies of most figured specimens of Jurassic bivalves along with locality and age information.

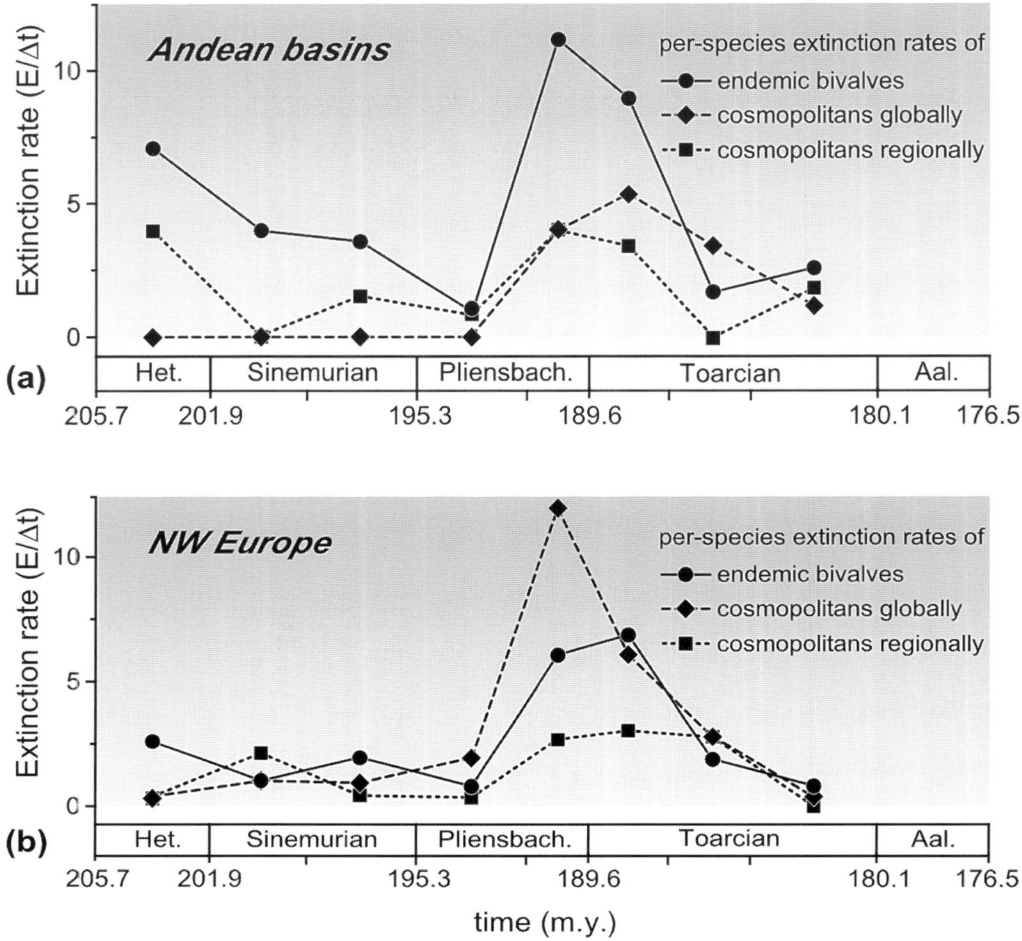

Fig. 3. Per-species extinction rates of bivalves per million years through Early Jurassic time. (**a**) Andean basins, (**b**) NW Europe. Key for time units as in Figure 1.

Within the temporal accuracy of the present study, several species (11 in South America and 14 in NW Europe) seem to have their first appearance simultaneously in different regions. For the regions concerned it remains unclear whether these species belong to the category of originating or that of immigrating species. When counting the numbers of originating and of immigrating species these taxa were omitted. Since their percentages in the various time intervals are relatively low (0% to 8.5% in South America and 1.6% to 6.1% in NW Europe), it is unlikely that this procedure strongly distorted the comparison of origination rates and immigration rates.

Patterns of diversity, extinction and immigration

Figure 1 shows the diversity of bivalves over the stages and substages of the Early Jurassic for the Andean basins and for NW Europe. Both regions show similar diversity trends with increasing values from the earliest Jurassic to a peak in the late Pliensbachian, followed by a sharp drop in the early Toarcian. In South America, this decline is continued, albeit less markedly, into the middle Toarcian and finally the late Toarcian, where the diversity reaches a relative minimum. Recovery from this Pliensbachian–Toarcian extinction did not take place before the Aalenian, but species richness remained distinctly below the Pliensbachian diversity maxima. In NW Europe, the diversity minimum is situated in the middle Toarcian. Recovery commenced in the late Toarcian, and by Aalenian times bivalve diversity surpassed even Pliensbachian diversity values.

Total extinction rates of endemics and cosmopolitans, and rates of the regional disappearance of cosmopolitans, are depicted in Figure 2. The most prominent features of the various curves are the marked extinction peaks in the late Pliensbachian. In NW Europe, they reflect the worldwide extinction of cosmopolitans and the extinction of endemics, whereas in the Andean basins, it was mainly the endemics that went extinct. Albeit less markedly, elevated extinction rates continue into the early Toarcian.

Late Pliensbachian extinction peaks also are evident in the graphs of per-species extinction rates (Fig. 3). In contrast to Figure 2, however, the extinction rate of NW European endemics, and that of cosmopolitans that went extinct in South America, reached their maxima in the early Toarcian. This indicates that a protracted interval of time, spanning the late Pliensbachian and early Toarcian, showed elevated extinction levels.

Figure 4 depicts the immigration rates for both regions, and emphasizes the portion of species that apparently spread through the Hispanic Corridor. Throughout the Early Jurassic, immigration rates were relatively low, ranging from zero to less than four species per million years. A tendency towards increasing immigration rates through time can be identified in NW Europe. An opposite, although repeatedly interrupted trend may be recognized in South America.

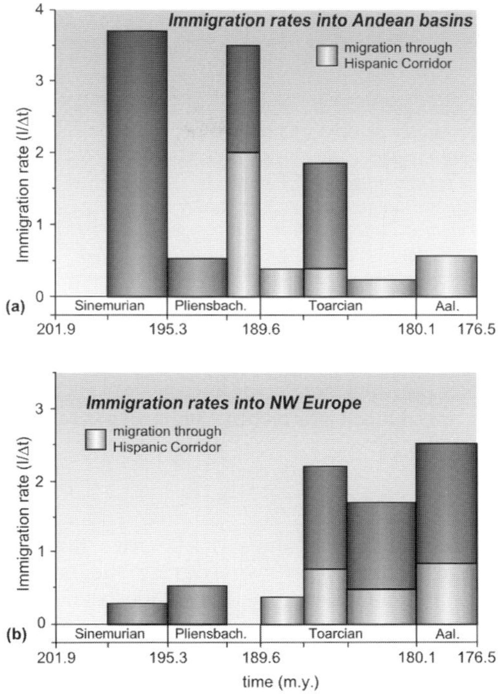

Fig. 4. Immigration rates of bivalve species per million years through time (Sinemurian to Aalenian). (**a**) Andean basins, (**b**) NW Europe.

Discussion

Dispersal through the Hispanic Corridor

As can be seen from Figure 4, none of the bivalve species seems to have migrated through the Hispanic Corridor before the Pliensbachian, and from Pliensbachian times onwards migration through the Corridor apparently was in both directions. These patterns argue for a Pliensbachian opening of the Corridor, providing a restricted faunal exchange between the eastern Pacific and western Tethys oceans for the remainder of Early Jurassic time. These results support the conclusions of Aberhan (2001), which were obtained by the same method (see the three criteria mentioned above). In contrast to Aberhan (2001), who concentrated on the distribution of a single group of bivalves in five regions, the present analysis considers the whole bivalve fauna, although the main focus is on only two regions. A slightly earlier, Sinemurian time for the effective operation of the Hispanic Corridor was proposed by Damborenea (2000), based on similarity coefficients at the generic level. However, with this approach it is difficult to preclude alternative dispersal routes. All in all, the Hispanic Corridor seems to be a viable feature of Early Jurassic Pangaea and, as a biogeographic link between two major oceans, had the potential to influence regional diversity.

Test of the extinction hypothesis

The extinction hypothesis implies that immigration of bivalves from Europe through the Hispanic Corridor contributed to the Pliensbachian–Toarcian bivalve extinction in South America (Aberhan & Fürsich 1997). Figure 4 shows that immigration rates into the Andean basins were extremly low in the early Pliensbachian and in the early Toarcian, whereas

Table 1. *Relative abundance of selected Early Jurassic bivalve species in northern Chile.*

Species	$N_{samples}$	$\%_{samples}$	Relative abundance (%)	Rel. abundance (mean)
Camptonectes auritus (Schlotheim)	7	3.5	0.5–3.7	2.0
Pseudopecten equivalvis (J. Sowerby)	4	2.0	0.5–3.5	1.4
Modiolus ventricosus (Roemer)	4	2.0	0.8–3.0	1.8
Plagiostoma giganteum J. Sowerby	–	–	–	–
Propeamusium pumilum (Lamarck)	(5)	(2.4)	(0.7–28.6)	(15.0)
Weyla alata (von Buch)	158	78.2	0.5–69.4	10.0

$N_{samples}$, number of samples in which species occurs (total number of samples = 202); $\%_{samples}$, percentage of Early Jurassic samples in which species occurs; relative abundance (%), range of relative abundances of species in samples; rel. abundance (mean), mean of relative abundances of species. Numbers in parentheses indicate data that include information from samples with less than 70 individuals. For comparison, the abundant bivalve *Weyla alata* has been included. Data based on Aberhan (1992) and unpublished information. For discussion see text.

moderately high values characterize the late Pliensbachian. However, during the critical time interval, from the early Pliensbachian to the early Toarcian, only five species possibly immigrated through the Corridor (Table 1). Compared to 57 species that went extinct during the same time interval, this group of five species seems to be too small to contribute significantly to the extinction.

It could be argued that a single, eurytopic and abundant species might have the potential to cause extinction of many other forms. Table 1 summarizes relative abundance data of species, including the five species in question. These data were obtained through detailed palaeoecological analysis of benthic associations from northern Chile (Aberhan 1992, 1993). It becomes evident that the first three species of Table 1 occurred in only a few samples (less than 4% of all quantitative samples) and in low abundances (on average 2% or less of the total fauna of a sample). Another species, *Plagiostoma giganteum*, occurred in very low numbers in a few samples, which were too small to be included in the statistical analysis of Aberhan (1992, 1993). For comparison, the ubiquitous bivalve species *Weyla alata* has been included in Table 1. It occurred in nearly 80% of the samples and is one order of magnitude more common than the other faunal elements (Table 1). In the Andean basins, the four species mentioned above can therefore be classified as rare and geographically restricted elements of Early Jurassic benthic communities, and it is indeed very unlikely that they caused competitive replacement.

An exception is the pectinoid bivalve *Propeamussium pumilum*, which becomes locally abundant (up to nearly 30%) in low-diversity, oxygen-controlled environments (Table 1; Aberhan 1993). Its occurrence in the early Toarcian coincided with the disappearance of the morphologically related species *Posidonotis semiplicata* (Hyatt), another low-oxygen-tolerant flat clam. *P. semiplicata* first appeared in the upper Sinemurian of several allochthonous terranes of the North American Cordillera and by late Pliensbachian times had spread to the Andean basins, where it persisted as a dominant benthic element of the dysaerobic biofacies into the early Toarcian (Aberhan & Pálfy 1996). The disappearance of *P. semiplicata* in the Andean basins, followed by the appearance of *P. pumilum* possibly represents a case of competitive replacement, but the South American age ranges of both species suggest that competition was pre-emptive rather than displacive in the sense of Hallam (1990). Also, competition with *P. pumilum* cannot explain the early Toarcian extinction of *P. semiplicata* in North America, since *P. pumilum* seems to be absent from that region.

It is beyond the scope of the present study to assess the immigration of predators through an opening Hispanic Corridor and their potential impact on the regional diversity of bivalves. Nevertheless it is worth mentioning that predation pressure apparently increased strikingly in the Late Cretaceous and Cenozoic following adaptive radiation of neogastropods, crabs and teleost fishes (Vermeij 1977), and in this respect post-dates the Early Jurassic diversity crisis by 100 million years and more. In summary, immigration into the Andean basins of bivalve species through the Hispanic Corridor, followed by competitive replacement, obviously did not contribute to the observed drop in diversity.

Test of the recovery hypothesis

In NW Europe immigration seems to have played a certain role in the recovery of the bivalve fauna (Fig. 4). Following the Pliensbachian–Toarcian extinction, immigration rates were moderately high in the middle Toarcian and reached a peak in the Aalenian. However, immigration from other regions, in particular from southern Europe, was more important than immigration from the eastern Pacific ocean through the Hispanic Corridor. In the middle Toarcian, all six species that immigrated into NW Europe (*Ctenostreon rugosum* (Smith)*, *Entolium corneolum* (Young & Bird)*, *Eopecten spondyloides* (Roemer), *Nuculana ovum* (J. de C. Sowerby)*, *Palaeonucula hammeri* (Defrance) and *Pseudopecten barbatus* (J. Sowerby)) also occurred in early Toarcian or older sedimentary rocks of southern Europe. Only three of them (those marked with an asterisk) are known from eastern Pacific areas, thus reducing the potential candidates that might have utilized the Hispanic Corridor for migration to a relatively low number. Moreover, the middle Toarcian increase in immigration rates did not set off a recovery since bivalve diversity fell to a minimum in the middle Toarcian (Fig. 1). Similarly, of the seven species immigrating into NW Europe during the late Toarcian (*Dacryomya lacryma* (J. de C. Sowerby), *Inoperna sowerbyana* (d'Orbigny), *Modiolus imbricatus* (J. Sowerby), *Nicaniella voltzii* (Hönighaus), *Parallelodon hirsonensis* (d'Archiac)*, *Plagiostoma rodburgensis* (Whidborne) and *Pseudolimea duplicata* (J. de C. Sowerby)*), only the two marked with an asterisk fulfil the criteria to rank as an immigrant through the Hispanic Corridor.

In Aalenian times, immigration again was largely from southern Europe rather than from eastern Pacific areas. From nine species immigrating into NW Europe (*Actinostreon gregareum* (J. Sowerby)*, *Arcomytilus pectinatus* (J. Sowerby), *Camptonectes laminatus* (J. Sowerby), *Ceratomya concentrica* (J. de C. Sowerby), *Gervillaria alaeformis* (J. Sowerby), *Gresslya peregrina* (Phillips)*, *Osteomya dilata* (Phillips), *Parainoceramus obliquus* (Morris & Lycett) and *Pholadomya fidicula* J. de C. Sowerby*), only three (those marked with an asterisk) apparently immigrated through the Hispanic Corridor. So there is little supporting evidence for the hypothesis that immigration of species through the Hispanic Corridor was responsible for the diversity increase observed in the late Toarcian and Aalenian. However, from Figure 4 alone the relative importance of immigration for recovery is not apparent.

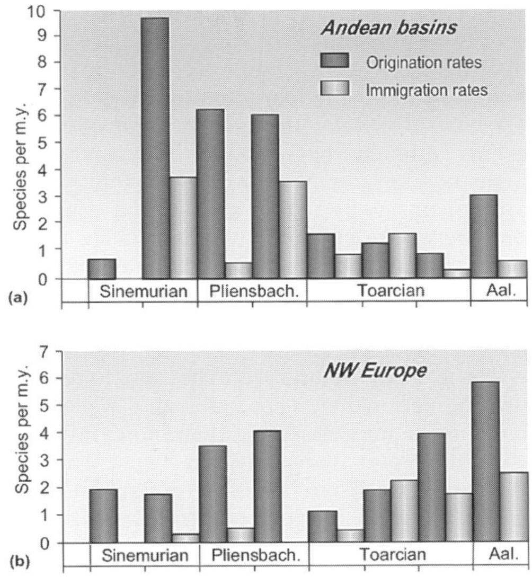

Fig. 5. Immigration and origination rates of bivalve species per million years for various Jurassic stages and substages. (**a**) Andean basins, (**b**) NW Europe.

The importance of immigration versus radiation for recovery

To put the contribution of immigration for biodiversity into perspective, I compared immigration rates of bivalve species to origination rates (Fig. 5). In both regions, Early Jurassic immigration rates usually remained distinctly below origination rates. Only in the time interval immediately following the Pliensbachian–Toarcian extinction was this pattern reversed, and the number of immigrant species exceeded that of newly originating species. However, middle Toarcian immigration rates were not significantly higher than at other times in the Early Jurassic and certainly were not high enough to cause an increase in total bivalve diversity (Fig. 1).

It remains possible that immigrants per se did not add significantly to the recovery in both regions, but that, after immigration, they evolved into new species and thus promoted a diversity increase. The latter group would then be counted in the category of originating species, and would have blurred the relative importance of immigrants. Precise knowledge of the phylogeny of a group is a prerequisite for recognizing such relationships, but this is rarely known for Jurassic bivalves. Nevertheless, some tentative statements can be made. Species immigrating into NW Europe from the middle

Toarcian to Aalenian belong to 22 genera, and species originating in the same time interval are distributed over 35 genera. Since both groups have only four genera in common (*Actinostreon*, *Gervillaria*, *Pholadomya* and *Plagiostoma*) it seems unlikely that immigrants gave rise to many new species. Moreover, these four genera are not unusually rich in species and comprise only between one and two species each. In the Andean basins, the corresponding numbers are seven genera for immigrating species, 15 genera for originating species, and the two groups do not share a single genus.

These results demonstrate that in both regions the origination of new species generally played a much more important role than immigration in controlling overall diversity. Recovery from the Pliensbachian–Toarcian mass extinction commenced when origination rates increased again, which, in the Andean basins, was in the Aalenian and in NW Europe started in the late Toarcian. Immigration in general, and immigration through the Hispanic Corridor in particular, cannot explain the observed recovery in diversity.

Conclusions

The aim of the study was to test two hypotheses, which relate extinction and recovery of Early Jurassic bivalves to immigration of species through the Hispanic Corridor. Both hypotheses failed the test: the Pliensbachian–Toarcian bivalve extinction in South America was not caused by the immigration of species through the Hispanic Corridor and subsequent competitive replacement; and recovery from the Pliensbachian–Toarcian extinction in NW Europe was not a consequence of the immigration of species from the eastern Pacific through the Hispanic Corridor.

Concerning the globally increased extinction rates of bivalves across the Pliensbachian–Toarcian boundary, the synchrony of the biotic crisis and voluminous continental flood basalt eruptions, sea-level highstand, widespread anoxia and possibly also massive release of methane from gas hydrates suggests causal links of these events (Pálfy & Smith 2000; Hesselbo *et al.* 2000). In addition to physicochemical changes, the drop in species diversity may have been amplified by biological factors. Thus, in the Andean basins, the extinction peak of endemic bivalves in the late Pliensbachian is preceded by an origination peak in the early Pliensbachian. The time span between these peaks (5.7 Ma) closely corresponds to the longevity peak of Jurassic endemic bivalves (4–6 Ma), and permits a causal relationship between these two features (Aberhan & Fürsich 2000).

Both analysed regions differ in the dynamics of originating and immigrating clades, and provide evidence for the biogeographic complexity of biotic recoveries. However, a common feature of both regions is the observation that recovery patterns of bivalves from the Pliensbachian–Toarcian mass extinction were largely controlled by increasing origination rates rather than immigration. Other studies showed that immigration intensity during the recovery phase may vary from basin to basin. For example, when analysing geographic variations in the molluscan recovery from the latest-Cretaceous extinction, Jablonski (1998) detected a significantly larger proportion of immigrants in the North American Gulf Coast biota as compared to three other biogeographic provinces. Future research should further investigate the relative importance of immigration versus radiation for the recovery of post-extinction biotas, and whether or not patterns differ between first-order and second-order mass extinctions.

I thank J. Pálfy, Berlin, for critically reading the manuscript and M. Gahr, Würzburg, for providing unpublished information on the stratigraphic distribution of Early Jurassic bivalves from Spain and Portugal. The comments by the journal referees C. Little and S. Damborenea are greatly appreciated. This study was financially supported by grants from the Deutsche Forschungsgemeinschaft (Ab 109/1–1, Ab 109/2–1), which are acknowledged with gratitude.

Appendix 1

Stratigraphic distribution of Hettangian to Aalenian bivalves from the Andean basins.

Taxon	Range
Palaeonucula cuevitana Aberhan	Het
Palaeoneilo elliptica (Goldfuss)	E.Plb – L.Plb
Nuculana ovum (J. de C. Sowerby)	L.Plb – L.Toa
Parallelodon groeberi Damborenea	E.Plb – M.Toa
Parallelodon aff. *groeberi* Damborenea	Het
Parallelodon hirsonensis (d'Archiac)	L.Sin – L.Plb
Paralleodon riccardii Damborenea	E.Plb – L.Plb
Parallelodon sp.	L.Plb – E.Toa
Grammatodon concinnus (Phillips)	E.Toa – L.Toa
Grammatodon costulatus (Leanza)	E.Plb – L.Plb
Grammatodon sulcatus Aberhan	L.Sin – L.Sin
Grammatodon sparsicosta (Gottsche)	Aal
Grammatodon cf. *toyorensis* Hayami	L.Plb
Cucullaea jaworskii Leanza	E.Plb – L.Plb
Cucullaea rothi Leanza	E.Plb – E.Toa
Cucullaea? sp.	L.Plb – L.Toa
Parainoceramus apollo (Leanza)	L.Sin – L.Plb

Parainoceramus? sp.	– Het
Pteria? sp.	Het
Pteroperna sp.	L.Sin – E.Plb
Aguilerella kobyi (de Loriol)	L.Sin – L.Plb
Bakevellia waltoni (Lycett)	L.Sin – L.Plb
Gervillaria alaeformis (J. Sowerby)	M.Toa – Aal
Gervillaria ashcroftensis (Crickmay)	L.Plb – E.Toa
Gervillaria hartmanni (Münster)	L.Sin – L.Toa
Gervillaria pallas (Leanza)	L.Sin – E.Toa
Gervillella araucana Damborenea	L.Sin – L.Plb
Gervillia cf. *dundriensis* Cox	L.Sin – E.Toa
Lithiotis cf. *gregarius* (Lupher & Packard)	M.Toa – M.Toa
Isognomon jupiter (Leanza)	L.Plb – Aal
Gervilleioperna turgida (Leanza)	L.Sin – E.Toa
Gervilleioperna aurita Aberhan & Hillebrandt	Aal
Pulvinites liasicus Damborenea	L.Plb – L.Plb
Bositra buchi (Roemer)	M.Toa – Aal
Posidonotis semiplicata (Hyatt)	L.Plb – E.Toa
Pinna cf. *folium* Young & Bird	L.Sin – E.Toa
Pinna cf. *radiata* Münster	L.Plb – L.Plb
Antiquilima cf. *nagatoensis* Hayami	L.Sin – E.Toa
Antiquilima sp.	Het
Ctenostreon paucicostatum Leanza	E.Plb – E.Toa
Ctenostreon cf. *rugosum* (Smith)	L.Sin – Aal
Plagiostoma giganteum J. Sowerby	L.Plb – L.Toa
Plagiostoma punctatum J. Sowerby	L.Plb – L.Plb
Plagiostoma sp. A	Het – L.Toa
Plagiostoma sp. B	L.Sin – L.Sin
Plagiostoma sp. C	Aal
Pseudolimea duplicata (J. de C. Sowerby)	L.Sin – Aal
Pseudolimea hettangiensis (Terquem)	Het
Plicatula spinosa (J. Sowerby)	L.Sin – L.Sin
Plicatula rapa Bayle & Coquand	L.Sin – E.Toa
Actinostreon costatum (J. de C. Sowerby)	L.Sin – E.Plb
Actinostreon longistriatum (Jaworski)	L.Sin – E.Toa
Actinostreon solitarium (J. de C. Sowerby)	E.Plb – Aal
Gryphaea darwini Forbes	Het – L.Sin
Gryphaea cf. *dumortieri* Joly	L.Sin – L.Plb
Gryphaea sp.	E.Toa – L.Toa
Gryphaea cf. *dilobotes* Duff	Aal
Gryphaea latior Steinmann	L.Sin – L.Plb
Gryphaea tricarinata Philippi	E.Sin – L.Sin
Nanogyra? cf. *auricularis (*Münster*)*	L.Sin – L.Plb
Otapiria neuquensis Damborenea	L.Sin – E.Plb
Otapiria pacifica Covacevich & Escobar	Het – E.Sin
Oxytoma inequivalvis (J. Sowerby)	Het – Aal
Oxytoma cf. *cygnipes* Young & Bird	Het
Oxytoma (Palmoxytoma) sp.	Het
Arctotis? *frenguellii* (Damborenea)	L.Toa – Aal
Meleagrinella cf. *echinata* (Smith)	Aal
Placunopsis radiata (Phillips)	L.Sin – L.Plb
Propeamussium pumilum (Lamarck)	E.Toa – Aal
Kolymonectes weaveri Damborenea	E.Plb – L.Plb
Kolymonectes sp.	L.Sin – L.Sin
Entolium corneolum (Young & Bird)	Het – Aal

Agerchlamys wunschae (Marwick)	Het – L.Plb
Camptonectes auritus (Schlotheim)	L.Plb – Aal
Chlamys textoria (Schlotheim)	Het – Aal
'*Chlamys*' cf. *tingensis* (Tilmann)	Het – L.Plb
Chlamys valoniensis (Defrance)	Het
Eopecten abjectus (Phillips)	L.Toa – Aal
Eopecten hartzi (Rosenkrantz)	E.Plb – E.Toa
Eopecten velatus (Goldfuss)	Het – Aal
Eopecten? sp.	E.Plb – E.Plb
Pseudopecten equivalvis (J. Sowerby)	L.Plb – E.Toa
Radulonectites sosneadoensis (Weaver)	E.Plb – L.Plb
Weyla alata (Buch)	L.Sin – E.Toa
Weyla bodenbenderi (Behrendsen)	L.Sin – E.Toa
Weyla (Weyla?) sp.	Het
Weyla unca (Philippi)	L.Sin – E.Toa
Weyla (Lywea?) sp.	Het
Lycettia hypertrigona Damborenea	L.Sin – E.Toa
Modiolus baylei (Philippi)	L.Sin – E.Toa
Modiolus chilensis Aberhan	E.Plb – L.Plb
Modiolus gerthi Damborenea	L.Plb – M.Toa
Modiolus giganteus Quenstedt	L.Sin – L.Plb
Modiolus imbricatus J. Sowerby	M.Toa – Aal
Modiolus lonsdalei (Morris & Lycett)	L.Plb – Aal
Modiolus ventricosus Roemer	L.Plb – L.Plb
"*Modiolus*" sp.	Aal
Palaeopharus? sp.	L.Plb – L.Plb
Groeberella neuquensis (Groeber)	L.Sin – Aal
Trigonia sp. A	M.Toa – M.Toa
Trigonia sp. B	M.Toa – M.Toa
Trigonia sp. C	Aal
Neuquenitrigonia huenickeni (Leanza & Garate)	Aal
Neuquenitrigonia sp. nov.	M.Toa – M.Toa
Prosogyrotrigonia sp. nov.	Het – E.Sin
Prosogyrotrigonia sp. A	E.Sin – E.Sin
Prosogyrotrigonia? sp. B	L.Sin – L.Sin
Prosogyrotrigonia? sp. C	Het – E.Sin
Frenguelliella chubutensis (Feruglio)	E.Plb – L.Plb
Frenguelliella inexspectata Jaworski	E.Plb – L.Plb
Frenguelliella poultoni Leanza	L.Sin – L.Plb
Frenguelliella tapiai Lambert	L.Plb – E.Toa
Jaworskiella burckhardti (Jaworski)	E.Plb – L.Plb
Jaworskiella gryphitica (Möricke)	L.Sin – L.Sin
Jaworskiella sp.	L.Sin – L.Sin
Quadratojaworskiella pustulata Reyes & Perez	E.Plb – L.Plb
Quadratojaworskiella sp. nov.	Het – L.Sin
Laevitrigoniinae gen. et sp. nov.	L.Sin – L.Sin
Psilotrigonia sp. nov.	L.Sin – E.Plb
Myophorella araucana (Leanza)	E.Plb – E.Toa
Myophorella catenifera (Hupé)	E.Plb – L.Plb
Myophorella cf. *tuberculata* (Agassiz)	E.Plb – L.Plb
Myophorella sp. nov. A	Aal
Myophorella sp. nov. B	M.Toa – L.Toa
Myophorella sp. A	Aal
Myophorella sp. B	L.Plb – L.Plb
Scaphorella sp. nov.	Aal
Vaugonia cf. *substriata* (Burmeister & Giebel)	M.Toa – M.Toa
Vaugonia sp. nov.	L.Sin – E.Toa
Vaugonia sp. A	M.Toa – M.Toa
Vaugonia sp. B	L.Plb – L.Plb

Vaugonia sp. C	L.Toa – L.Toa
'*Lucina*' *chubutensis* Wahnish	L.Sin – L.Plb
'*Lucina*' *feruglioli* Wahnish	E.Plb – L.Plb
Mesomiltha cf. *bellona* (D'Orbigny)	L.Sin – Aal
Mesomiltha huayquimili (Leanza)	E.Plb – L.Toa
Mesomiltha? sp.	Aal
Mactromya cardioideum (Phillips)	L.Sin – E.Plb
Kalentera sp.	L.Sin – E.Plb
Myoconcha neuquena Leanza	E.Plb – E.Toa
Myoconcha steinmanni (Jaworski)	Aal
"*Astarte*" *chubutensis* (Wahnish)	E.Plb – E.Toa
"*Astarte*" cf. *gregaria* Philippi	Aal
"*Astarte*" *keideli* Wahnish	E.Plb – L.Plb
"*Astarte*" cf. *subtetragona* Münster	Aal
Neocrassina andium (Gottsche)	Aal
Neocrassina aureliae (Feruglio)	E.Plb – Aal
Coelastarte excavata (J. Sowerby)	Aal
Opisoma excavatum Boehm	M.Toa – M.Toa
Trigonastarte? *steinmanni* (Mörike)	M.Toa – Aal
Astartidae gen. et sp. nov.	L.Plb – L.Plb
Cardinia andium (Giebel)	E.Sin – E.Toa
Cardinia cf. *listeri* (J. Sowerby)	L.Sin – L.Sin
Protocardia striatula (J. de C. Sowerby)	L.Plb – Aal
Protocardia sp.	L.Sin – L.Sin
Jurassicardium? *asaphum* (Leanza)	L.Sin – E.Plb
Quenstedtia? sp.	L.Sin – L.Plb
Corbicellopsis? sp.	L.Sin – L.Sin
Tancredia sp.	Aal
Anisocardia sp.	L.Sin – L.Plb
Isocyprina ancatruzi (Leanza)	L.Sin – L.Plb
Rollierella sp.	L.Toa – L.Toa
Arcticidae gen. et sp. nov.	L.Plb – L.Plb
Pseudisocardia cf. *cordata* (Buckmann)	Aal
Pseudisocardia? *liasina* Wahnish	E.Plb – E.Toa
"*Venus*" *peregrina* Gottsche	Aal
Pholadomya corrugata Koch & Dunker	L.Sin – M.Toa
Pholadomya aff. *corrugata* Koch & Dunker	Het
Pholadomya cf. *decorata* Hartmann	L.Sin – E.Plb
Pholadomya fidicula (J. Sowerby)	L.Sin – Aal
Pholadomya multilineata Gabb	E.Plb – E.Toa
Pholadomya sp. nov.	L.Sin – L.Sin
Goniomya cf. *proboscidea* Agassiz	E.Toa – Aal
Goniomya sp. nov. A	E.Plb – L.Plb
Goniomya sp. nov. B	Het
Homomya neuquena Leanza	L.Sin – E.Toa
Osteomya dilata (Phillips)	L.Sin – L.Sin
Pachymya rotundocaudata (Leanza)	E.Plb – L.Plb
Pachymya? sp.	L.Sin – L.Sin
Pachymya senckenbergi Möricke	Aal
Ceratomya sp.	E.Plb – L.Plb
Gresslya peregrina (Phillips)	L.Plb – Aal
Gresslya sp.	E.Sin – E.Plb
Pleuromya galathea Agassiz	Het
Pleuromya uniformis (J. Sowerby)	L.Sin – Aal
Cercomya undulata (J. de C. Sowerby)	E.Toa – Aal
Cercomya peruviana Cox	L.Toa – Aal
Platymyoidea? cf. *broliensis* (Buvignier)	L.Plb – L.Plb

Data based on evaluation of information in Hillebrandt (1980), Hillebrandt & Schmidt-Effing (1981), Pérez (1982), Hillebrandt & Westermann (1985), Damborenea (1987*a*,*b*, 1990*a*,*b*, 1992, 1993, 1996); Aberhan (1992, 1994), Damborenea & Manceñido (1992), Leanza (1993), Pérez & Reyes (1994) and unpublished information. *Range*: Het, Hettangian; Sin, Sinemurian; Plb, Pliensbachian; Toa, Toarcian; Aal, Aalenian; E, Early; M, Middle; L, Late. Note that 'Het' does not necessarily imply first appearance in the Hettangian, and 'Aal' does not necessarily imply last appearance in the Aalenian.

Appendix 2

Stratigraphic distribution of Hettangian to Aalenian bivalves from NW Europe.

Taxon	Range
Palaeonucula hammeri (Defrance)	M.Toa – Aal
Palaeonucula navis (Piette)	Het – E.Toa
Palaeoneilo elliptica (Goldfuss)	Het – L.Plb
Palaeoneilo galatea (d'Orbigny)	E.Plb – Aal
Nuculana ovum (J. de C. Sowerby)	M.Toa – M.Toa
Dacryomya gaveyi Cox	E.Plb – L.Plb
Dacryomya heberti (Martin)	Het – E.Plb
Dacryomya lacryma (J. de C. Sowerby)	L.Toa – Aal
Rollieria bronni (Andler)	Het – L.Plb
Ryderia doris (d'Orbigny)	Het – Aal
Ryderia texturata (Terquem & Piette)	Het – L.Plb
Isoarca subspirata (Münster)	Aal
Arca minuta J. de C. Sowerby	Aal
Parallelodon buckmani (Richardson)	Het – E.Toa
Parallelodon hirsonensis (d'Archiac)	L.Toa – Aal
Parallelodon trapezium Cox	E.Plb – E.Toa
Grammatodon chapuisi Arkell	M.Toa – M.Toa
Grammatodon inaequivalvis (Münster)	L.Toa – Aal
Grammatodon muensteri (Zieten)	E.Sin – Aal
Cucullaea oblonga J. Sowerby	L.Toa – Aal
Parainoceramus obliquus (Morris & Lycett)	Aal
Parainoceramus ventricosus (J. de C. Sowerby)	E.Plb – E.Plb
Pseudomytiloides dubius (J. de C. Sowerby)	E.Toa – M.Toa
Mytiloceramus polyplocus (Roemer)	M.Toa – Aal
Steinmannia bronni (Voltz)	E.Toa – M.Toa
Semuridia dorsetensis (Cox)	E.Sin – L.Sin
Pteria carixiensis Cox	E.Plb – E.Plb
Pteroperna deshayesi (Terquem)	Het – E.Plb
Pteroperna plana (Morris & Lycett)	Aal
Bakevellia binneyi (Brown)	Het
Bakevellia waltoni (Lycett)	Het – Aal
Gervillaria alaeformis (J. Sowerby)	Aal
Gervillaria hartmanni (Münster)	Het – Aal
Gervillaria tortuosa (J. de C. Sowerby)	L.Toa – Aal
Gervillella aviculoides (J. Sowerby)	Het – Aal
Gervillella siliqua	Aal

(Eudes-Deslongchamps)	
Gervillia lanceolata (J. de C. Sowerby)	Het – E.Toa
Gervillia metzertensis (Joly 1907)	Het
Gervillia northamptonensis Cox	Aal
Isognomon lugdunensis (Dumortier)	L.Sin – L.Plb
Isognomon mytiloides (Lamarck)	L.Toa – Aal
Linsdallia quadrata (J. de C. Sowerby)	Aal
Bositra buchi (Roemer)	E.Toa – Aal
Pinna cuneata Phillips	Aal
Pinna folium Young & Bird	Het – E.Toa
Antiquilima cubifera (Whidborne)	E.Toa – Aal
Antiquilima succincta (Schlotheim)	Het – L.Plb
Ctenostreon rugosum (Smith)	M.Toa – Aal
Ctenostreon tuberculatum (Terquem)	Het
Plagiostoma alticosta (Chapuis & Dewalque)	L.Toa – Aal
Plagiostoma cardiiformis J. Sowerby	Aal
Plagiostoma giganteum J. Sowerby	Het – L.Plb
Plagiostoma hersilia (d'Orbigny)	Aal
Plagiostoma punctatum J. Sowerby	L.Sin – E.Toa
Plagiostoma rodburgensis (Whidborne)	L.Toa – Aal
Plagiostoma toarcensis (Eudes-Deslongchamps)	M.Toa – Aal
Pseudolimea acuticostata (Münster)	E.Plb – E.Toa
Pseudolimea cristata (Dumortier)	E.Plb – L.Plb
Pseudolimea duplicata (J. de C. Sowerby)	L.Toa – Aal
Pseudolimea grettonensis Cox	E.Toa – E.Toa
Pseudolimea harburyiensis Cox	Het – E.Sin
Pseudolimea hettangiensis (Terquem)	Het – L.Plb
Pseudolimea koninckana (Chapuis & Dewalque)	Het – E.Sin
Pseudolimea pectinoides (J. Sowerby)	Het – E.Toa
Regalilima oolitica (Lycett)	Aal
Plicatula laevigata (d'Orbigny)	L.Plb – L.Plb
Plicatula spinosa (J. Sowerby)	E.Plb – E.Toa
Atreta intusstriata (Emmerich)	Het – E.Sin
Actinostreon marshii (J. Sowerby)	Aal
Actinostreon solitarium (J. de C. Sowerby)	Aal
Gryphaea arcuata Lamarck	Het – E.Sin
Gryphaea gigantea J. Sowerby	E.Plb – L.Plb
Gryphaea mccullochi J. de C. Sowerby	E.Sin – L.Sin
Gryphaea pictaviensis (Hébert)	L.Toa – L.Toa
Catinula crickleyensis Cox	Aal
Liostrea hisingeri (Nilsson)	Het – M.Toa
Nanogyra monoptera (Eudes-Deslongchamps)	E.Toa – E.Toa
Anningella faberi (Oppel)	E.Sin – L.Sin
Oxytoma calva (Schlönbach)	E.Sin – L.Plb
Oxytoma inequivalvis (J. Sowerby)	Het – Aal
Oxytoma cygnipes (Young & Bird)	E.Plb – L.Plb
Meleagrinella fallax (Pflücker)	Het
Meleagrinella papyria (Quenstedt)	L.Sin – L.Sin
Meleagrinella substriata (Münster)	E.Toa – M.Toa
Terquemia arietis (Quenstedt)	Het – L.Plb
Placunopsis radiata (Phillips)	Het – Aal
Propeamussium laeviradiatum (Waagen)	L.Plb – Aal
Propeamussium pumilum (Lamarck)	L.Plb – Aal
Entolium corneolum (Young & Bird)	M.Toa – Aal
Entolium lunare (Roemer)	Het – M.Toa
Camptonectes auritus (Schlotheim)	Het – Aal
Camptonectes laminatus (J. Sowerby)	Aal
Camptonectes subulatus (Münster)	Het – L.Toa
Chlamys pollux (d'Orbigny)	Het – E.Sin
Chlamys textoria (Schlotheim)	Het – Aal
Chlamys valoniensis (Defrance)	Het – E.Sin
Eopecten abjectus (Phillips)	L.Plb – Aal
Eopecten spondyloides (Roemer)	M.Toa – Aal
Eopecten velatus (Goldfuss)	Het – Aal
Pseudopecten barbatus (J. Sowerby)	M.Toa – Aal
Pseudopecten dentatus (J. de C. Sowerby)	Het – Aal
Pseudopecten equivalvis (J. Sowerby)	Het – E.Toa
Pseudopecten veyrasensis (Dumortier)	Het – L.Plb
Spondylopecten subspinosus (Schlotheim)	Het – Aal
Arcomytilus pectinatus (J. Sowerby)	Aal
Musculus subcancellata (Buvignier)	L.Sin – L.Plb
Inoperna lilliputensis Hodges	Het
Inoperna sowerbyana (d'Orbigny)	L.Toa – Aal
Modiolus hillanus (J. Sowerby)	Het – L.Sin
Modiolus imbricatus (J. Sowerby)	L.Toa – Aal
Modiolus minimus (J. Sowerby)	Het
Modiolus scalprum (J. Sowerby)	L.Sin – E.Toa
Modiolus ventricosus (Roemer)	Het – L.Plb
Liotrigonia lingonensis (Dumortier)	L.Plb – L.Plb
Trigonia costata J. Sowerby	L.Toa – Aal
Myophorella striata (J. Sowerby)	L.Toa – Aal
Scaphotrigonia navis (Lamarck)	M.Toa – Aal
Vaugonia flecta (Morris & Lycett)	Aal
Vaugonia literata (Young & Bird)	M.Toa – M.Toa
'*Lucina*' *plana* Zieten	L.Plb – L.Plb
Luciniola limbata (Terquem & Piette)	Het – L.Sin
Mesomiltha lirata (Phillips)	L.Toa – Aal
Mactromya cardioideum (Phillips)	Het – E.Toa
Sphaeriola subglobosa (Tate)	E.Plb – L.Toa
Myoconcha decorata (Münster)	Het – L.Plb
Myoconcha psilonoti Quenstedt	Het – E.Plb
Tutcheria cingulata (Goldfuss)	Het – E.Plb
Tutcheria submulticostata (d'Orbigny)	E.Plb – L.Plb
Pseudopis deslonchampsis (Tate)	E.Plb – L.Plb
Astarte guexi d'Orbigny	Het – L.Plb
Astarte obsoleta Dunker	Het – L.Plb
Neocrassina elegans (J. Sowerby)	M.Toa – Aal
Coelastarte excavata (J. Sowerby)	L.Toa – Aal
Nicaniella voltzii (Hönighaus)	L.Toa – Aal
Praeconia rhomboidalis (Phillips)	Aal
Coelopis lunulata (J. Sowerby)	L.Toa – Aal
Cardinia attenuata (Stutchbury)	L.Sin – E.Toa
Cardinia concinna (J. Sowerby)	Het – E.Toa
Cardinia listeri (J. Sowerby)	Het – L.Plb
Cardinia ovalis (Stutchbury)	Het – E.Sin
Nidarica slatteri (Wilson & Crick)	L.Plb – E.Toa

Hippopodium ponderosum J. Sowerby	E.Sin – L.Plb
Protocardia buckmani Morris & Lycett	L.Toa – Aal
Protocardia phillipiana (Dunker)	Het – L.Sin
Protocardia truncata (J. de C. Sowerby)	E.Plb – E.Toa
Quenstedtia laevigata (Phillips)	L.Toa – Aal
Sowerbya triangularis (Phillips)	L.Toa – Aal
Tancredia donaciformis Lycett	L.Toa – Aal
Tancredia securiformis Dunker	Het – E.Plb
Anisocardia minima (J. Sowerby)	Aal
Anisocardia truncata (Morris)	Aal
Isocyprina menkei (Dunker)	Het
Eotrapezium cucullatum (Münster)	E.Plb – E.Toa
Pronoella beneckei (Rollier)	Aal
Pronoella trigonellaris (Schlotheim)	L.Toa – Aal
Pseudotrapezium cordiforme (Deshayes)	L.Toa – Aal
Pseudisocardia cordata (Buckman)	Aal
Pholadomya ambigua (J. Sowerby)	Het – L.Plb
Pholadomya corrugata Koch & Dunker	Het – E.Plb
Pholadomya fidicula J. de C. Sowerby	Aal
Pholadomya lirata (J. Sowerby)	Aal
Pholadomya ovalis (J. Sowerby)	Aal
Goniomya hybrida (Münster)	E.Plb – E.Toa
Goniomya literata (J. Sowerby)	Aal
Homomya gibbosa (J. Sowerby)	L.Plb – Aal
Osteomya dilata (Phillips)	Aal
Pachymya vetusa (Phillips)	L.Sin – L.Plb
Ceratomya concentrica (J. de C. Sowerby)	Aal
Ceratomya petricosa (Simpson)	L.Sin – L.Plb
Ceratomya sublaevis (Tate)	L.Plb – L.Plb
Gresslya intermedia (Simpson)	L.Plb – Aal
Gresslya peregrina (Phillips)	Aal
Pteromya tatei (Richardson & Tutcher)	Het
Pleuromya alduini (Brongniart)	Aal
Pleuromya calceiformis (Phillips)	Aal
Pleuromya costata (Young & Bird)	E.Plb – E.Toa
Pleuromya galathea Agassiz	Het – E.Sin
Pleuromya uniformis (J. Sowerby)	Het – Aal
Platymyoidea broliensis (Buvignier)	L.Plb – L.Plb
Thracia depressa J. de C. Sowerby	Aal

Data based on evaluation of information in Hallam (1976, 1987), Johnson (1984), Hodges (1991, 2000), Muster (1995), Röhl (1998), Harries & Little (1999), and unpublished information. For abbreviations see Appendix 1. Note that 'Het' does not necessarily imply first appearance in the Hettangian, and 'Aal' does not necessarily imply last appearance in the Aalenian.

References

ABERHAN, M. 1992. Palökologie und zeitliche Verbreitung benthischer Faunengemeinschaften im Unterjura von Chile. *Beringeria*, **5**, 1–174.

ABERHAN, M. 1993. Benthic macroinvertebrate associations on a carbonate-clastic ramp in segments of the Early Jurassic back-arc basin of northern Chile (26–29°S). *Revista Geológica de Chile*, **20**, 105–136.

ABERHAN, M. 1994. Early Jurassic Bivalvia of northern Chile. Part I. Subclasses Palaeotaxodonta, Pteriomorphia, and Isofilibranchia. *Beringeria*, **13**, 3–115.

ABERHAN, M. 2001. Bivalve palaeobiogeography and the Hispanic Corridor: time of opening and effectiveness of a proto-Atlantic seaway. *Palaeogeography, Palaeoclimatology, Palaeoecology*, **165**, 375–394.

ABERHAN, M. & FÜRSICH, F. T. 1997. Diversity analysis of Lower Jurassic bivalves of the Andean Basin and the Pliensbachian–Toarcian mass extinction. *Lethaia*, **29**, 181–195.

ABERHAN, M. & FÜRSICH, F. T. 2000. Mass origination versus mass extinction: the biological contribution to the Pliensbachian–Toarcian extinction event. *Journal of the Geological Society, London*, **157**, 55–60.

ABERHAN, M. & PÁLFY, J. 1996. A low oxygen tolerant East Pacific flat clam (*Posidonotis semiplicata*) from the Lower Jurassic of the Canadian Cordillera. *Canadian Journal of Earth Sciences*, **33**, 993–1006.

DAMBORENEA, S. E. 1987*a*. Early Jurassic Bivalvia of Argentina. Part 1: Stratigraphical introduction and superfamilies Nuculanacea, Arcacea, Mytilacea and Pinnacea. *Palaeontographica (A)*, **199**, 23–111.

DAMBORENEA, S. E. 1987*b*. Early Jurassic Bivalvia of Argentina. Part 2: Superfamilies Pteriacea, Buchiacea and part of Pectinacea. *Palaeontographica (A)*, **199**, 113–216.

DAMBORENEA, S. E. 1990*a*. Bivalves. *In*: RICCARDI, A. C., DAMBORENEA, S. E. & MANCEÑIDO, M. O. (eds) Lower Jurassic of South America and Antarctic Peninsula. *Newsletters on Stratigraphy*, **21**, 85–90.

DAMBORENEA, S. E. 1990*b*. Bivalves. *In*: RICCARDI, A. C., WESTERMANN, G. E. G. & DAMBORENEA, S. E. (eds) Middle Jurassic of South America and Antarctic Peninsula. *Newsletters on Stratigraphy*, **21**, 114, 116–119.

DAMBORENEA, S. E. 1992. South American Jurassic bivalves. *In*: WESTERMANN, G. E. G. (ed.) *The Jurassic of the Circum-Pacific, Atlas*. Cambridge University Press, Cambridge, pls. 115–119.

DAMBORENEA, S. E. 1993. Early Jurassic South American pectinaceans and circum-Pacific palaeobiogeography. *Palaeogeography, Palaeoclimatology, Palaeoecology*, **100**, 109–123.

DAMBORENEA, S. E. 1996. Palaeobiogeography of Early Jurassic bivalves along the southeastern Pacific margin. *XIII Congreso Geológico Argentino y III Congreso de Exploración de Hidrocarburos, Actas*, **5**, 151–167.

DAMBORENEA, S. E. 2000. Hispanic Corridor: Its evolution and the biogeography of bivalve molluscs. *In*: HALL, R. L. & SMITH, P. L. (eds) *Advances in Jurassic Research 2000*. GeoResearch Forum, **6**, 369–380.

DAMBORENEA, S. E. & MANCEÑIDO, M. O. 1979. On the palaeogeographical distribution of the pectinid genus *Weyla* (Bivalvia, Lower Jurassic).

Palaeogeography, Palaeoclimatology, Palaeoecology, **27**, 85–102.
DAMBORENEA, S. E. & MANCEÑIDO, M. O. 1992. A comparison of Jurassic marine benthonic faunas from South America and New Zealand. *Journal of the Royal Society of New Zealand*, **22**, 131–152.
GRADSTEIN, F. M., AGTERBERG, F. P., OGG, J. G., HARDENBOL, J., VAN VEEN, P., THIERRY, J. & HUANG Z. 1994. A Mesozoic time scale. *Journal of Geophysical Research*, **99**, 24051–24074.
HALLAM, A. 1976. Stratigraphic distribution and ecology of European Jurassic bivalves. *Lethaia*, **9**, 245–259.
HALLAM, A. 1983. Early and mid-Jurassic molluscan biogeography and the establishment of the central Atlantic seaway. *Palaeogeography, Palaeoclimatology, Palaeoecology*, **43**, 181–193.
HALLAM, A. 1986. The Pliensbachian and Tithonian extinction events. *Nature*, **319**, 765–768.
HALLAM, A. 1987. Radiations and extinctions in relation to environmental change in the marine Lower Jurassic of northwest Europe. *Paleobiology*, **13**, 152–168.
HALLAM, A. 1990. Biotic and abiotic factors in the evolution of Early Mesozoic marine molluscs. *In*: ROSS, R. M. & ALLMON, W. D. (eds) *Causes of Evolution. A Paleontological Perspective*. The University of Chicago Press, Chicago, 249–260.
HALLAM, A. 1996. Recovery of the marine fauna in Europe after the end-Triassic and early Toarcian mass extinctions. *In*: HART, M. B. (ed.) *Biotic Recovery from Mass Extinction Events*. Geological Society, London, Special Publications, **102**, 231–236.
HALLAM, A. & WIGNALL, P. B. 1997. *Mass Extinctions and their Aftermath*. Oxford University Press, Oxford.
HARRIES, P. J. & LITTLE, C. T. S. 1999. The early Toarcian (Early Jurassic) and the Cenomanian–Turonian (Late Cretaceous) mass extinctions: similarities and contrasts. *Palaeogeography, Palaeoclimatology, Palaeoecology*, **154**, 39–66.
HESSELBO, S. P., GRÖCKE, D. R., JENKYNS, H. C., BJERRUM, C. J., FARRIMOND, P., MORGANS BELL, H. S. & GREEN, O. R. 2000. Massive dissociation of gas hydrate during Jurassic oceanic anoxic event. *Nature*, **406**, 392–395.
HILLEBRANDT, A. VON. 1980. Paleozoografía de Jurásico marino (Lías hasta Oxfordiana) en Suramérica. *In*: ZEIL, W. (ed.) *Nuevos Resultados de la Investigación Geocientífica Alemana en Lationoamérica*. Deutsche Forschungsgemeinschaft, Bonn, and Instituto de Colaboración Científica, Tübingen, 132–134.
HILLEBRANDT, A. VON & SCHMIDT-EFFING, R. 1981. Ammoniten aus dem Toarcium (Jura) von Chile (Südamerika). *Zitteliana*, **6**, 1–74.
HILLEBRANDT, A. VON. & WESTERMANN, G. E. G. 1985. Aalenian (Jurassic) ammonite faunas and zones of the Southern Andes. *Zitteliana*, **12**, 3–55.
HODGES, P. 1991. The relationship of the Mesozoic bivalve *Atreta* to the Dimyidae. *Palaeontology*, **34**, 963–970.
HODGES, P. 2000. The Early Jurassic Bivalvia from the Hettangian and Lower Sinemurian of south-west Britain. *Monograph of the Palaeontographical Society, London*, **154**, 1–64.
JABLONSKI, D. 1998. Geographic variation in the molluscan recovery from the end-Cretaceous extinction. *Science*, **279**, 1327–1330.
JOHNSON, A. L. A. 1984. The palaeobiology of the bivalve families Pectinidae and Propeamussiidae in the Jurassic of Europe. *Zitteliana*, **11**, 1–235.
LEANZA, H. A. 1993. Jurassic and Cretaceous trigoniid bivalves from west-central Argentina. *Bulletins of American Paleontology*, **105**, 5–95.
LITTLE, C. T. S. & BENTON, M. J. 1995. Early Jurassic mass extinction: a global long-term event. *Geology*, **23**, 495–498.
MUSTER, H. 1995. Taxonomie und Paläobiogeographie der Bakevelliidae (Bivalvia). *Beringeria*, **14**, 3–161.
PÁLFY, J. & SMITH, P. L. 2000. Synchrony between Early Jurassic extinction, oceanic anoxic event, and the Karoo-Ferrar flood basalt volcanism. *Geology*, **28**, 747–750.
PÉREZ, E. 1982. Bioestratigrafía del Jurásico de Quebrada Asientos, norte de Potrerillos, Región de Atacama. *Servicio Nacional de Geología y Minería – Chile, Boletin*, **37**, 1–149.
PÉREZ, E. & REYES, R. B. 1994. Catálogo de ejemplares tipo, conservados en la colección paleontológica del Sevicio Nacional de Geología y Minería, Chile, desde 1953 a 1993. *Servicio Nacional de Geología y Minería – Chile, Boletin*, **46**, 1–99.
RAUP, D. M. & SEPKOSKI, J. J., Jr. 1984. Periodicity of extinctions in the geologic past. *Proceedings of the National Academy of Sciences USA*, **81**, 801–805.
RÖHL, H.-J. 1998. Hochauflösende palökologische und sedimentologische Untersuchungen im Posidonienschiefer (Lias e) von SW-Deutschland. *Tübinger Geowissenschaftliche Arbeiten, Reihe A*, **47**, 1–170.
SMITH, P. L. 1983. The Pliensbachian ammonite *Dayiceras dayiceroides* and Early Jurassic paleogeography. *Canadian Journal of Earth Sciences*, **20**, 86–91.
SMITH, P. L. & TIPPER, H. W. 1986. Plate tectonics and paleobiogeography: Early Jurassic (Pliensbachian) endemism and diversity. *Palaios*, **1**, 399–412.
VERMEIJ, G. J. 1977. The Mesozoic marine revolution: Evidence from snails, predators and grazers. *Paleobiology*, **3**, 245–258.

Cretaceous patterns of floristic change in the Antarctic Peninsula

DAVID J. CANTRILL[1] & IMOGEN POOLE[2]

[1] *British Antarctic Survey, Natural Environment Research Council, High Cross, Madingley Road, Cambridge CB3 0ET, UK (e-mail:* D.Cantrill@bas.ac.uk*)*

[2] *Wood Anatomy Section, National Herbarium of the Netherlands, University of Utrecht branch, PO Box 80102, 3585 CS Utrecht, The Netherlands*

Abstract: Cretaceous radiation of angiosperms from low to high palaeolatitudes, coupled with the break-up of Gondwana, played a major role in establishing and maintaining biogeographic patterns across the southern hemisphere. Uncertainties in details of plate reconstructions provide conflicting hypotheses about area relationships of Gondwana fragments. This has led to a number of competing proposals concerning angiosperm migration across Gondwana. Central to this debate is the role of the Antarctic Peninsula, a region that is often envisaged as providing the main connection between east and west Gondwana. The initial radiation of angiosperms into the Antarctic Peninsula region, however, post-dates appearances elsewhere in east Gondwana (e.g. Australia), strongly suggesting that the Antarctic Peninsula was not the main gateway, at least in the early stages of Gondwana radiation. A steep climatic gradient in this part of the world probably acted as an effective barrier to angiosperm radiation. The peak of floristic replacement coincides with the peak of Cretaceous warmth (Turonian) which in turn suggests that climatic warming acted as a forcing mechanism by pushing latitudinal belts of vegetation southwards. Once into the southern high latitudes angiosperms diversified, and as climates cooled during the Late Cretaceous a number of important groups seem to have their origins here. Recent investigations of Antarctic macro- and microfloras indicate progressive floristic replacement through the Cretaceous. Bryophytes, hepatophytes, bennettites and other seed plants all show a rapid decline in diversity. In contrast, ferns initially decline then recover, while conifers remain relatively stable. The ecological preferences of the replaced groups imply that angiosperms initially occupied areas of disturbance and were understorey colonizers, only later replacing fern thickets and becoming important in the overstorey. This pattern is consistent with those observed elsewhere through the Cretaceous.

Two major events influenced Cretaceous to present-day biotic history of the southern hemisphere: the radiation of angiosperms and the concomitant break-up of Gondwana. During mid-Cretaceous times (130–90 Ma BP) terrestrial vegetation underwent rapid modernization, with an explosive radiation of crown group angiosperms (Lidgard & Crane 1990). In the northern hemisphere it is well established that angiosperms arose at low palaeolatitudes and spread polewards (Crane & Lidgard 1989; Crane *et al.* 1995). A paucity of knowledge about southern hemisphere floras, and in particular those of Antarctica, means that the pattern in the southern hemisphere is less well constrained (Drinnan & Crane 1990). However, a general pattern of earlier inception in low latitudes is evident from the limited data available (Drinnan & Crane 1990).

During Cretaceous times disintegration of Gondwana was important in maintaining, or creating, biotic gateways between what are now widely separated landmasses (Australia, Africa, India, New Zealand, South America and Antarctica) (Fig. 1). Consequently, the break-up of Gondwana underpins many biogeographic and diversity patterns seen in the southern hemisphere today. Antarctica is unique amongst fragments of Gondwana in that it was attached at some time to all other major components of the supercontinent (Fig. 1). For this reason understanding the geological evolution of Antarctica is essential for evaluating terrestrial connections across Gondwana through time. Although the timing and pattern of break-up is relatively well understood for the large continental fragments, this is less the case for smaller microplates (Storey *et al.* 1988). The present-day Antarctic region can be divided into two areas; East Antarctica, a craton; and West Antarctica, composed of a number of microcontinental fragments (e.g. Antarctic Peninsula, Thurston Island-Eights Land, Marie Byrd Land, Haag Nunataks, Ellsworth-Whitmore Mountains) (Fig. 2). These microcontinents have a long and complicated history (Storey *et al.* 1988) and represent one of the major uncertainties in Gondwana reconstructions. Nevertheless,

From: CRAME, J. A. & OWEN, A. W. (eds) 2002. *Palaeobiogeography and Biodiversity Change: the Ordovician and Mesozoic–Cenozoic Radiations*. Geological Society, London, Special Publications, **194**, 141–152.
0305-8719/02/$15.00

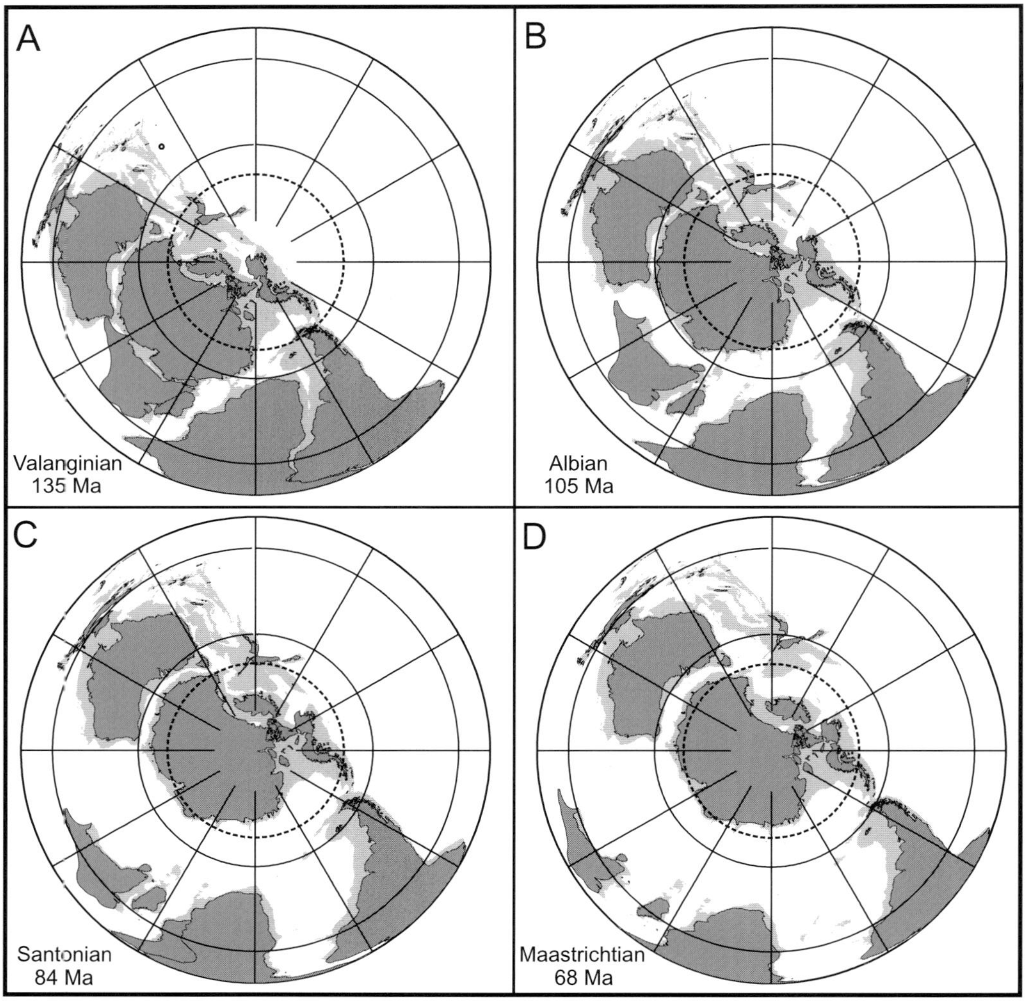

Fig. 1. View of Gondwana break-up for selected times in the Cretaceous. (**A**) Valanginian (135 Ma BP), illustrating the close proximity of Africa and India to the rest of Gondwana. (**B**) Albian (105 Ma BP). Note that India and South Africa have separated from the rest of Gondwana. (**C**) Santonian (84 Ma BP) showing extensive separation between India, Africa and the rest of Gondwana. Australia is beginning to rift from Antarctica along the southern margin. (**D**) Maastrichtian (68 Ma BP). New Zealand and Lord Howe rise are rifting from the eastern margin of Gondwana. Note the continued maintenance of connections between South America and east Gondwana (Australia, Antarctica, New Zealand). Latitude intervals are 30°; dashed black line represents palaeopolar circle. Reconstructions provided by R. A. Livermore, British Antarctic Survey.

developing a framework that reconstructs the position of these fragments through time is critical for understanding terrestrial connections between various landmasses and subsequent vicariant events.

The same is also true for other regions of Gondwana; for example, large uncertainties exist in our knowledge of the separation of India from Antarctica, due to a lack of seafloor data and age constraints on magnetic anomalies along the Antarctic margin. The development of marine conditions in present-day eastern Indian sedimentary basins suggests that the separation of India was diachronous from west to east (Truswell *et al.* 1999). However, reconstructions based on seafloor magnetic anomalies indicate

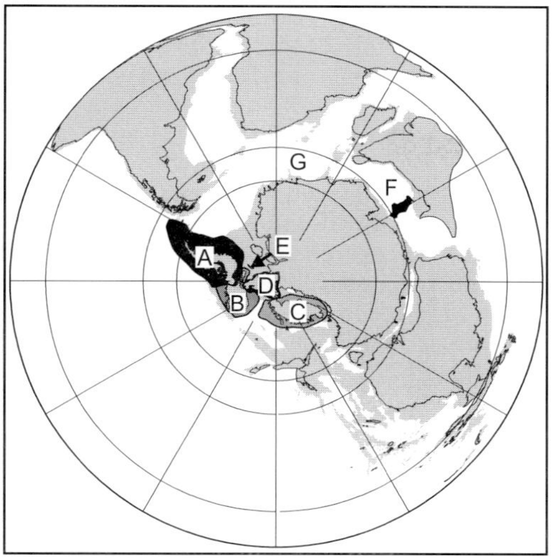

Fig. 2. Albian reconstruction of Gondwana courtesy of R. A. Livermore, British Antarctic Survey. Polar view illustrating the importance of the West Antarctic microcontinents (A, Antarctic Peninsula; B, Thurston Island; C, Marie Byrd Land; D, Ellsworth-Whitmore Mountains; E, Haag Nunataks) in maintaining connections between east and west Gondwana. Note also the potential importance of the Kergeulen Plateau (F) and the Mozambique Rise (G) in the early stages of rifting.

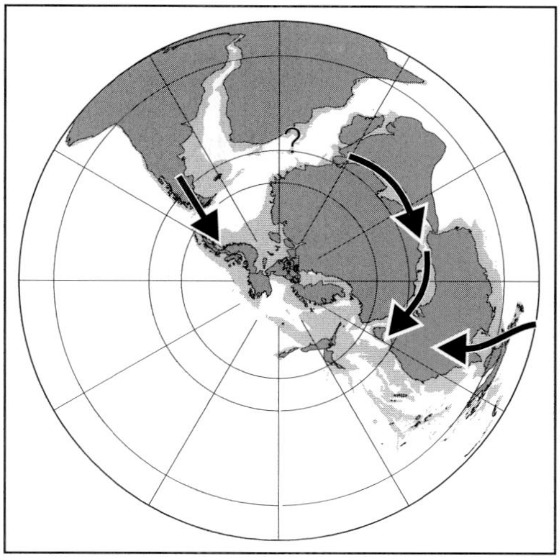

Fig. 3. Alternative scenarios for the expansion of angiosperms across Gondwana plotted on an Aptian reconstruction provided by R. A. Livermore, British Antarctic Survey.

opening from east to west (Veevers *et al.* 1991; Royer & Coffin 1992). Although seafloor spreading was well established by Aptian times (*c.* 120 Ma BP), microcontinental fragments such as the Kerguelen Plateau (continental crust) (Fig. 2) probably maintained terrestrial connections for a considerable period of time. Recent work has identified terrestrial deposits of Turonian age (*c.* 90 Ma) on the Kerguelen Plateau (Shipboard Scientific Party 2000), implying that the microcontinent was above sea level at this time.

The reconstruction of Africa/Antarctica and the subsequent early stages of break-up are also problematic due to a paucity of magnetic anomaly data (Marks & Tikku 2001). Reconstructions such as those of Lawver *et al.* (1997), Muller *et al.* (1998) and Roeser *et al.* (1996) differ fundamentally from those of Segoufin & Patriat (1980), Livermore & Hunter (1996) and Marks & Tikku (2001), and consequently have markedly different implications for the timing of terrestrial separation. Sea floor spreading had started by 155 Ma BP in, for example, the Mozambique Basin (Segoufin & Patriat 1980), and off the Dronning Maud Land Coast (Roeser *et al.* 1996). However, much of the initial African/Antarctic motion had a large transform component that probably allowed landmasses to be juxtaposed (e.g. Mozambique Ridge), and recent reconstructions suggest terrestrial connections were present at least until 120 Ma BP and perhaps later (Reeves & de Wit 2000). The recent discovery of Late Cretaceous dinosaurs in Madagascar (Sampson *et al.* 1998), and their occurrence in South America and India, also suggests that these landmasses were still connected (via Antarctica) in the Late Cretaceous (Sampson *et al.* 1998; Krause *et al.* 1999; Hay *et al.* 1999). The uncertainties in the timing of terrestrial separation between the component continents of Gondwana remain a major constraint in explaining present-day disjunct distribution patterns.

Despite the present uncertainty, conventional reconstructions of Gondwana break-up imply separation of Africa/South America from east Gondwana by at least 155 Ma BP (early Late Jurassic times), and therefore substantially before angiosperm radiation and diversification. The only connection between east (Africa, South America) and west (India, Australia,

Antarctica, New Zealand) Gondwana is via the Antarctic Peninsula, a region that has been postulated as providing the main connecting link during the radiation of angiosperms (Fig. 3). Alternative scenarios have also been postulated; palaeogeographic reconstructions of SE Asia, Australia and east Antarctica during the Cretaceous indicate land connections between these areas (Metcalfe 1990), and point to a possible invasion from Asia to the north (Burger 1981; Truswell *et al.* 1987; Hill & Scriven 1995, fig. 5) (Fig. 3). Other workers have postulated that angiosperms arrived in Gondwana via India (Burger 1990), while more recently Dettmann (1989, 1994) suggested that angiosperms migrated along rift valleys, as these provided highly disturbed environments that favoured early angiosperms (Fig. 3).

To date, relatively little is known of Cretaceous angiosperm floras from Antarctica (Truswell 1991), yet they are critical for testing the validity of hypothesized radiation routes. If the Antarctic Peninsula acted as the main gateway, then Cretaceous floras from this region should have earlier records of angiosperms than elsewhere in eastern Gondwana (i.e. Australia, New Zealand, India). Alternatively, if angiosperms radiated into Gondwana via another route then they should appear later in Antarctic floras than elsewhere in Gondwana.

Floristic change in the Antarctic Peninsula

Over the last ten years our knowledge of the floristic composition of the Antarctic Peninsula macro- and microfloras has increased dramatically. Well documented macrofloras are now known from Aptian (Cantrill 1997, 2000; Césari *et al.* 1998, 1999; Torres *et al.* 1997), Albian (Cantrill 1995, 1996; Cantrill & Nichols 1996; Falcon-Lang & Cantrill 2000; Cantrill & Falcon-Lang 2001), Coniacian (Hayes 1999), Santonian (Hayes 1999), Campanian and Maastrichtian (Poole & Francis 1999, 2000; Poole *et al.* 2000*a,b,c;* Poole & Gottwald 2001) strata. However, the Maastrichtian record of leaf floras is sparse and could not be used in this study. In contrast, microfloras have been recorded from all of the major stages of the Cretaceous (excluding the Hauterivian and Barremian) (Askin 1990, 1994; Duane 1996; Riding *et al.* 1998; Cantrill 2000), and so provide a record for the latest Cretaceous (Maastrichtian).

Patterns of floristic replacement through the Cretaceous were examined using within-flora diversity plots (Lidgard & Crane 1990). These were calculated as a percentage of the total species diversity for each macroflora and microflora. Within-flora diversity moderates sample size effects (Spicer 1988) but the diversity measure for groups is not independent, as the values of a single flora must sum 100%. Comparisons of absolute within-flora species richness with within-flora diversity measures demonstrate that within-flora diversity captures patterns seen in absolute diversity and does not induce spurious trends (Lidgard & Crane 1990). Despite taxonomic biases in both the macrofloral and microfloral record that make it difficult to discriminate certain groups (e.g. bennettites in the microflora, lycopods in the macroflora), the datasets show similar patterns through time. This gives us some confidence that the observed patterns are a real reflection of changes in vegetation composition.

Bennettitales and other non-coniferous seed plants

Bennettites were initially reasonably diverse (Aptian, 14%), but then rapidly disappear from the floras becoming extinct by the Campanian (Fig. 4A). Unfortunately bennettites do not have a good pollen record; their monosulcate grains lack distinguishing features at the light-microscope level, which makes discrimination of species difficult. Thus the pollen record appears to be less diverse than the macrofloral record (see Dettmann 1986*b*). Like the bennettites, other non-coniferous seed plants (e.g. *Pachypteris*, Ginkgoales) all show a progressive decline from the Aptian to Albian, then recover slightly in Cenomanian times before declining in importance through the later part of the Cretaceous (Fig. 4B).

The decline in bennettites is also reflected in their abundance through the Cretaceous. They are one of the most frequently encountered foliage taxa in the Aptian floras, but by Albian times, although still accounting for 3.5% of the floristic diversity, they are very rare. Only a handful of specimens are known from the Albian and one from the Coniacian. This suggests that not only did they become less diverse but that they also became less important ecologically within the vegetation. A similar pattern is seen elsewhere globally. For example, macrofloras in Australia have abundant Bennettitales in early Aptian floras (e.g. Douglas 1969) but they disappear by Aptian–Albian times (Drinnan & Chambers 1986). Recent work on dispersed cuticles indicates that bennettites remained a minor constituent of the Australian vegetation at least until the Cenomanian (Pole & Douglas 1999).

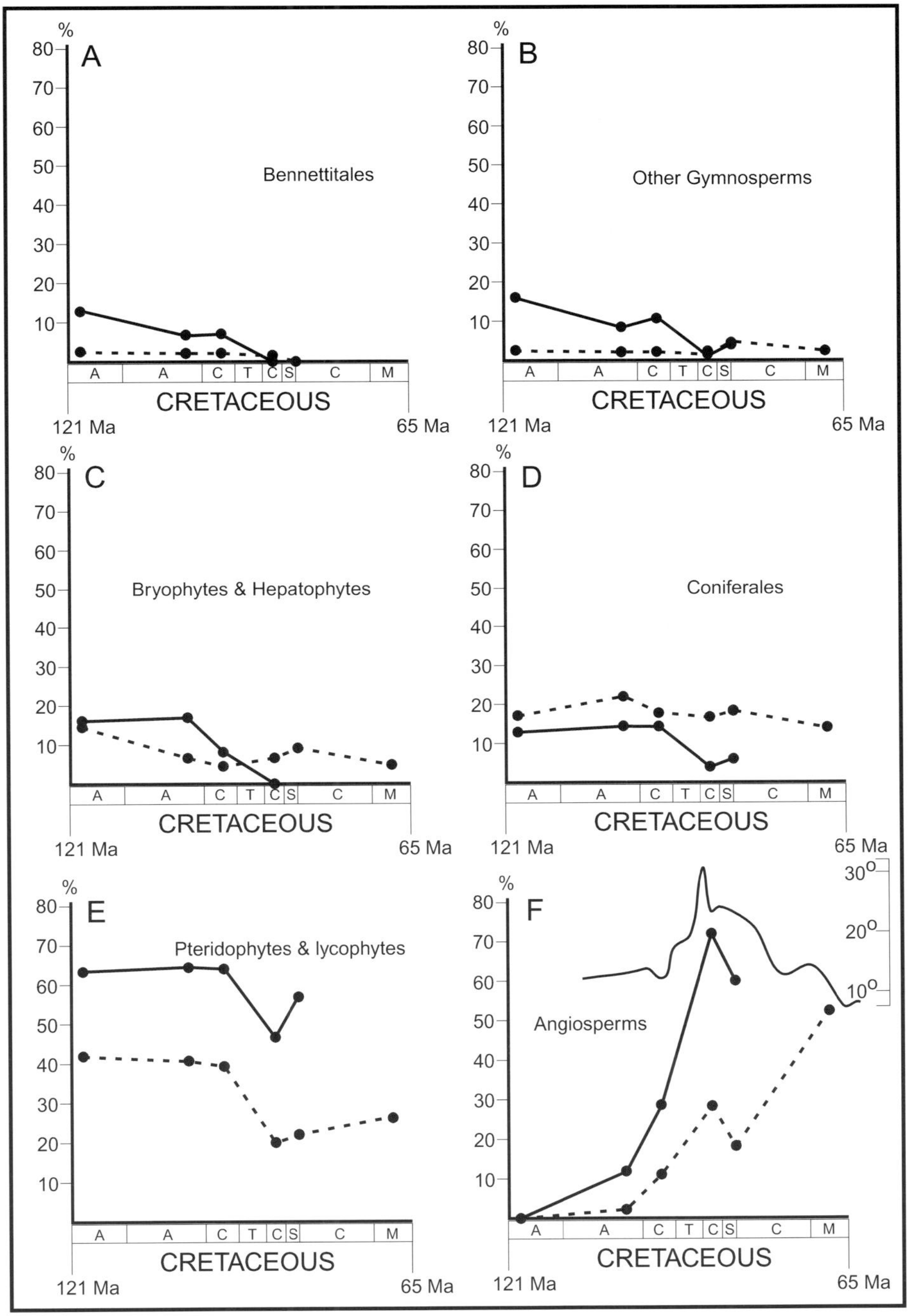

Fig. 4. Diversity trends through the Cretaceous for macrofloral (solid line) and microfloral (dashed line) data. Note the close correspondence between angiosperm diversity and palaeotemperature estimates (F) for the Southern Ocean. Temperature curve modified and smoothed from Huber (1998). Macrofloral taxic diversity compiled from Orlando (1968), Hernández & Azcárate (1971), Lacey & Lucas (1981), Banerji & Lemoigne (1987), Rees & Smellie (1989), Barale *et al.* (1995), Cesari *et al.* (1999), Hayes (1999), Cantrill (2000) and Falcon-Lang *et al.* (2001). Microfloral taxic diversity compiled from Dettmann & Thomson (1987), Askin (1990, 1994), Dolding (1992), Keating (1992), Keating *et al.* (1992), Riding *et al.* (1998), Barreda *et al.* (1999), Hathway *et al.* (1999) and Cantrill (2000).

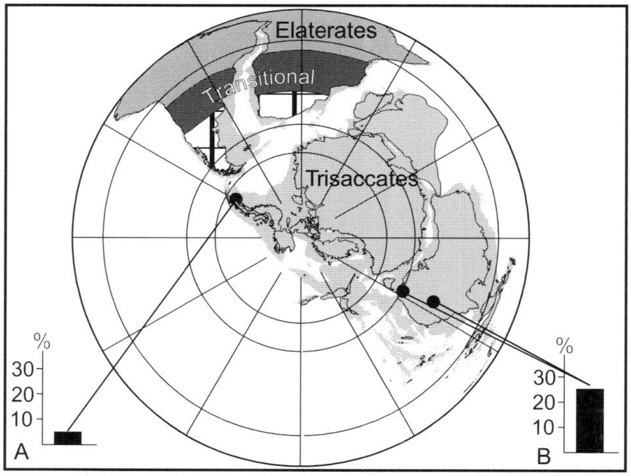

Fig. 5. Floristic provincialism across Gondwana derived from Herngreen *et al.* (1996). Note the regional differences in abundance of key trisaccate species such as *Microcachyridites antarcticus* within the Trisaccate Province. Histograms show, percentage abundance of *M. antarcticus* in Lower Cretaceous strata: (**A**) from Dettmann & Thomson (1987); (**B**) southeastern Australia from Dettmann (1986*b*), central Australia from Burger (1980).

Bryophytes/Hepatophytes

In the earliest Cretaceous, bryophytes and hepatophytes form an important and characteristic component of the floristic diversity. Not only are they diverse, accounting for up to 20% of the species (Fig. 4C), but they are also ecologically abundant ranging from colonizers of fresh sediment to components of established fern thickets and forests (Cantrill 1997). Indeed, many localities are characterized by assemblages comprising just hepatophytes, where they can cover individual bedding surfaces for hundreds of metres. This is also true for other southern high-latitude sites (e.g. southeastern Australia; Douglas 1973) and appears to be a phenomenon of high latitude floras in the early part of the Mesozoic (Cantrill 1997). The highest hepatophyte diversity occurs in the Early Cretaceous but they rapidly become a minor constituent of the vegetation by the Late Cretaceous. This pattern is also reflected in the microfloral diversity (Fig. 4C).

Coniferales

Conifers generally maintain between 10 and 20% of the within-flora diversity through the Cretaceous interval (Fig. 4D). This is apparent from the microfloral record but is less clear in the macroflora. The macrofloral record shows a distinct drop in within-flora diversity in the mid-Cretaceous. The differences may be due to wind-dispersed conifer pollen representing a wider range of plant communities than those seen in the macrofloras.

Pteridophytes and lycophytes

Macrofloral remains are most diverse in the Early Cretaceous but decline towards mid-Cretaceous times (Fig. 4E). Following this decline ferns recover through the later part of the Cretaceous (Fig. 4E). This pattern is supported by the microfloral record, which suggests that the recovery continues through the Late Cretaceous (Fig. 4E). Although a number of groups show a marked decline in diversity through the Cretaceous (e.g. Dipteridaceae), other families remain relatively stable (e.g. Gleicheniaceae). In terms of abundance, ferns show a similar pattern to the bennettites: they are extremely common in Aptian and Albian floras but are less frequently encountered in Late Cretaceous floras. So, although diverse in the later part of the Cretaceous, they are probably less important ecologically.

Angiosperms

Macrofloral remains of angiosperms are not recorded until the Late Albian (Cantrill & Nichols 1996) where they make up to 12% of the flora. This includes a diversity of habit ranging

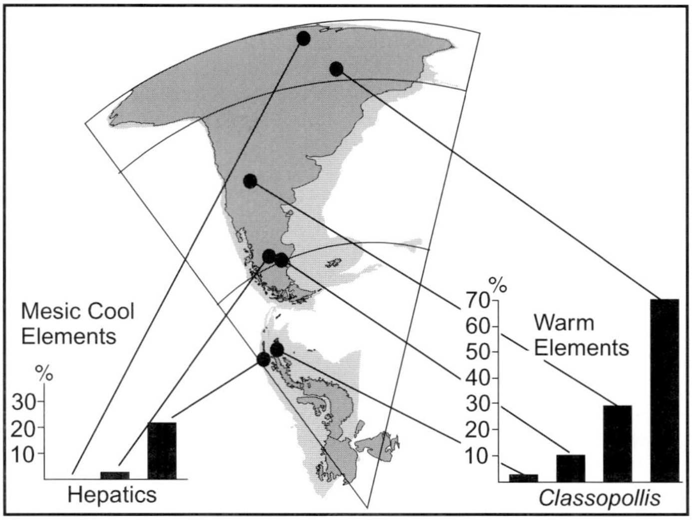

Fig. 6. Palaeolatitudinal gradients through South America and the Antarctic Peninsula highlighting the increase in diversity of mesic elements (hepatics) and the decrease in abundance of thermophilic species (e.g. *Classopollis*) from low to high latitudes. Diversity data for hepatics derived from Pons (1988), Riccardi (1988) and Falcon-Lang *et al.* (2001). Abundance data for *Classopollis* derived from Herngreen (1975), Dettmann & Thomson (1987), Herngreen *et al.* (1996) and Pramparo & Volkheimer (1999).

from herbaceous plants to probable trees (Cantrill & Nichols 1996). In the macrofloras angiosperms rapidly increase in importance rising to a peak in the Coniacian (72%); this is followed by a decline into the Santonian. In the microfloras angiosperms appear in the early Albian (Dettmann & Thomson 1987) and show a rapid diversification, reaching up to 75% within-flora diversity by Turonian times (Fig. 4F). From this peak of diversity they initially decline, before rapidly recovering in the Campanian and Maastrichtian.

Summary

Angiosperm within-flora diversity increases at the expense of bryophytes/hepatophytes, bennettites and other gymnosperms and, initially, ferns and lycophytes. This suggests that the angiosperm invasion took place in the niches occupied by these groups. Within the Antarctic Cretaceous ecosystem liverworts were colonizers of bare sediment and an understorey component of the vegetation (Cantrill 1997), whilst ferns and lycophytes were either understorey or thicket-forming plants (Cantrill 1996). So it seems most likely that the early angiosperms occupied colonization and understorey niches. Further support for this idea can be seen in the overstorey elements: conifers, the main tree species, maintained a relatively constant diversity throughout the Cretaceous. Indeed, angiosperm wood does not become common until Coniacian times, and the early angiosperm floras of the Antarctic Peninsula (Albian) are dominated by herbaceous leafy forms, although some leaf morphologies are suggestive of trees (Cantrill & Nichols 1996). The lack of wood also suggests that these early angiosperms occupied under/middle-storey niches. We conclude then that the angiosperms were invading the understorey and colonization niches within the ecosystem as those are the groups that were being replaced. Angiosperms later expanded into the overstorey, perhaps by filling gaps in the canopy left by conifers. The recovery of fern diversity (Fig. 4E) suggests that pteridophytes were able to colonize new niches created by the angiosperms.

Antarctic Peninsula: biotic gateway or biological backwater?

In order to evaluate whether the Antarctic Peninsula acted as a biotic gateway during the mid-Cretaceous radiation of angiosperms, the floral succession needs to be densely sampled and well age-constrained. A pilot study (Dettmann & Thomson 1987) identified angiosperm pollen in early Albian samples from the James Ross Basin. Unfortunately these were the oldest samples examined, but ?Barremian to Aptian strata in the South Shetlands lacked angiospermous grains (Askin 1983). Since these studies over 250 pre-Albian (e.g. Keating *et al.* 1992; Duane 1996; Riding *et al.* 1998; Hathway *et al.* 1999), as well as numerous Albian and younger samples (e.g. Barreda *et al.* 1999) have been examined from various parts of the Antarctic Peninsula. Despite the intensive sampling of Aptian and older strata, angiosperms do not appear in the Antarctic Peninsula until early Albian times. This event clearly post-dates their occurrence elsewhere in eastern Gondwana (Burger 1990, 1993). For example, angiospermous pollen grains (*Clavatipollenites hughesii*) are known from the basal *Cyclosporites hughesi* Palynozone (early Aptian) of Australia (Burger 1988), and macrofossils with attached flowers of Aptian age occur in south-eastern Australia (Taylor & Hickey 1990). Since these unequivocal records predate the appearance of angiosperms in Antarctica it seems unlikely that the Antarctic Peninsula could have acted as the main biotic gateway for dispersal across Gondwana, at least in the initial stages of the radiation. Although it seems certain that terrestrial land connections occurred between southern South America and the Antarctic Peninsula during the Cretaceous, other factors must have been acting as a barrier to radiation.

The Early to mid-Cretaceous in the southern hemisphere is marked by strong floral provincialism (Herngreen *et al.* 1996). A distinct floral province dominated by trisaccate pollen grains characterized microfloras of Australia, India, New Zealand, southern Africa and southern South America (Herngreen *et al.* 1996, and references therein) (Fig. 5). In particular, there is a steep floristic gradient through South America that has been recognized from microfloras. For example, in Aptian strata *Classopollis* can account for 30–60% of certain palynofloras by abundance in low latitudes, but drops sharply in high latitudes, a feature also noted by Dettmann & Thomson (1987). In the Antarctic Peninsula region, *Classopollis* rarely accounts for more than 5% of the residue and is often absent (Fig. 6). This type of pattern is repeated for a number of groups including Elaterate (Gnetales) and angiosperm pollen. Furthermore the reverse pattern is true, with some taxa being more diverse in the high latitudes. For example, hepatophytes make up to 18% of the within-flora diversity in high-latitude sites, yet they are almost absent from mid-latitude floras (Cantrill 1997) (Fig. 6). Similar but less extreme examples include the ferns and lycopsids.

Equally, within the large Trisaccate Province that covers Australia, India, New Zealand, Antarctica, southern Africa and southern South America there is floristic regionalism. This phytogeographic floral province can be subdivided into two regions: the *Cyclusphaera-Classopollis* subprovince (Africa plus South America), and the *Murospora* subprovince (Australia, India) (Herngreen *et al.* 1996). Within these two provinces there are important variations. *Microcachyridites antarcticus*, one of the elements that defines the province, typically constitutes up to 25% by abundance of palynoresidues at Australian localities (Fig. 5). In contrast, *M. antarcticus* is never more than 5–6% by abundance in the Antarctic Peninsula microfloras (Fig. 5). These regional differences also extend to the macrofloras and provide indirect support for a late arrival of angiosperms in the Antarctic Peninsula. One group that is apparently replaced by the angiosperms is the bennettites. In Australia, bennettites are common in Valanginian to early Aptian macrofloras of southeastern Australia but disappear by the Albian (Douglas 1969). In the Antarctic Peninsula macrofloras, bennettites are common in the Aptian, and to a lesser extent the Albian, but they persist until Coniacian times. This time lag in regional extinction supports the idea that angiosperms arrived later in the Antarctic Peninsula region than elsewhere in Gondwana.

One explanation for the steep floristic gradients observed through South America and into the Antarctic Peninsula is a strong climatic gradient (Dettmann 1986*a*). Recent climatic modelling suggests that the thermal and effective precipitation gradients through this latitudinal transect were extreme (Valdes *et al.* 1996), and this is supported by the plant fossils. A comparison of Aptian macrofloras between the Antarctic Peninsula (Cerro Negro Formation) and southern South America (Baquero Formation) indicates important differences. Antarctic Peninsula floras are high in cool- and moisture-loving forms (e.g. hepatophytes, ferns, etc.). In contrast, the Baquero Formation is dominated by desiccation-tolerant, warm-loving plants (e.g. bennettites and other seed-plants). This can be seen in the cuticular adaptations of the Baquero Formation plants with strongly sunken stoma, highly papillate leaves and thick cuticles (Archangelsky *et al.* 1995). Although some of the plant groups are represented in both floras, those such as the bennettites from the higher latitude sites of the Antarctic Peninsula have thin cuticles, and appear to lack the adaptations to cope with warm, desiccating conditions. It could well be that this extreme thermal and moisture gradient acted as an effective barrier to migration into the Antarctic Peninsula region, and that the initial radiation of angiosperms across Gondwana was not via the Antarctic Peninsula but by some other route.

Peak of Cretaceous warmth as a forcing mechanism

It is probably no coincidence that the arrival of angiosperms and their subsequent rise to dominance in the high southern latitudes correspond to a global climatic warming trend. The peak of the Cretaceous greenhouse was in the Turonian (Huber 1998) and this coincides with the peak of angiosperm diversity in the Antarctic Peninsula (Fig. 4E). Certainly if the thermal regime that existed in the Early Cretaceous was a barrier to migration, then global warming would have resulted in latitudinal range expansion as climatic zones widened. This would have pushed floristic boundaries southwards, and ultimately have led to angiosperm arrival in the Antarctic Peninsula region.

Following the peak of Cretaceous warmth in the Turonian, the high southern latitudes remained warm until late Santonian to early Campanian times when cooling started to occur (Huber 1998). This latest Cretaceous period saw the origin of a number of groups that today characterize temperate southern floras (e.g. *Nothofagus*, certain groups in the Proteaceae and Myrtaceae; Dettmann 1989). Furthermore, events such as the uplift of the Andes (*c.* 80 Ma BP) and a general cooling trend through the latest Cretaceous provided opportunities for these taxa to expand northwards. As the globe cooled the tropical biomes contracted equatorwards creating space for temperate biota to expand. A good example of this is the fern family Lophosoriaceae, a group that arose in the Antarctic Peninsula regions during the earliest Cretaceous (Berriasian) (Dettmann 1986*a*; Cantrill 1998), appeared later in southern South America (Aptian–early Albian), and later still in Colombia (Pleistocene). Although today it occupies an extreme latitudinal range along the Andean chain, it is clear that the closer to the tropics it occurs, then the higher (cooler) the altitudinal setting (Cantrill 1998). Clearly for this group, migration is incumbent upon a suitable habitat either provided by cooling climates or the creation of high-altitude sites. Similar patterns of later northward inceptions are seen in the earliest appearance of *Nothofagus* in the Antarctic Peninsula (early Campanian) and subsequent spread to southern South America (Maastrichtian).

Summary

Patterns of floristic replacement through the Cretaceous in the high southern latitudes occur amongst the understorey and colonizer elements of the vegetation, suggesting that this was the niche that the angiosperms initially occupied. This is a pattern similar to that seen elsewhere in the world (Lidgard & Crane 1988), and is supported by the lack of angiosperm wood in these early deposits, despite the presence of pollen. Later stage invasion occurred in the overstorey with the appearance of tree species as attested to by abundant angiosperm wood in the Late Cretaceous sequences (Chapman & Smellie 1992; Poole & Francis 1999, 2000; Poole *et al.* 2000*a*,*b*,*c;* Poole & Cantrill 2001; Poole & Gottwald 2001).

The timing of the radiation into the Antarctica Peninsula implies that this region was not the initial gateway for invasion of Gondwana by angiosperms. It seems most likely that the initial invasion occurred through Africa or India but further work is needed to refine not only the plate tectonic reconstructions but also the floristic history of these areas. The macro- and microfloral record is not well known for southern Africa, and is poorly age-constrained for India, making it difficult to discriminate between these hypotheses. Despite the terrestrial connection between South America and the Antarctic Peninsula, it is suggested that a steep climatic gradient was probably responsible for blocking the early radiation through this region. However, mid-Cretaceous (Turonian) warming resulted in latitudinal range expansion of climatic belts and appears to have been the forcing mechanism for the radiation of angiosperms into the Peninsula.

References

ARCHANGELSKY, A., ANDREIS, R. R., ARCHANGELSKY, S. & ARTABE, A. 1995. Cuticular characters adapted to volcanic stress in a new Cretaceous cycad leaf from Patagonia, Argentina. Considerations on the stratigraphy and depositional history of the Baquero Formation. *Review of Palaeobotany and Palynology*, **89**, 213–233.

ASKIN, R. A. 1983. Tithonian (uppermost Jurassic) – Barremian (Lower Cretaceous) spore, pollen and microplankton from the South Shetland Islands, Antarctica. *In:* OLIVER, R. L., JAMES, P. R. & JAGO, J. B. (eds) *Antarctic Earth Science*. Australian Academy of Sciences, Canberra, 295–297.

ASKIN, R. A. 1990. Cryptogam spores from the upper Campanian and Maastrichtian of Seymour Island, Antarctica. *Micropaleontology*, **36**, 141–156.

ASKIN, R. A. 1994. Monosulcate angiosperm pollen from the Lopez de Bertodano Formation (upper

Campanian-Maastrichtian-Danian) of Seymour Island, Antarctica. *Review of Palaeobotany and Palynology*, **81**, 151–164.

BANERJI, J. & LEMOIGNE, Y. 1987. Significant additions to the Upper Triassic flora of Williams Point, Livingston Island, South Shetlands (Antarctica). *Geobios*, **20**, 469–487.

BARALE, G., PHILIPPE, M., TORRES, T. & THEVENARD, F. 1995. Reappraisal of the Triassic flora from Williams Point, Livingston Island (South Shetland Islands, Antarctica): systematical, biostratigraphical and paleogeographical implications. *Serie Científica del Instituto Antárctico Chileno*, **45**, 9–38.

BARREDA, V., PALAMARCZUK, S. & MEDINA, F. 1999. Palinología de la Formación Hidden Lake (Coniianciano-Santoniano), Isla James Ross, Antártida. *Revista Espanõla de Micropalaeontología*, **31**, 53–72.

BURGER, D. 1980. Palynology of the lower Cretaceous in the Surat Basin. *Bureau of Mineral Resources, Geology and Geophysics Bulletin*, **189**, 1–106.

BURGER, D. 1981. Observations on the earliest angiosperms development with special reference to Australia. *Proceedings of the 4th International Palynological Conference, Lucknow*, **3**, 418–428.

BURGER, D. 1988. Early Cretaceous environments in the Eromanga Basin: palynological evidence from GSQ Wyandra 1 corehole. *Australasian Association of Palaeontologists Memoir*, **5**, 173–186.

BURGER, D. 1990. Early Cretaceous angiosperms from Queensland, Australia. *Review of Palaeobotany and Palynology*, **65**, 153–163.

BURGER, D. 1993. Early and middle Cretaceous angiosperm pollen grains from Australia. *Review of Palaeobotany and Palynology*, **78**, 183–234.

CANTRILL, D. J. 1995. The occurrence of the fern *Hausmannia* Dunker (Dipteridaceae) in the Cretaceous of Alexander Island, Antarctica. *Alcheringa*, **19**, 243–254.

CANTRILL, D. J. 1996. Fern thickets from the Cretaceous of Alexander Island, Antarctica containing *Alamatus bifarius* Douglas and *Aculea acicularis* sp. nov. *Cretaceous Research*, **17**, 169–182.

CANTRILL, D. J. 1997. Hepatophytes from the Early Cretaceous of Alexander Island, Antarctica: systematics and palaeoecology. *International Journal of Plant Sciences*, **158**, 476–488.

CANTRILL, D. J. 1998. Early Cretaceous fern foliage from President Head, Snow Island, Antarctica. *Alcheringa*, **22**, 241–258.

CANTRILL, D. J. 2000. A Cretaceous (Aptian) flora from President Head, Snow Island, Antarctica. *Palaeontographica B*, **253**, 153–191.

CANTRILL, D. J. & FALCON-LANG, H. J. 2001. Cretaceous (Late Albian) Coniferales of Alexander Island, Antarctica. Part 2: Foliage, reproductive structures and roots. *Review of Palaeobotany and Palynology*, **115**, 119–145.

CANTRILL, D. J. & NICHOLS, G. J. 1996. Taxonomy and palaeoecology of Early Cretaceous (Late Albian) angiosperm leaves from Alexander Island, Antarctica. *Review of Palaeobotany and Palynology*, **92**, 1–28.

CÉSARI, S., PARICA, C., REMESAL, M. & SALANI, F. 1998. First evidence of Pentoxylales in Antarctica. *Cretaceous Research*, **19**, 733–743.

CÉSARI, S., PARICA, C., REMESAL, M. & SALANI, F. 1999. Palaeoflora del Cretácico Inferior de peninsula Byers, islas Shetland del Sur, Antártida. *Ameghiniana*, **36**, 3–22.

CHAPMAN, J. L. & SMELLIE, J. L. 1992. Cretaceous fossil wood and palynomorphs from Williams Point, Livingston Island, Antarctic Peninsula. *Review of Palaeobotany and Palynology*, **74**, 163–192.

CRANE, P. R. & LIDGARD, S. 1989. Angiosperm diversification and paleolatitudinal gradients in Cretaceous floristic diversity. *Science*, **246**, 675–678.

CRANE, P. R., FRIIS, E. M. & PEDERSEN, K. R. 1995. The origin and early diversification of angiosperms. *Nature*, **374**, 27–33.

DETTMANN, M. E. 1986*a*. Significance of the Cretaceous-Tertiary spore genus *Cyatheacidites* in tracing the origin and migration of *Lophosoria* (Filicopsida). *Special Papers in Paleontology*, **35**, 63–94.

DETTMANN, M. E. 1986*b*. Early Cretaceous palynoflora of subsurface strata correlative with the Koonwarra Fossil Bed, Victoria. *Australiasian Association of Palaeontologists Memoir*, **3**, 79–110.

DETTMANN, M. E. 1989. Antarctica: Cretaceous cradle of austral temperate rainforests? *In*: CRAME J. A. (ed.) *Origins and Evolution of the Antarctic Biota*. Geological Society, London, Special Publications, **47**, 89–105.

DETTMANN, M. E. 1994. Cretaceous vegetation: the microfloral record. *In*: HILL, R. S. (ed.) *History of the Australian Vegetation, Cretaceous to Recent*. Cambridge University Press, Cambridge, 143–170.

DETTMANN, M. E. & THOMSON, M. R. A. 1987. Cretaceous palynomorphs from the James Ross Island area – a pilot study. *British Antarctic Survey Bulletin*, **77**, 13–59.

DOLDING, P. J. D. 1992. Palynology of the Marambio Group (Upper Cretaceous) of northern Humps Island. *Antarctic Science*, **4**, 311–326.

DOUGLAS, J. G. 1969. The Mesozoic floras of Victoria. Parts 1 & 2. *Geological Survey of Victoria Memoir*, **28**, 1–310.

DOUGLAS, J. G. 1973. The Mesozoic floras of Victoria. Part 3. *Geological Survey of Victoria Memoir*, **29**, 1–185.

DRINNAN, A. N. & CHAMBERS, T. C. 1986. Flora of the Lower Cretaceous Koonwarra Fossil Bed (Korumburra Group), South Gippsland, Victoria. *Association of Australasian Palaeontologists Memoir*, **3**, 1–77.

DRINNAN, A. N. & CRANE, P. R. 1990. Cretaceous paleobotany and its bearing on the biogeography of austral angiosperms. *In*: TAYLOR, T. N. & TAYLOR, E. L. (eds) *Antarctic Paleobiology: its Role in the Reconstruction of Gondwana*. Springer Verlag, New York, 192–219.

DUANE, A. M. 1996. Palynology of the Byers Group (late Jurassic–Early Cretaceous) of Livingston and Snow Island, Antarctic Peninsula: its biostratigraphic and palaeoenvironmental significance: *Review of Palaeobotany and Palynology*, **91**, 241–281.

FALCON-LANG, H. J. & CANTRILL, D. J. 2000. Cretaceous (Late Albian) Coniferales of Alexander Island, Antarctica. Part 1: Wood taxonomy: a quantitative approach. *Review of Palaeobotany and Palynology*, **111**, 1–17.

FALCON-LANG, H. J., CANTRILL, D. J. & NICHOLS, G. J. 2001. Biodiversity and terrestrial ecology of a mid-Cretaceous, high-latitude floodplain, Alexander Island, Antarctica. *Journal of the Geological Society, London*, **150**, 709–724.

HATHWAY, B., DUANE, A. M., CANTRILL, D. J. & KELLEY, S. P. 1999. $^{40}Ar/^{39}Ar$ geochronology and palynology of the Cerro Negro Formation, South Shetland Islands, Antarctica: a new radiometric tie for Cretaceous terrestrial biostratigraphy in the Southern Hemisphere. *Australian Journal of Earth Sciences*, **46**, 593–606.

HAY, W. W., DE CONTO, R. M., WOLD, C. N., *ET AL.* 1999. An alternative global Cretaceous palaeogeography. *In*: BARRERA, E. & JOHNSON, C. (eds) *Evolution of the Cretaceous Ocean–Climate System*. Geological Society of America, Special Papers, **332**, 1–48.

HAYES, P. 1999. *Cretaceous Angiosperm Floras of Antarctica*. PhD Thesis, University of Leeds.

HERNÁNDEZ, P. J. & AZCÁRATE, V. 1971. Estudio paleobotanico preliminar sobre restos de una tafoflora de la Peninsula Byers (Cerro Negro), Isla Livingston, Islas Shetland del Sur, Antartica. *Serie Científica del Instituto Antárctico Chileno*, **2**, 15–50.

HERNGREEN, G. F. W. 1975. Palynology of Middle and Upper Cretaceous strata in Brazil. *Mededelingen Rijks Geologische Dienst Nieuwe Serie*, **26**, 39–90.

HERNGREEN, G. F. W., KEDVES, M., ROVINA, L. V. & SMIRNOVA, S. B. 1996. Cretaceous palynofloral provinces: a review. *In*: JANSONIUS, J. & MCGREGOR, D.C. (eds) *Palynology: Principles, and Applications*. American Association of Stratigraphic Palynologists Foundation, **3**, 1157–1188.

HILL, R. S. & SCRIVEN, L. J. 1995. The angiosperm-dominated woody vegetation of Antarctica: a review. *Review of Palaeobotany and Palynology*, **86**, 175–198.

HUBER, B. T. 1998. Tropical paradise at the Cretaceous poles? *Science*, **282**, 2199–2200.

KEATING, J. M. 1992. Palynology of the Lachmans Crags Member, Santa Marta Formation (Upper Cretaceous) of north-west James Ross Island. *Antarctic Science*, **4**, 293–304

KEATING, J. M., SPENCER-JONES, M. & NEWHAM, S. 1992. The stratigraphical palynology of the Kotick Point and Whisky Bay formations, Gustav Group (Cretaceous), James Ross Island. *Antarctic Science*, **4**, 279–292.

KRAUSE, D. W., ROGER, R. R., FORSTER, C. A., HARTMAN, J. H. & BUCKLEY, G. A. 1999. The Late Cretaceous vertebrate fauna of Madagascar: implications for gondwanan palaeobiogeography. *GSA Today*, **9**(8), 1–7.

LACEY, W. S. & LUCAS, R. C. 1981. The Triassic flora of Livingston Island, South Shetland Islands. *British Antarctic Survey Bulletin*, **53**, 157–173.

LAWVER, L. A., GAHAGAN, L. M. & DALZIEL, I. W. D. 1997. A tight fit early Mesozoic Gondwana, a plate reconstruction perspective. *Memoirs of the National Institute of Polar Research*, **53**, 214–229.

LIDGARD, S. & CRANE, P. R. 1988. Quantitative analyses of the early angiosperm radiation. *Nature*, **331**, 344–346.

LIDGARD, S. & CRANE, P. R. 1990. Angiosperm diversification and Cretaceous floristic trends: a comparison of palynofloras and leaf macrofloras. *Paleobiology*, **16**, 77–93.

LIVERMORE, R. A. & HUNTER, R. J. 1996. Mesozoic seafloor spreading in the southern Weddell Sea. *In*: STOREY, B. C., KING, E. C. & LIVERMORE, R. A. (eds) *Weddell Sea Tectonics and Gondwana Break-up*. Geological Society, London, Special Publications, **108**, 227–241.

MARKS, K. M. & TIKKU, A. A. 2001. Cretaceous reconstructions of East Antarctica, Africa and Madagascar. *Earth and Planetary Science Letters*, **186**, 479–495.

METCALFE, I. 1990. Allochthonous terrane processes in Southeast Asia. *Philosophical Transactions of the Royal Society London A*, **331**, 625–640.

MULLER, R. D., MIHUT, D. & BALDWIN, S. 1998. A new kinematic model for the formation and evolution of the west and northwest Australian margin. *In*: PURCELL, P. G. & PURCELL, P. R. (eds) *The Sedimentary Basins of Western Australia*. Petroleum Exploration Society of Australia, Perth, 55–71.

ORLANDO, H. A. 1968. A new Triassic flora from Livingston Island, South Shetland Islands. *British Antarctic Survey Bulletin*, **16**, 1–13.

POLE, M. S. & DOUGLAS, J. G. 1999. Bennettitales, Cycadales and Ginkgoales from the mid-Cretaceous of the Eromanga Basin, Queensland, Australia. *Cretaceous Research*, **20**, 523–538.

PONS, D. 1988. *Le Mésozoïque de Colombie macroflores et microflores*. CNRS, Paris.

POOLE, I. & CANTRILL, D. J. 2001. Fossil woods from Williams Point Beds, Livingston Island, Antarctica: a Late Cretaceous southern high latitude flora. *Palaeontology* **44**, 1081–1112.

POOLE, I. & FRANCIS, J. E. 1999. The first record of fossil atherospermataceous wood from the Upper Cretaceous of Antarctica. *Review of Palaeobotany and Palynology*, **107**, 97–107.

POOLE, I. & FRANCIS, J. E. 2000. The first record of fossil wood of Winteraceae from the Upper Cretaceous of Antarctica. *Annals of Botany*, **85**, 307–315.

POOLE, I. & GOTTWALD, H. 2001. Monimiaceae *sensu lato*, an element of gondwanan polar forests: evidence from the Late Cretaceous–Early Tertiary wood flora of Antarctica. *Australian Systematic Botany*, **14**, 207–230.

POOLE, I., CANTRILL, D. J., HAYES, P. & FRANCIS, J. E. 2000*a*. The fossil record of Cunoniaceae: new evidence from Late Cretaceous wood of Antarctica? *Review of Palaeobotany and Palynology*, **111**, 127–144.

POOLE, I., GOTTWALD, H. & FRANCIS, J. E. 2000*b*. *Illicioxylon*, an element of Gondwanan Polar forests? Late Cretaceous and early Tertiary woods of Antarctica. *Annals of Botany*, **86**, 421–432.

POOLE, I., RICHTER, H. G. & FRANCIS, J. E. 2000*c*. Evidence for Gondwanan origins for Sassafras (Lauraceae)? Late Cretaceous fossil wood of Antarctica. *International Association of Wood Anatomists Journal*, **21**, 463–475.

PRAMPARO, M. B. & VOLKHEIMER, W. 1999. Palinología del Miembro Avilé (Formación Agrio, Cretácico Inferior) en la cerro de la Parva, Neuquén. *Ameghiniana*, **36**, 217–227.

REES, P. M. & SMELLIE, J. L. 1989. Cretaceous angiosperms from an allegedly Triassic flora at Williams Point, Livingston Island, South Shetland Islands. *Antarctic Science*, **1**, 239–248.

REEVES, C. V. & DE WIT, M. J. 2000. Making ends meet in Gondwana: retracing the transforms of the Indian Ocean and reconnecting continental shear zones. *Terra Nova*, **12**, 272–280.

RICCARDI, A. C. 1988. The Cretaceous system of southern South America. *Geological Society of America Memoir*, **168**, 1–161.

RIDING, J. B., CRAME, J. A., DETTMANN, M. E. & CANTRILL, D. J. 1998. The age of the base of the Gustav Group in the James Ross Basin, Antarctica. *Cretaceous Research*, **19**, 87–105.

ROESER, H. A., FRITSCH, J. & HINZ, K. 1996. The development of the crust of Dronning Maud Land, East Antarctica. *In*: STOREY, B. C., KING, E. C. & LIVERMORE, R. A. (eds) *Weddell Sea Tectonics and Gondwana Break-up*. Geological Society, London, Special Publications, **108**, 243–264.

ROYER, J-V. & COFFIN, M. F. 1992. Jurassic to Eocene plate tectonic reconstructions in the Kerguelen Plateau region. *Proceedings of the Ocean Drilling Program, Scientific Results*, **120**, 917–928.

SAMPSON, S. D., WITMER, L. M., FORSTER, C. A., KRAUSE, D. W., O'CONNOR, P. M., DODSON, P. & RAVOAVY, F. 1998. Predatory dinosaur remains from Madagascar: implications for the Cretaceous biogeography of Gondwana. *Science*, **280**, 1048–1051.

SEGOUFIN, J. & PATRIAT, P. 1980. Existence d'anomalies mesozoiques dans le bassin de Somalie. Implications pour les relations Afrique-Antarctique-Madagascar. *Compte Rendu Academie Sciences Paris*, **291**, 85–88.

SHIPBOARD SCIENTIFIC PARTY. 2000. Leg 183 summary: Kerguelen Plateau–Broken Ridge—A large igneous province. *In*: COFFIN, M. F., FREY, F. A., WALLACE, P. J., *ET AL.* (eds) *Proceedings Ocean Drilling Program, Initial Reports*, **183**, 1–101.

SPICER, R. A. 1988. Quantitative sampling of plant megafossil assemblages. Paleontological Society Special Publication, **3**, 29–51.

STOREY, B. C., DALZIEL, I. W. D., GARRETT, S. W., GRUNOW, A. M., PANKHURST, R. J. & VENNUM, W. R. 1988. West Antarctic in Gondwanaland: crustal blocks, reconstruction and break-up processes. *Tectonophysics*, **155**, 381–390.

TAYLOR, D. W. & HICKEY, L. J. 1990. An Aptian plant with attached leaves and flowers: implication for angiosperm origin. *Science*, **247**, 702–704.

TORRES, T., BARALE, G., THÉVERNARD, F., PHILIPPE, M. & GALLEGUILLOS, H. 1997. Morfología y sistemática de la flora del Cretácico Inferior de President Head, isla Snow, archipélago de las Shetland del Sur, Antártica. *Serie Científica del Instituto Antárctico Chileno*, **2**, 15–50.

TRUSWELL, E. M. 1991. Antarctica: a history of terrestrial vegetation. *In*: TINGEY, R. J. (ed.) *The Geology of Antarctica*. Oxford Monographs on Geology and Geophysics, **17**, Clarendon Press, Oxford, 499–537.

TRUSWELL, E. M., KERSHAW, A. P. & SLUITER, I. R. 1987. The Australian–south-east Asian connection: evidence from the palaeobotanical record. *In*: WHITEMORE, T. C. (ed.) *Biogeographical Evolution of the Malay Archipelago* Oxford University Press, Oxford, 32–49.

TRUSWELL, E. M., DETTMANN, M. E. & O'BRIEN, P. E. 1999. Mesozoic palynofloras from the Mac. Robertson Shelf, East Antarctica: geological and phytogeographic implications. *Antarctic Science*, **11**, 239–255.

VALDES, P. J., SELLWOOD, B. W. & PRICE, G. D. 1996. Evaluating concepts of Cretaceous equability. *Palaeoclimates*, **2**, 139–158.

VEEVERS, J. J., POWELL, C. MCA. & ROOTS, S. R. 1991. Review of seafloor spreading around Australia. 1. Synthesis of the patterns of spreading. *Australian Journal of Earth Sciences*, **38**, 373–389.

Cenozoic palaeogeography and the rise of modern biodiversity patterns

J.A. CRAME[1] & B.R. ROSEN[2]

[1] *British Antarctic Survey, High Cross, Madingley Road, Cambridge CB3 0ET, UK (e-mail:* A.Crame@bas.ac.uk*)*

[2] *Department of Palaeontology, The Natural History Museum, Cromwell Road, London SW7 5BD, UK (e-mail:* B.Rosen@nhm.ac.uk*)*

Abstract: The steepest latitudinal and longitudinal gradients in taxonomic diversity at the present day are those associated with tropical high diversity foci. Although there has been a tendency in the past to regard these features as either evolutionary 'cradles' or 'museums' of considerable antiquity, this may not be the case. Within the marine realm, a uniform, pan-tropical fauna was progressively disrupted by a series of plate tectonic events, the most important of which were the Early Miocene (*c.* 20 Ma) collisions of Africa/Arabia with Europe and Australia/New Guinea with Indonesia, and the Middle Miocene–latest Pliocene rise of the Central American Isthmus. This had the net effect of establishing two main tropical high diversity foci: the Indo-West Pacific and the Atlantic–Caribbean–East Pacific. Similar foci were also established at the same time in the terrestrial realm.

Together with the physical isolation of Antarctica, these same tectonic events contributed significantly to global cooling throughout the Cenozoic Era. This in turn led to the imposition of a series of thermally defined provinces, and thus a considerable degree of biotic differentiation on a regional scale. However, something else seems to have been involved in the creation of very steep tropical diversity peaks. This could in part be a coincidental radiation of a series of unrelated taxa, or some sort of evolutionary feedback mechanism between interacting clades. Alternatively, Late Cenozoic rates of origination may have been enhanced by an external forcing mechanism such as changes in Orbital Range Dynamics.

Despite the enormous increase in biodiversity research in recent years we are still a long way from understanding the true nature of the larger-scale patterns of life on Earth. There is a general appreciation that, for many groups of plants and animals, there are simply more of them in the tropical and low-latitude regions than in the high-latitude and polar ones, and latitudinal gradients in taxonomic diversity constitute one of the most striking macroecological patterns at the present day (e.g. Gaston & Blackburn 2000, and references therein). In addition we are also beginning to understand that tropical high diversity is distinctly heterogeneous. Two tropical hotspots or high diversity foci are often distinguished in the marine realm, one in the Indo-West Pacific (IWP) and the other in the Atlantic, Caribbean and East Pacific region (ACEP) (*sensu* Ellison *et al.* 1999). Analogous patterns have now been established in the terrestrial realm too, and in a global study of the family richness of seed plants, amphibians, reptiles and mammals, Williams *et al.* (1997) found hotspots of maximum richness in Colombia, Nicaragua, Oaxaca (Mexico) and Malaysia. Tropical longitudinal gradients in taxonomic diversity are equally impressive contemporary biogeographical phenomena (Crame 2000*a*,*b*, and references therein). They are of particular interest as they are far less likely to co-vary with any known environmental parameter than latitudinal gradients (but see Bellwood & Hughes 2001).

We also lack a convincing integrated theory or model to account for the origins of large-scale biodiversity patterns in both time and space (Gaston 2000). Are they, as some have argued, the product of the 'ecological moment', representing no more than the adjustment of regional biotas to climates and habitats that have developed since the Last Glacial Maximum (i.e. some 18 ka ago)? Or are there longer-term processes involved too, stretching back over a geological timescale of millions or even tens of millions of years? There has recently been a considerable development in our understanding of the ways in which geological processes could influence the formation of large-scale biodiversity patterns, and this has led to an intense debate on the relative roles of contemporary versus historical factors in the creation of taxonomic diversity gradients (e.g. Turner *et al.* 1996; Francis &

From: CRAME, J. A. & OWEN, A. W. (eds) 2002. *Palaeobiogeography and Biodiversity Change: the Ordovician and Mesozoic–Cenozoic Radiations*. Geological Society, London, Special Publications, **194**, 153–168.
0305-8719/02/$15.00

Currie 1998; Huston 1999; Ricklefs 1999; Qian & Ricklefs 2000).

It is the intention of this study to further this line of enquiry by focusing on a series of major tectonic and climatic events that occurred throughout the Cenozoic Era (i.e. the last 65 Ma of Earth history). It will be suggested that this was a time of major expansion in the size and scale of taxonomic diversity gradients and the question asked whether this can be attributed directly to the relative isolation of the continents, fundamental reorganization of oceanic circulation systems and progressive imposition of a steep latitudinal temperature gradient. The potential role of biotic factors in promoting the massive Cenozoic radiation of selected tropical taxa will not be ignored, but the emphasis here will essentially be on evolving Cenozoic palaeogeographies and how they affected such key processes as climate change, the establishment of new colonization and dispersal routes, and, possibly, pronounced changes in origination (speciation) and extinction rates.

Cenozoic global diversification

There is general agreement that there was a very substantial rise in global biodiversity through the Cenozoic Era. This was particularly so at the lower taxonomic levels of species and genus where the rise was the steepest and most prolonged of the entire Phanerozoic interval (Signor 1990; Benton 1999; Foote 2000). We could perhaps be looking at an order of magnitude increase in species diversity at this time (Signor 1990; but see also Alroy *et al.* 2001). Equally striking is the fact that this radiation is recorded across such a broad range of organisms in both the marine and terrestrial realms. This was the time of widespread proliferation of groups such as the neogastropods, heteroconch bivalves, cheilostome bryozoans, decapod crustaceans and teleost fish in shallow seas, and amphibians, reptiles, birds, mammals, flowering plants and insects on land (e.g. Hallam 1994). Parallel temporal trends in so many different groups has been taken as a strong indication that there may be a single, basic mechanism underpinning the process of Cenozoic biodiversification (Signor 1990; Benton 1999).

The presence of very clear latitudinal gradients at the present day strongly suggests that this pulse of Cenozoic diversification was concentrated in low-latitude and tropical regions. Unfortunately, we cannot yet demonstrate this by reconstructing diversity gradients directly from the fossil record, but a number of indirect lines of evidence suggest that this was indeed the case. For example, in a recent study of latest Mesozoic–Cenozoic bivalve diversification, Crame (2000*a*) demonstrated that the steepest latitudinal gradients occurred in the youngest clades (and especially in the very large heteroconch clade). He went on to suggest that the same phenomenon may occur within the marine gastropods, where the youngest major clade, the neogastropods, shows particularly strong latitudinal gradients (Crame 2001). In the terrestrial realm the three largest core eudicot angiosperm clades are amongst the youngest, and very similar trends are shown by various insects, birds, mammals, fish and reptiles. There is growing evidence to suggest that the age of certain major clades and the strength of their latitudinal gradients are consistently negatively correlated (Crame 2001).

Cenozoic global cooling

One of the most striking and puzzling aspects of the Cenozoic biodiversification event is that it took place against a backdrop of global climatic deterioration. From an Early Eocene climatic optimum (55 Ma BP), global climates deteriorated in a series of pronounced steps to their present-day values. Although the Eocene seems to have been an epoch of relative global warmth, a 5°C drop in oceanic surface water temperatures at the Eocene–Oligocene boundary (33 Ma BP) is widely interpreted to mark the onset of permanent Antarctic glaciation. Further abrupt cooling phases occurred in the Middle–Late Miocene (approx. 15–13 Ma BP), latest Miocene–earliest Pliocene (approx. 6.2–4.8 Ma BP), Middle–Late Pliocene (3.6–2.4 Ma BP), and Late Pliocene–Recent. Even during the Neogene climatic optimum (early Middle Miocene, 17–15 Ma BP) average marine global temperature values only briefly returned to those of the Late Eocene (Clarke & Crame 1992; Frakes *et al.* 1992; Zachos *et al.* 2001).

The principal effect of these changes within the tropical and subtropical regions was not so much a reduction in temperature as one in area. It has been estimated that, since the Early–Middle Eocene (i.e. the last 45 Ma), the tropical biome has contracted some 40–50% in size (Adams *et al.* 1990). In other words, the marked Cenozoic diversification event took place as the tropics actually contracted in size and the temperate and polar regions expanded. The principal steps in the physical control of this process are outlined below.

One of the crucial early Cenozoic palaeogeographic events to affect global oceanographic circulation patterns and climates was the

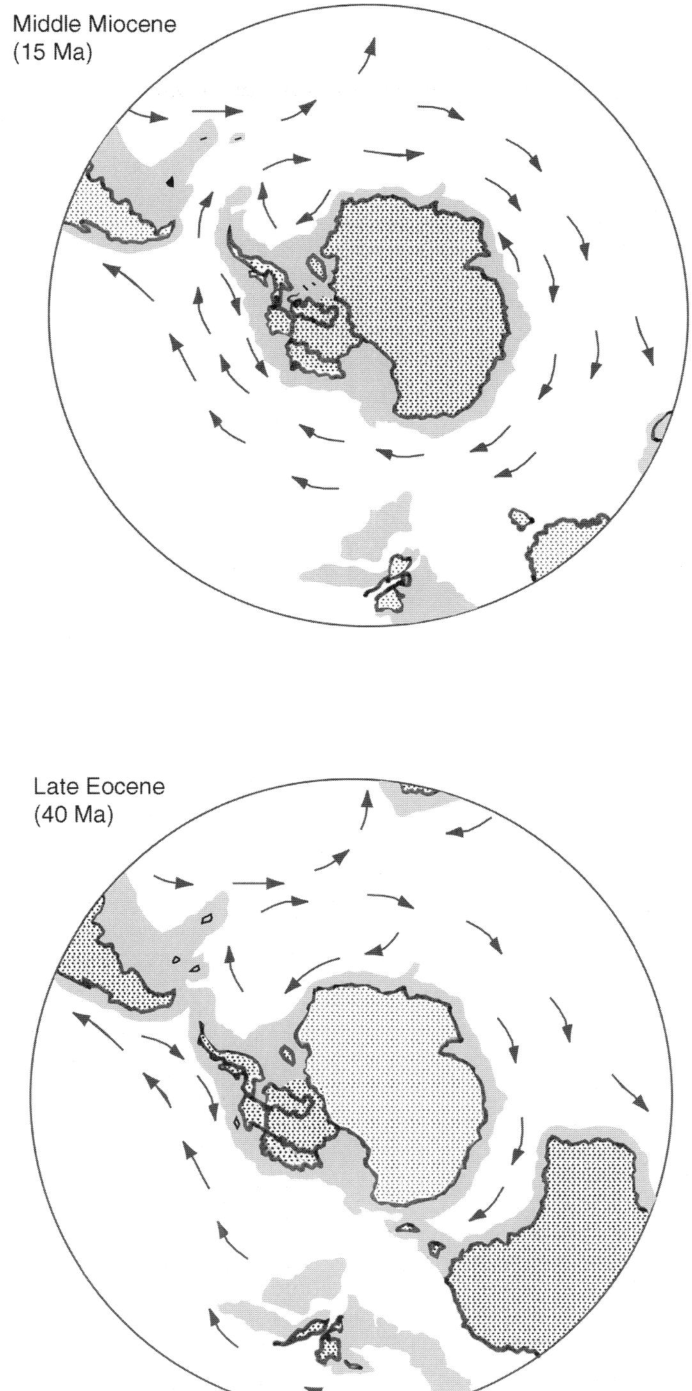

Fig. 1. The physical isolation of Antarctica. Late Eocene and Middle Miocene south polar projections redrawn and simplified from Lazarus & Caulet (1993, figs 19 and 22), with some additional detail from maps contained in Lawver *et al.* (1992). Key: stipple, continents; grey shading, continental shelf; arrows, principal current directions.

physical isolation of Antarctica. Although largely separated from the other southern continents by the end of the Cretaceous Period, there were still intermittent terrestrial and shallow-water connections between both eastern Antarctica and southeasternmost Australia, and the Antarctic Peninsula and southernmost South America (Crame 1999, fig. 1B) (Fig. 1). Full, deep-water separation of the South Tasman Rise from Antarctica was probably not achieved until the late Early Eocene (50 Ma BP), by which time the continent must have been almost completely isolated by deep water. The first indications of global cooling occured in the interval 50–40 Ma BP, and there were significant 2°C temperature falls in both the late Middle Eocene and Middle–Late Eocene boundary (Stott *et al.* 1990). These were followed by a major cooling event at approximately the Eocene–Oligocene boundary (37 Ma BP) where both surface and bottom waters may have decreased by as much as 5°C in 75–100 ka (Lazarus & Caulet 1993; Crame 1999; and references therein).

The precise time of opening of the Drake Passage is still uncertain. Whereas marine geophysical evidence indicates that seafloor spreading began in the early Late Oligocene (28 Ma BP), it may have been as late as 23.5 ± 2.5 Ma (Oligocene–Miocene boundary) before full deep-water circulation was established (Barker *et al.* 1991). On the other hand, there is some micropalaeontological evidence to suggest that there was an opening, at least at shallow and intermediate depths, as early as 36 Ma BP (Diester-Haas & Zahn 1996), and a taxonomic assessment of mid-Eocene mammal faunas from Seymour Island (northern tip of the Antarctic Peninsula) shows that they had been isolated from a South American ancestral stock since at least the Early Eocene (54–51 Ma BP) (Woodburne & Case 1996). In any event it is apparent that a full deep-water, circum-Antarctic current system and polar frontal zone had been established by the Middle Miocene (15 Ma BP) (Lazarus & Caulet 1993) (Fig. 1). This was the time of development of a latitudinal temperature gradient very similar to that seen today (Loutit *et al.* 1983; Kennett *et al.* 1985).

Nevertheless, it should also be emphasized that other Cenozoic tectonic events almost certainly contributed to significant global cooling. Of particular importance was the progressive northward movement of the Africa/Arabia landmass that led to the constriction of the Tethyan Ocean and its eventual closure in the Early Miocene (20 Ma BP), followed by the Early–Middle Miocene (20–15 Ma BP) collision of the Australian plate with Indonesia (Kennett 1977; Kennett *et al.* 1985; see below). The net effect of these changes was to shift oceanic circulation from predominantly equatorial to strongly meridional (i.e. north–south) or gyral; this was especially so in the Pacific Ocean (Kennett *et al.* 1985; Grigg 1988). The transport of warm waters into high-latitude regions is thought to have led to increased levels of precipitation there which in turn contributed to the gradual build-up of glacial ice (see also below).

Even Late Neogene palaeogeographical changes are thought to have had profound climatic implications. For example, the gradual uplift of the Central American Isthmus (CAI, i.e. the Isthmus of Panama), which occurred over the interval of 13–1.9 Ma BP, began to have a major effect on oceanographic circulation patterns by about 4–6 Ma BP (Coates *et al.* 1992; Haug & Tiedemann 1998). At this time the Central American Seaway had shallowed to <100 m and the Gulf Stream was beginning to deflect warm, shallow waters northward along the eastern US seaboard (Fig. 2). Although this undoubtedly led to some warming in northern hemisphere mid- to high latitudes, it is apparent that, as a result of evaporation of surface waters in the relatively narrow North Atlantic Ocean by trade winds, this water would have been slightly hypersaline. When it reached the northern high latitudes it began to descend in both the Norwegian and Labrador seas to form North Atlantic Deep Water (NADW). This then spread into both the South Atlantic and central Pacific to initiate a major 'conveyor belt' of deep-ocean circulation (Haug & Tiedemann 1998) (Fig. 2).

Further closure of the CAI saw the North Atlantic thermohaline circulation system intensify by 3.6 Ma BP, and the Arctic Ocean effectively isolated from the warm Atlantic waters. It is this isolation from the oceans to the south that led to the drastic temperature decline in the north polar regions and the eventual onset of glaciation at 2.5 Ma BP (Stanley 1995). Changes in the Earth's obliquity amplitude fluctuations between 3.1 and 2.5 Ma BP may also have been an important contributory factor (Haug & Tiedemann 1998). A further important effect of the closure of the CAI was to reverse the flow of water through the Bering Strait, and this in turn influenced Pliocene–Pleistocene patterns of North Atlantic thermohaline circulation (Shaffer & Bendtsen 1994; Marincovich 2000).

It can be concluded that, throughout the Cenozoic Era, a series of tectonic changes occurred that led directly to global cooling. First

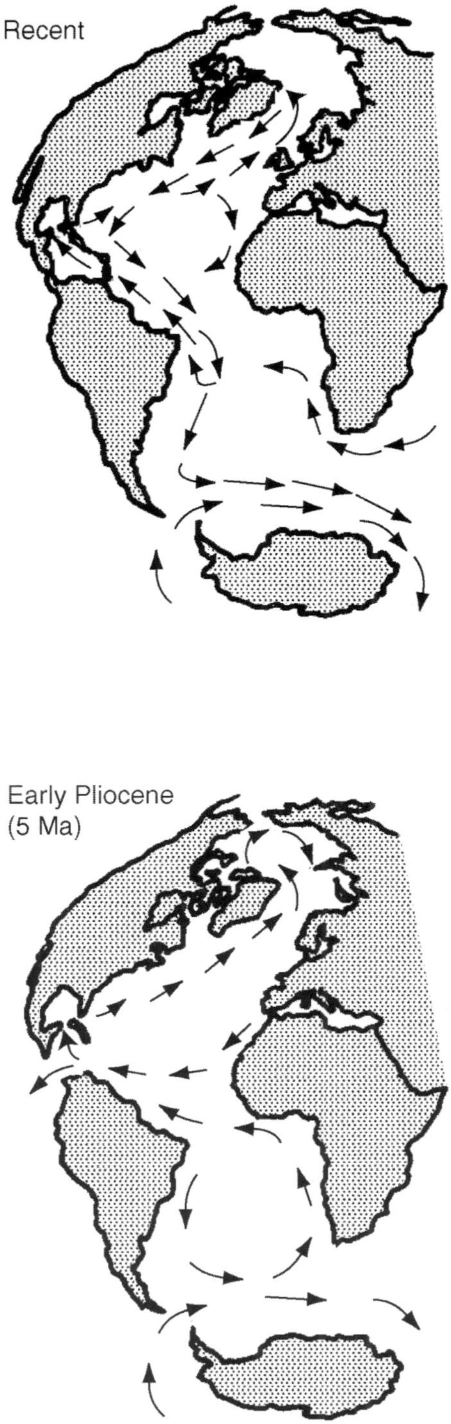

Fig. 2. Pliocene closure of the Central American Seaway. Maps redrawn and simplified from Stanley (1995, fig. 4). Arrows indicate principal current directions.

there was the tectonic isolation of Antarctica, then the progressive disruption of the circum-equatorial current system, and finally the thermal isolation of the Arctic Ocean by the formation of NADW. Of course, there may well have been other contributory factors to global cooling too, such as the uplift of the Himalayas and changes in atmospheric CO_2 levels (Raymo & Ruddiman 1992), but there is an undeniable link throughout the Cenozoic between changes in palaeogeography and global cooling.

The net effects of global cooling on biodiversification

Although an intuitive reaction might be to link global cooling with a progressive loss of biodiversity, this is not necessarily the case. The reason for this lies in the phenomenon of differentiation diversity, or how many communities can be packed within a province/region (beta diversity), and how many provinces can be packed within the biosphere (delta diversity) (*sensu* Magurran 1988). It is thought that, throughout the Cenozoic Era, a series of thermal provinces was gradually imposed on the now widely separated continents. The present-day north–south barriers formed by a unique configuration of continents and ocean basins has led to the development of an unprecedented number of latitude-parallel provincial chains in both the marine and terrestrial realms (Valentine *et al.* 1978; Jablonski *et al.* 1985; Niklas *et al.* 1985). Independent evidence gleaned from the Palaeozoic fossil record has recently attested to the strong control of differentiation processes (i.e. beta diversification) on the formation of global diversity patterns (Adrain *et al.* 2000).

However, we also have to consider whether an increase in provinciality alone would have been sufficient to account for an order of magnitude increase in global species diversity. Some have argued that it would not, and indicated that we must also take into account significant Cenozoic increases in alpha (within-habitat) diversity (Bambach 1977, 1990; Niklas *et al.* 1985; Vermeij 1987). In his seminal study of Palaeozoic diversification, Sepkoski (1988) estimated that alpha diversity might account for some 50–70% of the taxa produced in the Ordovician radiations. There is some evidence to suggest that alpha and beta diversity co-vary widely in nature; at the present day increased regional diversity is accompanied by both increased local diversity (i.e. alpha) and increased provincialism (i.e. beta) (Ricklefs & Schluter 1993; Gaston 2000).

Causes of Cenozoic biodiversification

It would appear then that if we are to find a satisfactory explanation for the global Cenozoic biodiversification event, it must be one that involves a synchronous increase at all four basic levels (i.e. alpha, beta, gamma and delta diversity). In other words this must be a process that allows both an increase in the numbers of taxa within individual habitats as well as increased differentiation between these habitats on a variety of geographical scales. For example, at one of the very simplest levels it has been postulated that the Cenozoic rise in biodiversity could be linked to a global rise in trophic resources provided by the radiation of the angiosperms (Bambach 1977). Even in the marine realm the increased flux of plant detritus through river systems may have raised benthic productivity. Similarly, it has been suggested that oceanic nutrient cycling may have been influenced by changes in the global thermohaline circulation system brought about by the formation of Antarctic deep bottom waters at the Eocene–Oligocene boundary (Bambach 1977).

Such ideas are indeed appealing, and we should not lose sight of the fact that there may well have been times of abrupt increase in both terrestrial and marine productivity through Earth history. Nevertheless it is by now well established that there is not necessarily a direct link between increased productivity and increased biodiversity. An increase in productivity can trigger an increase in biomass, but there is no reason why this should not, in turn, just be within one or a small number of species. Something else is required to generate a large number of new taxa from a rise in productivity (Blackburn & Gaston 1996). We also have to bear in mind that any potential mechanism for Cenozoic global biodiversification must explain the greater relative production of new taxa in the low- than in the high-latitude regions. In their discussion of the origin and diversification of major taxonomic groups, Jablonski & Bottjer (1990) identified five basic explanations for global radiations. We might rationalize these into two main types: intrinsic and extrinsic.

Intrinsic mechanisms of diversification

One of the simplest explanations for global Cenozoic diversification would be that it represents the coincidental expansion of a series of unrelated clades. Although key adaptive breakthroughs, such as mantle fusion and the development of posterior siphons in bivalves, or the development of plants with flowers, undoubtedly occurred well before the Cenozoic Era, it may have taken periods of tens of millions of years for such clades to build up substantial numbers of taxa (Jablonski & Bottjer 1990). Once certain groups became established in the latest Mesozoic/earliest Cenozoic they may in turn have triggered the automatic radiation of others. Angiosperm assemblages would have provided a variety of novel terrestrial habitats, and the rapid development of coral reefs, sea-grass beds and mangroves at this time could have promoted diversification in shallow, tropical seas (Vermeij 1977).

It is also possible that intense biological interactions between various taxonomic groups were a driving force behind diversification. In particular, if biological hazards due to competitors and predators have increased through time, then so may the responses of various prey taxa to them. The net result is a sort of 'evolutionary arms race', with first one group gaining a numerical ascendancy, and then the other (Vermeij 1987, 1994). This special form of co-evolutionary relationship, known as escalation, emphasizes the importance of enemies as agents of natural selection: over periods of geological time enemy-related adaptations bring about long-term evolutionary trends in the morphology, ecology or behaviour of other organisms. It is essentially a tropical phenomenon and is particularly well represented in species-rich groups, such as angiosperms, arthropods, vertebrates and molluscs, that have highly developed competitive and defensive capabilities (Vermeij 1987). Striking examples would appear to be the co-radiation of angiosperms and pollinating insects in the terrestrial realm, and the link between durophagous predators and both deeper-burrowing infauna and more heavily ornamented epifauna in the marine one.

Although escalation seems to be an intuitively satisfying explanation for certain tropical radiations, some doubts have been raised as to its widespread efficacy. For example, Vermeij (1987) himself drew attention to the low incidence of damage repair within many infaunal bivalve groups, and the high-resolution fossil record in general has comparatively few examples of gradual morphological change induced within a prey lineage (Hansen *et al.* 1999; Dietl *et al.* 2000; Miller 2000). Many biotic transitions within the fossil record could equally well be interpreted as the replacement of incumbent taxa by some form of fundamental environmental change (Jablonski 2000). Non-competitive expansions may be just as common through geological time (Benton 1999).

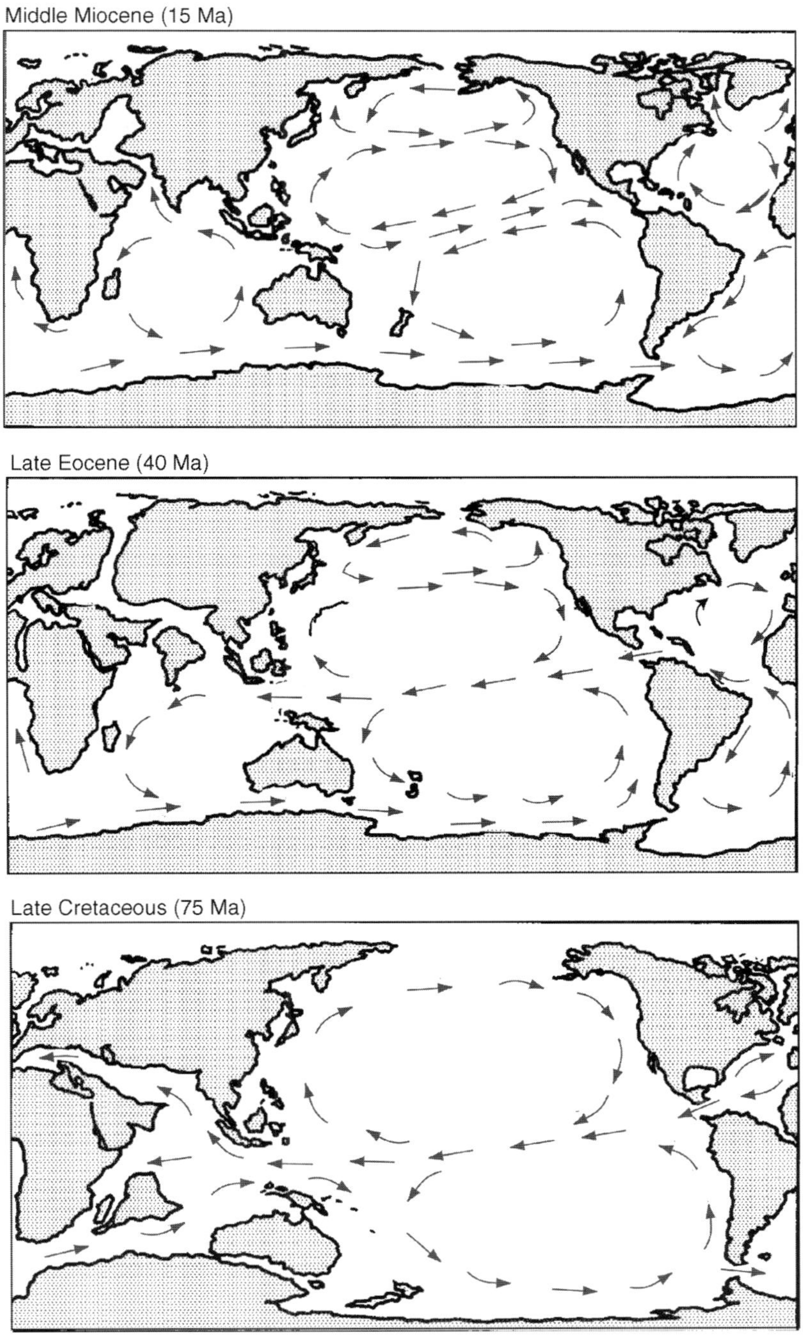

Fig. 3. Disruption of a circum-equatorial seaway by Cenozoic plate tectonic movements. Late Cretaceous (75 Ma BP): equatorial circulation dominated by a strong westerly flowing current system; no major biotic differentiation. Late Eocene (40 Ma BP): partial initiation of a circum-Antarctic current and some intensification of Pacific and Atlantic oceanic gyres. Middle Miocene (15 Ma BP): closure of Tethys in the Mediterranean region, closure of the Indonesian seaway, and earliest beginnings of the uplift of the Central American Isthmus. Circum-Antarctic current fully formed and Pacific and Atlantic oceanic gyres intensified. This was the time of initiation of the IWP and ACEP high-diversity foci. Arrows indicate principal current directions. Redrawn and simplified from Veron (1995, figs 28–30).

Extrinsic mechanisms of diversification

Closure of the Tethyan Ocean in the Middle East. At the end of the Cretaceous period a vast, circum-equatorial tropical ocean (Tethys) could be traced westwards from the Indonesian region, through the Middle East and southern Europe, to the proto-Atlantic Ocean (Fig. 3). There is thought to have been a predominantly westward current flow in the ocean and this continued through the Central American Seaway to the eastern Pacific Ocean (Panthalassa). Although there are some signs of faunal differentiation within the Late Cretaceous Tethyan Ocean, this does not appear to have been such as to create high-diversity foci in any way equivalent to those seen at the present day (Kauffman 1973; Hallam 1994).

The same picture is essentially true for the early part of the Cenozoic Era (i.e. the Paleogene). There was one very large, homogeneous tropical marine realm, with only minor evidence of physical differentiation in areas such as the Caribbean, Middle East and deep-water eastern Pacific (Newell 1971; Kay 1984; Rosen 1988) (Fig. 3). A major pulse in coral reef development occurred at the end of the Oligocene epoch, but even then there was marked similarity, at the generic level, between western Tethys (i.e. Mediterranean) and Caribbean/Gulf of Mexico corals (Frost 1981; Veron 1995). It was not in fact until the Early Miocene (*c.* 20 Ma BP) that the Tethyan Ocean was closed in its central region by the northward movement of the Africa/Arabia landmass. This dramatically curtailed the westward-flowing tropical current and led, eventually, to the Mediterranean region being excluded from the coral reef belt (Fig. 3). At the same time the eastern Atlantic became significantly cooler and this accentuated the physical isolation of the Caribbean–eastern Pacific region (Rosen 1988; Veron 1995). It is to this time that we can trace the origins of the distinct IWP and ACEP foci of high tropical marine diversity.

The Early Miocene fossil record suggests that there was a reduced coral fauna in the Caribbean region that was transitional in compositional terms between a Late Oligocene cosmopolitan one and a later Miocene one containing a number of endemics (Veron 1995). It would appear that the overall trend through the Neogene (i.e. the last 23 Ma) was one of relative impoverishment of ACEP coral reef faunas and relative enrichment of IWP ones. For example, whereas there are some 24 coral genera in the ACEP region at the present day, there are 87 in the IWP; a similar comparison at the species level gives a ratio of approximately 62: 450 (Rosen 1988; Veron 1995). Briggs (1995) estimated that the vast IWP region contains more than 6000 species of molluscs, 800 echinoderms, 500 hermatypic corals (probably an overestimate; Veron 1995) and 4000 species of fishes. This equates to a total species richness that is approximately 2.5 times that of the western Atlantic, 3.5 times that of the eastern Pacific, and 7.3 times that of the eastern Atlantic (Briggs 1996). Ellison *et al.* (1999) estimated that the species richness of trees, shrubs and ferns in mangrove forests is an order of magnitude higher in the IWP than the ACEP.

Collision of Australia–New Guinea with SE Asia. At the beginning of the Cenozoic Era the Australia–New Guinea block was separated from mainland SE Asia by a deep-water Indo-Pacific gateway measuring some 3000 km across. This feature was progressively reduced in width through the Paleogene (i.e. 65–23 Ma BP), chiefly by the northward subduction of Indian–Australian lithosphere beneath the Sunda–Java–Sulawesi arcs (Hall 1998). By the mid-Oligocene (30 Ma BP) the gap had narrowed substantially but there was still a clear deep-water passage floored by oceanic crust of the Indian and Pacific plates (Fig. 4). Major changes in plate boundaries occurred at approximately 25 Ma BP when the New Guinea passive margin collided with the leading edge of the east Philippines–Halmahera–New Guinea arc system and the Australian margin, in the Bird's Head region, was very close to collision with the Eurasian margin in West Sulawesi (Hall 1998). By 20 Ma BP the continent–arc collision had closed the deep-water passage between the Pacific and Indian oceans, resulting in a major reorganization of tropical oceanic current systems (Kennett *et al.* 1985; Grigg 1988). Continued northward movement of Australia from 20 to 10 Ma BP caused the rotation of several plate boundaries and the formation of tectonic provinces that are recognizable at the present day (Fig. 4).

In comparison with both the Mediterranean and Caribbean regions, Eocene and Lower Oligocene hermatypic coral faunas appear to have been genuinely sparse in the central IWP. However, this situation was dramatically reversed in the Early Miocene when there was a four-fold increase in the number of coral genera in the IWP focus (Rosen 1988; Wilson & Rosen 1998). The Early–Middle Miocene was a time of widespread deposition of coral-dominated carbonates in northern Australia, New Guinea and throughout SE Asia; reefs proliferated and

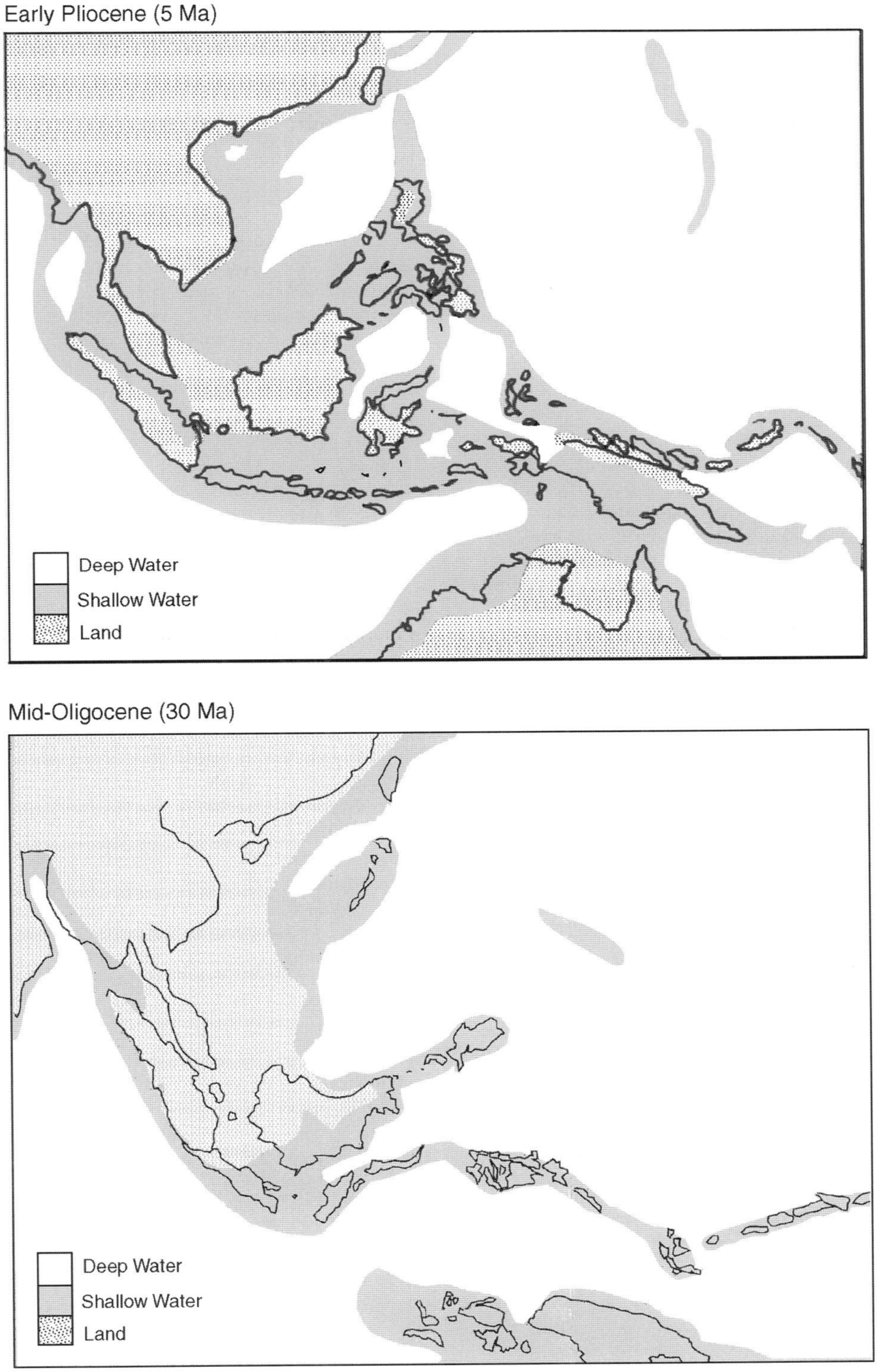

Fig. 4. Closure of the Indonesian seaway. Redrawn and simplified from Hall (1998, figs 12 and 16). See text for further explanation.

many modern genera and species evolved (Veron 1995; Crame 2001, and references therein). At the present day some 55% (by area) of the world's coral reefs occur in the SE Asia–New Guinea–Australia region (Wilson & Rosen 1998).

There can be no doubt that this dramatic rise in the numbers of both reef and reef-associated taxa in the IWP region through the Neogene was linked to a huge increase in the availability of tropical shallow-water habitats. Besides the northward movement of Australia and New Guinea into the coral reef belt, collision-related uplift led to the provision of more islands and carbonate shelves in SE Asia (Fig. 4). This in turn must have greatly increased the amount of local habitat heterogeneity and the potential for allopatric speciation between fragmented shallow-water areas. As a result of the synchronous closure of Tethys in the Middle East and progressive westward movement of the Pacific archipelagos associated with the Darwin Rise/Superswell, SE Asia also became a 'crossroads' for tropical shallow-marine organisms (Rosen 1988; Pandolfi 1992).

Finally, it should be emphasized that the Australian–New Guinea block continued to move northwards throughout the Neogene. It has been suggested that at approximately 4 Ma BP (Early Pliocene) it reached a critical point when it came into close contact with the rapidly growing island of Halmahera. This had the effect of deflecting warm south Pacific waters eastwards at the Halmahera eddy to form the Northern Equatorial Countercurrent (Cane & Molnar 2001). Thus warm waters in the Indonesian throughflow were replaced by relatively cold ones from the north Pacific, leading to a drop in sea surface temperatures in the Indian Ocean and the aridification of East Africa. These changes were the catalyst for a shift in the relative heat balance between the east and west Pacific, which in turn may have helped trigger the onset of northern hemisphere glaciation (Cane & Molnar 2001).

Neogene climate change and biodiversification

Following the lines of evidence presented above it could be maintained that Cenozoic palaeogeographic changes were very largely responsible for the evolution of some of the major patterns of life on Earth. A once-homogeneous tropical biota was disrupted by vicariant events such as the closure of Tethys in the Middle East, the collision of Australia–New Guinea with SE Asia, and the rise of the CAI. The net effect, in the marine realm, was to isolate an ACEP centre of high tropical diversity from an IWP one. Even in the terrestrial realm, which is complicated to some extent by pockets of high diversity in both central and southern Africa, the effects were to produce not dissimilar Palaeotropical and Neotropical realms (e.g. Barthlott *et al.* 1997). In both the marine and terrestrial realms the steepest latitudinal gradients at the present day are associated with the western margins of the Neotropics/ACEP and eastern margins of the Palaeotropics/IWP, respectively (Crame 2000*a*,*b*).

Perhaps the marked heterogeneity observed in the tropical biota at the present day can be attributed simply to the range in sizes of the various subregions imposed by Cenozoic tectonics. For example, in the shallow marine realm the Indo-West Pacific province is approximately four times the area of the Western Atlantic and Eastern Pacific provinces combined (Briggs 1996). If we assume that the greater species richness of the tropics is a time-invariant feature, caused perhaps by the greater size of the tropics in comparison with all other biomes (e.g. Rosenzweig 1995), or some form of species–energy hypothesis (e.g. Wright *et al.* 1993), then what we see at the present day may be due as much to tectonic as to biological factors.

Nevertheless, important as these processes undoubtedly are to the generation and maintenance of large-scale diversity patterns, there is a distinct impression that something else must have been involved too. As our knowledge of the tropical fossil record slowly improves it is becoming apparent that much of the very pronounced tropical Cenozoic diversification event actually occurred in the mid- to late Neogene (i.e. last 10–15 Ma) In the marine realm this is certainly the case for zooxanthellate corals (Veron 1995; Wilson & Rosen 1998), as well as certain reef-associated molluscan taxa (Crame 2001, appendix 2). There is also some palaeontological evidence to demonstrate that certain major eudicot angiosperm clades are of essentially Neogene origin (Magallón *et al* 1999).

Some of this Neogene rise could well be attributed to differentiation diversity, with communities and provinces being distinguished as much in a longitudinal sense as a latitudinal one. Nevertheless, there has long been a suspicion that global climate change was an important driver of diversification too, for this was a time of marked intensification of Milankovitch cyclicity (Bennett 1997). These cycles, which are based on the obliquity of the Earth's axis and eccentricity of orbit, have periods of 41 ka and 100 ka, respectively; they are complemented by the annual timing of the minimum Earth–Sun distance which varies with a 21 ka period. These orbital oscillations lead to variations in insolation and thus ambient temperatures. Such temperature variations would have been greater towards the poles, but led to eustatic sea-level changes that were global in effect (Bennett 1997; Cronin & Raymo 1997).

In a series of previous papers, Rosen (1981, 1984, 1988) has already suggested how an intensification of glacioeustatic cyclicity may have promoted diversification within the IWP focus. In essence, a type of species diversity pump was envisaged whereby taxa created in the outer islands of the western Pacific and eastern Indian oceans during sea level lowstands were subsequently 'pumped' into the central Indonesian focus during highstands. In time, the latter region became a form of refugium for numerous sympatric species. A similar mechanism of species production by repeated cycles of allopatry and secondary sympatry over at least a 10 Ma period has recently been suggested for the temperate zone floras of eastern Asia (Qian & Ricklefs 2000). Such floras have twice as many species as their counterparts in eastern North America, and it is thought that this is the product of repeated climatic and sea-level changes over a geographically much more heterogeneous terrain (Qian & Ricklefs 2000).

We are also beginning to appreciate more fully the effect of Milankovitch oscillations on species ranges; such orbitally forced range dynamics (ORD) may well be the basis of a range of macroecological phenomena (Gaston & Blackburn 2000). In a wide-ranging review, Dynesius & Jansson (2000) demonstrated that ORD tended to be larger and more pronounced in high- than low-latitude regions. Although this might be taken as a strong indication that the formation of ecological isolates, a necessary first step in the process of allopatric speciation, would be more likely in the high latitudes, these authors have argued that this is not in fact the case. Instead, they contend that the short, stable periods between high-latitude oscillations were not long enough for the process of gradual speciation to be completed; isolates were either brought back together again or they became extinct. Such a process is essentially reversed in the tropics, where much smaller ORD promotes the formation of isolates and drives the process of allopatric speciation (Dynesius & Jansson 2000).

In an alternative model, Chown & Gaston (2000) have pointed out that many tropical taxa are stenotopic and can, in theory, range large distances within a more or less constant temperature belt defined by 25°N and 25°S. High-latitude taxa can achieve large ranges too, but these tend to be eurytopes buffered to withstand a considerable amount of climate change. Thus it is the tropical taxa that may well be the most vulnerable to temperature change; repeated temperature oscillations through the late Neogene could have caused far more range disjunctions in the low- than high-latitude regions (Chown & Gaston 2000).

Although the precise mechanisms have yet to be substantiated, it is likely that ORD played a key role in the Neogene diversification event (Bennett 1997). It should also be emphasized that their effects were not necessarily confined to just tropical regions. Some high-latitude and polar clades obviously radiated through the Neogene too (e.g. Crame & Clarke 1997), where both sea-level and temperature oscillations would have been magnified in certain regions. It is perhaps the superimposition of ORD on other factors, such as greater habitat heterogeneity and productivity, that makes their presence most keenly felt in the tropics.

In formulating his diversity pump model, Rosen (1984) supposed that much the same sort of process had operated in the ACEP focus as the IWP one, though on a somewhat smaller scale. From a comprehensive stratigraphical study of Neogene molluscs in the Caribbean region, we know that diversity did indeed rise steadily from approximately 14 to 5 Ma BP, but then levelled off substantially (Jackson & Johnson 2000). As there is good evidence to show that over the last 5 Ma extinction rates were higher in this region than anywhere else in the tropics, there must also have been an equivalent burst of originations (Allmon *et al.* 1996; Jackson *et al.* 1999). For Caribbean corals, we know that diversification increased from approximately 16 to 4 Ma BP, but then from 4 to 1 Ma BP a peak of originations preceded a peak of extinctions (Budd *et al.* 1996; Allmon 2001). Between 9 and 1 Ma BP there were >120 coral species in the Caribbean, as compared with 62 at the present day. Patterns of faunal change in the Western Atlantic Neogene are obviously complex and so too must be the environmental changes that underpin them. Nevertheless it is likely that origination and extinction patterns in this region were strongly affected by the progressive closure of the Central American Seaway and rise of the CAI (see above). Although deep-water circulation through the seaway was blocked at 3.6 Ma BP, and shallow-water circulation at 3.0 Ma BP, the slowly rising

CAI may have had an effect since 15–11 Ma BP (Cronin & Dowsett 1996).

Even though temperature changes did occur in the Western Atlantic, they were almost certainly of secondary importance in the formation of diversity patterns. Instead it is likely that the major reorganization of ocean currents brought about by the full emplacement of the CAI led to a reduction in the areas of upwelling and thus primary productivity (Allmon 2001). At first a decline in productivity may have actually promoted the formation of isolates, and thus local speciation, but in time it would have led to a marked decrease in the rate of isolate persistence, decreased origination and increased extinction (Allmon 2001). The striking turnover in East Pacific coral faunas since the Pliocene closure of the CAI can be related to changes in temperature, salinity and nutrient levels (Budd *et al.* 1996; Budd & Johnson 1999). Today there is only a very depauperate coral fauna in this region but a surprisingly rich molluscan one.

Discussion

Taxonomic diversity gradients can be traced back in the fossil record to the Late Palaeozoic Era, and perhaps even farther than that (Crame 2001, and references therein). Despite the fact that it is often difficult to make meaningful comparisons between these gradients, there is a distinct impression that they were not nearly so steep as those seen at the present day. Dramatic steepening appears to have occurred through the Cenozoic, and we now suspect that this was especially so during the Neogene (i.e. the last 23 Ma). Indeed it is unlikely that, in the marine realm, the present-day IWP and ACEP foci of high taxonomic diversity are any more than 20 Ma old. This was the time of crucial plate tectonic movements that essentially gave the tropics their modern form.

It is also striking how the imposition of steep diversity gradients coincided with marked global climatic cooling, and tempting to postulate a link between the two. Whilst some form of large-scale differentiation diversity undoubtedly contributed to the increase in global biodiversity at this time, it is not easy to see how it could have specifically promoted radiations in two or more tropical foci. There may have been some form of longitudinal as well as latitudinal biotic differentiation, but it is unlikely that this would have been on a scale sufficient to generate the features seen at the present day.

Within the tropical marine realm, the richest Paleogene (i.e. 65–23 Ma BP) coral faunas occurred in Europe and the Caribbean (Rosen 1988, table 2; Wilson & Rosen 1998, fig. 9). In the Early Miocene, there is evidence to suggest that at least 12 common IWP genera were formerly more widespread (Veron 1995). Of particular interest in this context is the fossil record of the diverse branching genus, *Acropora*. The oldest occurrence is in the Paleocene of Somalia, and there are also numerous European records from the Eocene–Miocene. However, *Acropora* does not occur in either the Caribbean or the IWP until the Oligocene (Wilson & Rosen 1998, and references therein), and a very marked radiation in the IWP only is dated as latest Miocene or Pliocene (McCall *et al.* 1994). Similarly, fungiid corals, which show their richest diversity at the present day in the IWP centre, can be traced back to records of *Cycloseris* in the Paleocene of Somalia and western Pakistan, as well as other occurrences in the Eocene of SE Asia and Europe (Wilson & Rosen 1998).

Similar patterns to those of certain coral genera are shown by mangroves. The fossil record strongly supports a Tethyan origin both for mangroves and their associated gastropod fauna in the Maastrichtian/Paleocene (Ellison *et al.* 1999). A pan-tropical distribution of many key taxa had been achieved by the Middle Eocene, with, for example, *Bruguiera* being recorded from the London Clay. Range retractions first occurred during a phase of global cooling at the Eocene–Oligocene boundary and a cosmopolitan biota was then split by a series of vicariant events. Subsequent in-situ speciation within both the IWP and ACEP foci occurred from the mid- to late Miocene onwards (Ellison *et al.* 1999).

If such trends are repeated in other tropical taxa then it will be extremely difficult to classify tropical high-diversity foci as either evolutionary 'cradles' or 'museums' (Briggs 1992; Jablonski 1993, and references therein); in a sense they could be seen as both. A seminal study by McCoy & Heck (1976) first emphasized the importance of tectonic and climatic events in producing disjunct tropical distributions in the marine realm and this has since been substantiated by a number of other workers (Rosen & Smith 1988; Pandolfi 1992; Veron 1995; Paulay 1997). In essence, we would appear to be looking at a Paleogene phase of range retractions and disjunctions followed by a Neogene phase of in-situ radiations from various foci.

It is important to determine whether the mid to late Neogene tropical radiations represent an acceleration of evolutionary rates, and, if so, what caused them. Are the various tropical diversity patterns that we see at the present day simply a function of province/region size, or is something else involved? It is again tempting to link a phase of rapid diversification to one of rapid climate change, and several mechanisms have been proposed whereby enhanced tropical speciation could have been caused by changes in ORD. These all seem entirely plausible, but have yet to be subjected to any form of rigorous examination.

There are also counter-arguments available in the literature to suggest that at least Late Pliocene–Pleistocene climatic cycles were detrimental to the process of speciation. For example, in the marine realm Potts (1984) presented a model whereby Plio-Pleistocene disturbances (principally sea-level changes) were so frequent in the central IWP region that evolutionary-scale processes such as speciation would rarely be seen through to completion. This is, in effect, precisely the opposite of the prediction made by the Dynesius & Jansson (2000) isolate formation model. However, it would seem likely that most living marine invertebrate taxa associated with coral reefs originated well before the Pleistocene and we should perhaps be looking at origination events that occurred in the mid-Miocene to mid-Pliocene time frame (i.e. 15–3 Ma BP) (Jackson & Johnson 2000).

Evidence for a direct link between enhanced Neogene diversification and climatic change may be even harder to find in the terrestrial realm. Intensive stratigraphic study of North American Cenozoic mammal faunas has shown very little obvious correlation with short-term climate change (e.g. Alroy 1998; Prothero 1999); for many prominent turnover events there is no obvious external trigger. Palynological evidence suggests that the vast Amazonian lowland forests were stable throughout much of the Cenozoic (Colinvaux & de Oliveira 2001), and molecular phylogenetic evidence to date does not point to any great Pleistocene radiation of rainforest vertebrate faunas (Moritz *et al.* 2000). Nevertheless, a recent synthesis of phylogenetic and geological data suggests that the species-rich Cape Flora of South Africa is no more than 7–8 Ma old (Richardson *et al.* 2001). In this particular instance speciation seems to have been driven by a combination of aridification of a subtropical flora and imposition of regular climate cycles.

Clearly these are complex issues that will not be resolved either quickly or easily. However, what is becoming clearer is the major role played by Cenozoic plate tectonics over the last 15–20 Ma. Only when we know much more about complex plate interactions in a number of key areas, and how these affected parameters such as oceanic circulation, climate change and organism dispersal, will we be able to understand more fully the nature and origin of our major extant biodiversity patterns.

We are grateful to M. J. Benton, A. Brandt, A. Clarke, A. Hallam and A. W. Owen for reading the manuscript and suggesting various ways in which it could be improved. Contribution to British Antarctic Survey Project, 'Antarctic marine biodiversity: a historical perspective'.

References

ADAMS, C. G., LEE, D. E. & ROSEN, B. R. 1990. Conflicting isotope and biotic evidence for tropical sea-surface temperatures during the Tertiary. *Palaeogeography, Palaeoclimatology, Palaeoecology,* **77**, 289–313.

ADRAIN, J. M., WESTROP, S. R., CHATTERTON, B. D. E. & RAMSKÖLD, L. 2000. Silurian trilobite alpha diversity and the end-Ordovician mass extinction. *Paleobiology*, **26**, 625–646.

ALLMON, W. D. 2001. Nutrients, temperature, disturbance, and evolution: a model for the late Cenozoic marine record of the western Atlantic. *Palaeogeography, Palaeoclimatology, Palaeoecology*, **166**, 7–26.

ALLMON, W. D., ROSENBERG, G., PORTELL, R. W. & SCHINDLER, K. 1996. Diversity of Pliocene–Recent mollusks in the Western Atlantic: extinction, origination, and environmental change. *In*: JACKSON, J. B. C., BUDD, A. F. & COATES, A. G. (eds) *Evolution and Environment in Tropical America.* University of Chicago Press, Chicago, 271–302.

ALROY, J. 1998. Long-term equilibrium in North American mammalian diversity. *In*: MCKINNEY, M. L. (ed.) *Biodiversity Dynamics: Turnover of Populations, Taxa and Communities*. Columbia University Press, New York, 272–309.

ALROY, J., MARSHALL, C. R. & BAMBACH, R. K. 2001. Effects of sampling standardization on estimates of Phanerozoic marine diversification. *Proceedings of the National Academy of Sciences USA*, **98**, 6261–6266.

BAMBACH, R. K. 1977. Species richness in marine benthic habitats through the Phanerozoic. *Paleobiology*, **3**, 152–167.

BAMBACH, R. K. 1990. Late Palaeozoic provinciality in the marine realm. *In*: MCKERROW, W. S. & SCOTESE, C. R. (eds) *Palaeozoic Palaeogeography and Biogeography*. Geological Society, London, Memoirs, **12**, 307–323.

BARKER, P. F., DALZIEL, I. W. D. & STOREY, B. C. 1991. Tectonic development of the Scotia arc region. *In*: TINGEY, R. J. (ed.) *The Geology of Antarctica*. Clarendon Press, Oxford, 215–248.

BARTHLOTT, W., BIEDINGER N., BRAUN, G., FEIG, F. KIER, G., LAUER, W. & MUTKE, J. 1997. Global biodiversity: species numbers of vascular plants. Map published on the internet at: http: //www.botanik.uni-bonn.de/system/biomaps.htm

BELLWOOD, D. R. & HUGHES, T. P. 2001. Regional-scale assembly rules and biodiversity of coral reefs. *Science*, **292**, 1532–1534.

BENNETT, K. D. 1997. *Evolution and Ecology: the Pace of Life*. Cambridge University Press, Cambridge.

BENTON, M. J. 1999. The history of life: large databases in palaeontology. *In*: HARPER, D. A. T. (ed.) *Numerical Palaeobiology*. Wiley, Chichester, 249–283.

BLACKBURN, T. M. & GASTON, K. J. 1996. A sideways look at patterns in species richness, or why there are so few species outside the tropics. *Biodiversity Letters*, **3**, 44–53.

BRIGGS, J. C. 1992. The marine East Indies: centre of origin? *Global Ecology and Biogeography Letters*, **2**, 149–156.

BRIGGS, J. C. 1995. *Global Biogeography*, Elsevier, Amsterdam.

BRIGGS, J. C. 1996. Tropical diversity and conservation. *Conservation Biology*, **10**, 713–718.

BUDD. A. F. & JOHNSON, K. G. 1999. Origination preceding extinction during late Cenozoic turnover of Caribbean reefs. *Paleobiology*, **25**, 188–200.

BUDD, A. F., JOHNSON, K. G. & STENMANN, T. A. 1996. Plio-Pleistocene turnover and extinctions in the Caribbean reef-coral fauna. *In*: JACKSON, J. B. C., BUDD, A. F. & COATES, A. G. (eds) *Evolution and Environment in Tropical America*. University of Chicago Press, Chicago, 168–204.

CANE. M. A. & MOLNAR, P. 2001. Closing of the Indonesian seaway as a precursor to east African aridification around 3–4 million years ago. *Nature*, **411**, 157–162.

CHOWN, S. L. & GASTON, K. J. 2000. Areas, cradles and museums: the latitudinal gradient in species richness. *Trends in Ecology & Evolution*, **15**, 311–315.

CLARKE, A. & CRAME, J. A. 1992. The Southern Ocean benthic fauna and climate change: a historical perspective. *Philosophical Transactions of the Royal Society of London, Series B*, **338**, 299–309.

COATES, A. G., JACKSON, J. B. C., COLLINS, L. S., ET AL. 1992. Closure of the Isthmus of Panama: The near-shore marine record of Costa Rica and Panama. *Geological Society of America Bulletin*, **104**, 814–828.

COLINVAUX, P. A. & DE OLIVEIRA, P. E. 2001. Amazon plant diversity and climate through the Cenozoic. *Palaeogeography, Palaeoclimatology, Palaeoecology*, **166**, 51–63.

CRAME, J. A. 1999. An evolutionary perspective on marine faunal connections between southernmost South America and Antarctica. *Scientia Marina*, **63**, 1–14.

CRAME, J. A. 2000*a*. Evolution of taxonomic diversity gradients in the marine realm: evidence from the composition of Recent bivalve faunas. *Paleobiology*, **26**, 188–214.

CRAME, J. A. 2000*b*. The nature and origin of taxonomic diversity gradients in marine bivalves. *In*: HARPER, E. M., TAYLOR, J. D. & CRAME, J. A. (eds) *The Evolutionary Biology of the Bivalvia*. Geological Society, London, Special Publications, **177**, 347–360.

CRAME, J. A. 2001. Taxonomic diversity gradients through geological time. *Diversity and Distributions*, **7**, 175–189.

CRAME, J. A. & CLARKE, A. 1997. The historical component of marine taxonomic diversity gradients. *In*: ORMOND, R. F. G., GAGE, J. D. & ANGEL, M. V. (eds) *Marine Biodiversity: Patterns and Processes*. Cambridge University Press, Cambridge, 258–273.

CRONIN, T. M. & DOWSETT, H. J. 1996. Biotic and oceanographic response to Pliocene closing of the Central American Isthmus. *In*: JACKSON, J. B. C., BUDD, A. F. & COATES, A. G. (eds) *Evolution and Environment in Tropical America*. University of Chicago Press, Chicago, 76–104.

CRONIN, T. M. & RAYMO, M. E. 1997. Orbital forcing of deep-sea benthic species diversity. *Nature*, **385**, 624–627.

DIESTER-HAAS, L. & ZAHN, R. 1996. Eocene–Oligocene transition in the Southern Ocean: History of water mass circulation and biological productivity. *Geology*, **24**, 163–166.

DIETL, G. P., ALEXANDER, R. R. & BIEN, W. F. 2000. Escalation in Late Cretaceous–early Paleocene oysters (Gryphaeidae) from the Atlantic Coastal Plain. *Paleobiology*, **26**, 215–237.

DYNESIUS, M. & JANSSON, R. 2000. Evolutionary consequences of changes in species' geographical distributions driven by Milankovitch climate oscillations. *Proceedings of the National Academy of Sciences USA*, **97**, 9115–9120.

ELLISON, A. M., FARNSWORTH, E. J. & MERKT, R. E. 1999. Origins of mangrove ecosystems and the mangrove biodiversity anomaly. *Global Ecology and Biogeography*, **8**, 95–115.

FOOTE, M. 2000. Origination and extinction components of taxonomic diversity: general problems. *In*: ERWIN, D. H. & WING, S. L. (eds) *Deep Time: Paleobiology's Perspective. Paleobiology*, Supplement, **26**(4), 74–102.

FRAKES, L. A., FRANCIS, J. E. & SYKTUS, J. I. 1992. *Climate Modes of the Phanerozoic*. Cambridge University Press, Cambridge.

FRANCIS, A. P. & CURRIE, D. J. 1998. Global patterns of tree species richness in forests: another look. *Oikos*, **81**, 598–602.

FROST, S. H. 1981. Oligocene reef coral biofacies of the Vicentin, northeast Italy. *In*: TOOMEY, D. H. (ed.) *European Fossil Reef Models*. Society of Economic Paleontologists and Mineralogists, Special Publications, **30**, 483–539.

GASTON, K. J. 2000. Global patterns in biodiversity. *Nature*, **405**, 220–227.

GASTON, K. J. & BLACKBURN, T. M. 2000. *Pattern and Process in Macroecology*. Blackwell Science, Oxford.

GRIGG, R. W. 1988. Paleoceanography of coral reefs in the Hawaiian-Emperor Chain. *Science*, **240**, 1737–1743.

HALL, R. 1998. The plate tectonics of Cenozoic SE Asia and the distribution of land and sea. *In*: HALL, R. & HOLLOWAY, J. D. (eds) *Biogeography and Geological Evolution of SE Asia*. Backhuys, Leiden, 99–131.

HALLAM, A. 1994. *An Outline of Phanerozoic Biogeography*. Oxford University Press.

HANSEN, T. A., KELLEY, P. H., MELLAND, V. D. & GRAHAM, S. E. 1999. Effect of mass extinctions on escalation in molluscs. *Geology*, **27**, 1139–1142.

HAUG, G. H. & TIEDEMANN, R. 1998. Effect of the formation of the Isthmus of Panama on Atlantic thermohaline circulation. *Nature*, **393**, 673–676.

HUSTON, M. A. 1999. Local processes and regional patterns: appropriate scales for understanding variation in the diversity of plants and animals. *Oikos*, **86**, 393–401.

JABLONSKI, D. 1993. The tropics as a source of novelty through geological time. *Nature*, **364**, 142–144.

JABLONSKI, D. 2000. Micro- and macroevolution: scale and hierarchy in evolutionary biology and paleobiology. *In*: ERWIN, D. H. & WING, S. L. (eds) *Deep Time: Paleobiology's Perspective Paleobiology*, Supplement, **26**(4), 15–52.

JABLONSKI, D. & BOTTJER, D. J. 1990. The origin and diversification of major groups: environmental patterns and macroevolutionary lags. *In*: TAYLOR, P. D. & LARWOOD, G. P. (eds) *Major Evolutionary Radiations*. Systematics Association, Special Volume **42**, Clarendon Press, Oxford, 17–57.

JABLONSKI, D., FLESSA, K. W. & VALENTINE, J. W. 1985. Biogeography and paleobiology. *Paleobiology*, **11**, 75–90.

JACKSON, J. B. C. & JOHNSON, K. G. 2000. Life in the last few million years. *In*: ERWIN, D. H. & WING, S. L. (eds) *Deep Time: Paleobiology's Perspective. Paleobiology*, Supplement, **26**(4), 221–235.

JACKSON, J. B. C., TODD, J. A., FORTUNATO, H. & JUNG, P. 1999. Diversity and assemblages of Neogene Caribbean Mollusca of lower Central America. *In*: COLLINS, L. S. & COATES, A. G. (eds) *The Neogene of the Isthmus of Panama: a Paleobiotic Survey of the Caribbean Coast*. Bulletins of American Paleontology, **357**, 193–230.

KAUFFMAN, E. G. 1973. Cretaceous Bivalvia. *In*: HALLAM, A. (ed.) *Atlas of Palaeobiogeography*. Elsevier, Amsterdam, 353–383.

KAY, E. A. 1984. Patterns of speciation in the Indo-West Pacific. *Special Publications of the Bernice Pauahi Bishop Museum*, **72**, 15–31.

KENNETT, J. P. 1977. Cenozoic evolution of Antarctic glaciation, the circum-Antarctic ocean, and their impact on global paleoceanography. *Journal of Geophysical Research*, **82**, 3843–3860.

KENNETT, J. P., KELLER, G. & SRINIVASEN, M. S. 1985. Miocene planktonic foraminiferal biogeography and paleoceanographic development of the Indo-Pacific region. *In*: KENNETT, J. P. (ed.) *The Miocene Ocean: Paleoceanography and Biogeography*. Geological Society of America, Memoirs, **163**, 197–236.

LAWVER, L. A., GAHAGAN, L. M. & COFFIN, M. F. 1992. The development of paleoseaways around Antarctica. *In*: KENNETT, J. P. & WARNKE, D. A. (eds) *The Antarctic Paleoenvironment: A Perspective on Global Change*. Antarctic Research Series, **56**, 7–30.

LAZARUS, D. & CAULET, J. P. 1993. Cenozoic Southern Ocean reconstructions from sedimentologic, radiolarian and other microfossil data. *In*: KENNETT, J. P. & WARNKE, D. A. (eds) *The Antarctic Paleoenvironment: A Perspective on Global Change. Part Two*. Antarctic Research Series, **60**, 145–174.

LOUTIT, T. S., KENNETT, J. P. & SAVIN, S. M. 1983. Miocene equatorial and Southwest Pacific paleoceanography from stable isotope evidence. *Marine Micropalaeontology*, **8**, 215–233.

MCCALL, J., ROSEN, B. R. & DARRELL, J. G. 1994. Carbonate deposition in accretionary prism settings: Early Miocene coral limestones and corals of the Makran mountain range in southern Iran. *Facies*, **31**, 141–178.

MCCOY, E. D. & HECK, K. L. 1976. Biogeography of corals, seagrasses, and mangroves: an alternative to the center of origin concept. *Systematic Zoology*, **25**, 201–210.

MAGALLÓN, S., CRANE, P. R. & HERENDEEN, P. S. 1999. Phylogenetic pattern, diversity, and diversification of eudicots. *Annals of the Missouri Botanic Garden*, **86**, 297–372.

MAGURRAN, A. E. 1988. *Ecological Diversity and its Measurement*. Croom Helm, London.

MARINCOVICH, L., JR. 2000. Central American paleogeography controlled Pliocene Arctic Ocean molluscan migrations. *Geology*, **28**, 551–554.

MILLER, A. I. 2000. Conversations about Phanerozoic global diversity. *In*: ERWIN, D. H. & WING, S. L. (eds) *Deep Time: Paleobiology's Perspective. Paleobiology*, Supplement, **26**(4), 53–73.

MORITZ, C., PATTON, J. L., SCHNEIDER, C. J. & SMITH, T. B. 2000. Diversification of rainforest faunas: An integrated molecular approach. *Annual Reviews of Ecology and Systematics*, **31**, 533–563.

NEWELL, N. D. 1971. An outline history of tropical organic reefs. *American Museum Novitates*, **2465**, 1–37.

NIKLAS, K. J., TIFFNEY, B. H. & KNOLL, A. H. 1985. Patterns in vascular land diversification: An analysis at the species level. *In*: VALENTINE, J. W. (ed.) *Phanerozoic Diversity Patterns: Profiles in Macroevolution*. Princeton University Press, Princeton, NJ, 97–128.

PANDOLFI, J. M. 1992. Successive isolation rather than evolutionary centres for the origination of Indo-Pacific reef corals. *Journal of Biogeography*, **19**, 593–609.

PAULAY, G. 1997. Diversity and distribution of reef organisms. *In*: BIRKLAND, C. E. (ed.) *Life and Death of Coral Reefs*. Chapman and Hall, London, 298–352.

POTTS, D. C. 1984. Generation times and the Quaternary evolution of reef-building corals. *Paleobiology*, **10**, 48–58.

PROTHERO, D. R. 1999. Does climatic change drive mammalian evolution? *GSA Today*, **9**, 1–7.

QIAN, H. & RICKLEFS, R. E. 2000. Large-scale processes and the Asian bias in species diversity of temperate plants. *Nature*, **407**, 180–182.

RAYMO, R. E. & RUDDIMAN, W. F. 1992. Tectonic forcing of late Cenozoic climate. *Nature*, **359**, 117–122.

RICHARDSON, J. E., WELTZ, F. M., FAY, M. F., CRONK, Q. C. B., LINDER, H. P., REEVES, G. & CHASE, M. W. 2001. Rapid and recent origin of species richness in the Cape flora of South Africa. *Nature*, **412**, 181–183.

RICKLEFS, R. E. 1999. Global patterns of tree species richness in moist forests: distinguishing ecological influences and historical contingency. *Oikos*, **86**, 369–373.

RICKLEFS, R. E. & SCHLUTER, D. 1993. Species diversity: regional and historical influences. *In*: RICKLEFS, R. E. & SCHLUTER, D. (eds) *Species Diversity in Ecological Communities: Historical and Geographical Perspectives*. University of Chicago Press, Chicago, 350–363.

ROSEN, B. R. 1981. The tropical high diversity enigma – the corals'-eye view. *In*: FOREY, P. L. (ed.) *The Evolving Biosphere*. British Museum (Natural History), London, and Cambridge University Press, Cambridge, 103–129.

ROSEN, B. R. 1984. Reef coral biogeography and climate through the late Cainozoic: just islands in the sun or a critical pattern of islands? *In*: BRENCHLEY, P. J. (ed.) *Fossils and Climate*. Wiley, Chichester, 210–262.

ROSEN, B. R. 1988. Progress, problems and patterns in the biogeography of reef corals and other tropical marine organisms. *Helgoländer Meeresuntersuchungen*, **42**, 269–301.

ROSEN, B. R. & SMITH, A. B. 1988. Tectonics from fossils? Analysis of reef-coral and sea-urchin distributions from the late Cretaceous to Recent, using a new method. *In*: HALLAM, A. & AUDLEY-CHARLES, M. G. (eds) *Gondwana and Tethys*. Geological Society, London, Special Publications, **37**, 275–306.

ROSENZWEIG, M. L. 1995. *Species Diversity in Space and Time*. Cambridge University Press, Cambridge.

SEPKOSKI, J. J. JR. 1988. Alpha, beta, or gamma: where does all the diversity go? *Paleobiology*, **14**, 221–234.

SHAFFER, G. & BENDTSEN, J. 1994. Role of the Bering Strait in controlling North Atlantic ocean circulation and climate. *Nature*, **367**, 354–357.

SIGNOR, P. W. 1990. The geologic history of diversity. *Annual Reviews of Ecology and Systematics*, **21**, 509–539.

STANLEY, S. M. 1995. New horizons for paleontology, with two examples: the rise and fall of the Cretaceous Supertethys and the cause of the modern ice age. *Journal of Paleontology*, **69**, 999–1007.

STOTT, L. D., KENNETT, J. P., SHACKLETON, N. J. & CORFIELD, R. M. 1990. The evolution of Antarctic surface waters during the Paleogene: Inferences from the stable isotopic composition of planktonic foraminifers, ODP Leg 113. *Proceedings of the Ocean Drilling Program Scientific Results*, **113**, 849–863.

TURNER, J. R. G., LENNON, J. J. & GREENWOOD, J. J. D. 1996. Does climate cause the global biodiversity gradient? *In*: HOCHBERG, M. E., CLOBERT, J. & BARBAULT, B. (eds) *Aspects of the Genesis and Maintenance of Biological Diversity*. Oxford University Press, Oxford, 199–220.

VALENTINE, J. W., FOIN, T. C. & PEART, D. 1978. A provincial model of Phanerozoic marine diversity. *Paleobiology*, **4**, 55–66.

VERMEIJ, G. J. 1977. The Mesozoic marine revolution: evidence from snails, predators grazers. *Paleobiology*, **3**, 245–258.

VERMEIJ, G. J. 1987. *Evolution and Escalation. An Ecological History of Life*. Princeton University Press, Princeton, NJ.

VERMEIJ, G. J. 1994. The evolutionary interaction among species. Selection, escalation, and coevolution. *Annual Reviews of Ecology and Systematics*, **25**, 219–236.

VERON, J. E. N. 1995. *Corals in Space and Time. The Biogeography and Evolution of the Scleractinia*. Comstock/Cornell, London.

WILLIAMS, P. H., GASTON, K. J. & HUMPHRIES, C. J. 1997. Mapping biodiversity values worldwide: combining higher-taxon richness from different groups. *Proceedings of the Royal Society of London*, **B264**, 141–148.

WILSON, M. E. J. & ROSEN, B. R. 1998. Implications of paucity of corals in the Paleogene of SE Asia: plate tectonics or Centre of Origin ? *In*: HALL, R. & HOLLOWAY, J. D. (eds) *Biogeography and Geological Evolution of SE Asia*. Backhuys, Leiden, 165–195.

WOODBURNE, M. O. & CASE, J. A. 1996. Dispersal, vicariance, and Late Cretaceous to Early Tertiary land mammal biogeography from South America to Australia. *Journal of Mammalian Evolution*, **3**, 121–161.

WRIGHT, D. H., CURRIE, D. J. & MAURER, B. A. 1993. Energy supply and patterns of species richness on local and regional scale. *In*: RICKLEFS, R. E. & SCHLUTER, D. (eds) *Species Diversity in Ecological Communities: Historical and Geographical Perspectives*. University of Chicago Press, Chicago, 66–74.

ZACHOS, J., PAGANI, M., SLOAN, L., THOMAS, E. & BILLUPS, K. 2001. Trends, rhythms, and aberrations in global climate 65 Ma to present. *Science*, **292**, 686–693.

Palaeontological databases for palaeobiogeography, palaeoecology and biodiversity: a question of scale

PAUL J. MARKWICK[1] & RICHARD LUPIA[2]

[1] *Robertson Research International Limited. Llandudno, Conwy, LL30 1SA, UK (e-mail:* paul.markwick@which.net*)*

[2] *Sam Noble Oklahoma Museum of Natural History and School of Geology and Geophysics, University of Oklahoma, 2401 Chautauqua Avenue, Norman OK 73072, USA*

Abstract: Computerized databases provide an essential tool for investigating large-scale spatial and temporal palaeontological problems. Although advances in both software and hardware have made the logistics of building a database much easier, fundamental problems remain concerning the representation and qualification of the data. Data from the fossil record are highly heterogeneous. Databases must be designed to account for variations in scale (grain, resolution), inconsistency in the data, and potential errors (inaccuracy). These issues vary with the scope of the study (extent), the biological group, and the nature and scale-dependence of supplementary, non-biological datasets (e.g. climate and ocean parameters). With the application of desktop geographic information systems (GIS) to global Earth systems science, and the ability to efficiently integrate and query large, diverse datasets, the need to ensure robust qualification of data, especially scale, has become all the more essential. This chapter examines some of the issues involved, defines terminology and offers pragmatic solutions.

The fossil record is vast, despite its inherent incompleteness, and computerized databases provide the only practical means for investigating large-scale palaeobiological patterns and the processes responsible (e.g. Sepkoski 1982; Raup & Sepkoski 1986; Boulter *et al.* 1991; Benton 1993; Damuth 1993; Labandeira & Sepkoski 1993; Alroy 1995; Krebs *et al.* 1996; Markwick 1996; Lupia 1999; Lupia *et al.* 1999; Alroy *et al.* 2001). But a database is only as good as the data it contains and the questions asked of it, and palaeontological data are more complex than most. Palaeontological databases must be designed to take account of heterogeneities in scale (grain, resolution), inconsistencies in the data, and potential errors (accuracy). As more studies examine the interplay of diverse datasets (e.g. climate, soil, biodiversity), qualifying these inherent scaling differences becomes critical, as mixing of incongruent datasets may lead to erroneous results. This has become an important issue in landscape ecology (see Levin 1992) from which we draw examples, but the problems are exacerbated for palaeontologists by the addition of a temporal dimension in the data (see Kidwell & Behrensmeyer 1993). This paper reviews some of these issues, examines the potential consequences of ignoring scale, and suggests pragmatic solutions that are applicable to the design and implementation of palaeontological databases. For many these concepts will be familiar, but in the absence of a manual for building palaeontological databases, this is aimed at helping researchers just beginning to construct their own databases.

Databases and basic database structure for palaeontological data

The earliest and most basic form of computer database is the 'flat file database' in which data are stored as a single set of records of the same kind (Fig.1a). Conceptually this is similar to a card index system, and suffers from the same weaknesses in that data of only one kind can be queried, and each record must be edited individually (e.g. Sepkoski 1982). 'Relational databases' (Fig.1b) can be viewed as multiple 'flat file databases' or tables (relations) linked together (related), such that complex queries can be made integrating varied and diverse data. The advantages of separating data in this way are that they only need to be entered or updated once in one table, but can be utilized by many different records in other tables of the database.

Geographic information systems (GIS) couple the power of relational databases with the visual efficiency of geographic maps (Fig.1c). In GIS, a record can be represented by

From: CRAME, J. A. & OWEN, A. W. (eds) 2002. *Palaeobiogeography and Biodiversity Change: the Ordovician and Mesozoic–Cenozoic Radiations*. Geological Society, London, Special Publications, **194,** 169–178.
0305-8719/02/$15.00

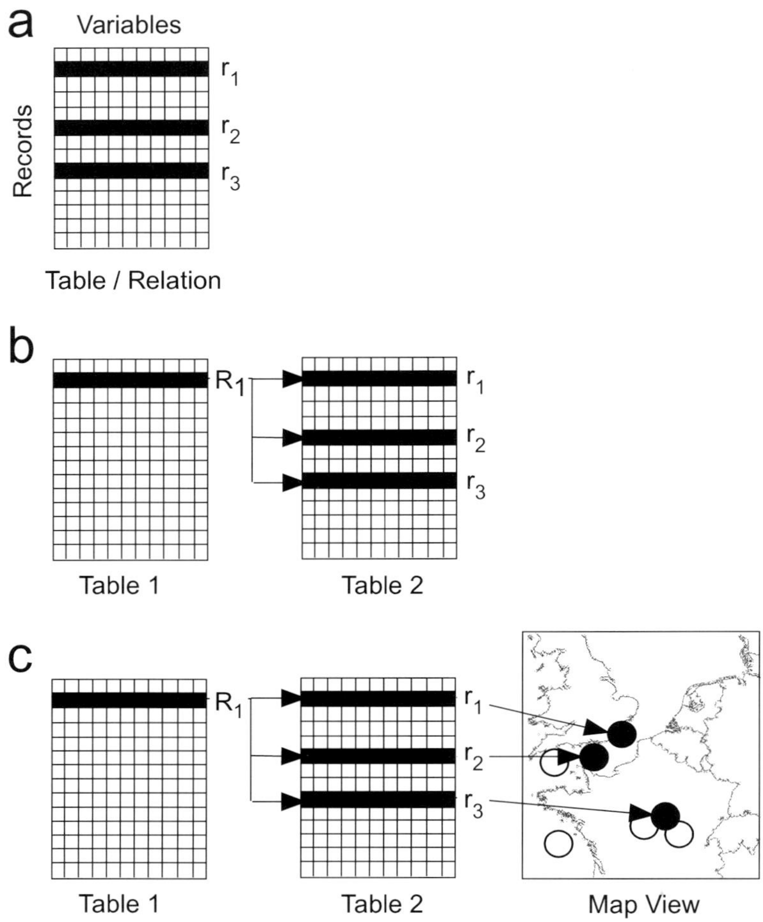

Fig. 1. Schematic representation of different database types. (**a**) Flat file format: r_1, r_2 and r_3 represent three records in a single table. (**b**) Relational database: different kinds of data can be represented in separate tables, e.g. table 1 and table 2; data in record R_1 can be linked to relevant records in table 2, r_1, r_2 and r_3. (**c**) GIS: relational database linked and queried using map views.

a point, area (polygon), linear feature (line or 'arc') or grid (raster data). This has been particularly important to Earth scientists over the last few years investigating the complex interactions of different and diverse elements of the Earth system using different types and scales of data (e.g. points, grids). For Earth scientists geography need not be the modern geography but any reconstructed map of the world (palaeogeography).

The basic structure of most published palaeontological databases is quite similar, reflecting the nature of the fossil record (Fig.2): an individual fossil represents a record (occurrence table) of a particular organic group (taxonomy table) at a specific time and place (locality table). However, the inclusion of temporal and spatial information in the same table means that potentially the same geographic location might be represented by more than one record – one for each different stratigraphic level (e.g. samples from a core). A more efficient structure, which removes this duplication, is one in which the geographical and age data are placed in separate tables (space and time tables, respectively; Fig.3), with the two linked together in what was the locality table of Figure 2. A 'locality' (in time and space) can then vary according to requirements: a chronostratigraphic interval; depth range in a well; a lithological or palaeoenvironmental associated interval; a single point (depth or time) such as a

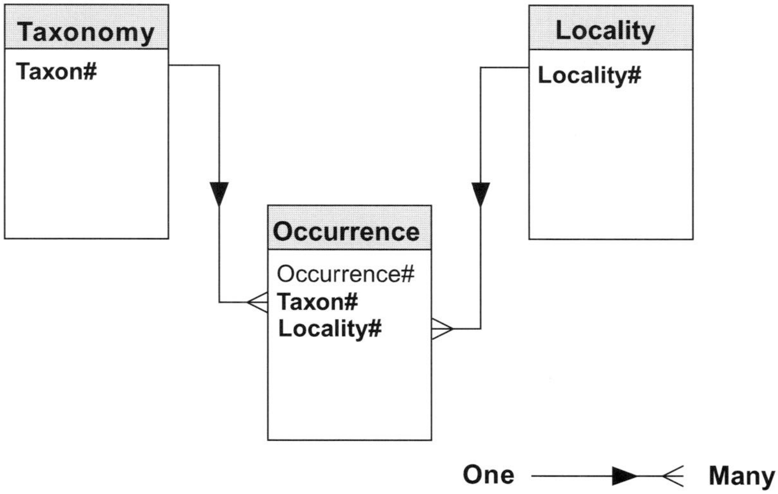

Fig. 2. Basic database structure for palaeontological databases. Fields in bold (e.g. 'Taxon#') indicate principal links between tables, although any field can be linked and queried. The 'Taxonomy' table can include any information on the individual taxon, e.g. taxonomy, ecology, habitat. The 'locality' table can comprise information such as geographic and stratigraphic data. The 'occurrence' table consists of data appropriate to that unique occurrence of the taxon at the specified time and place, such as specimen information or abundance.

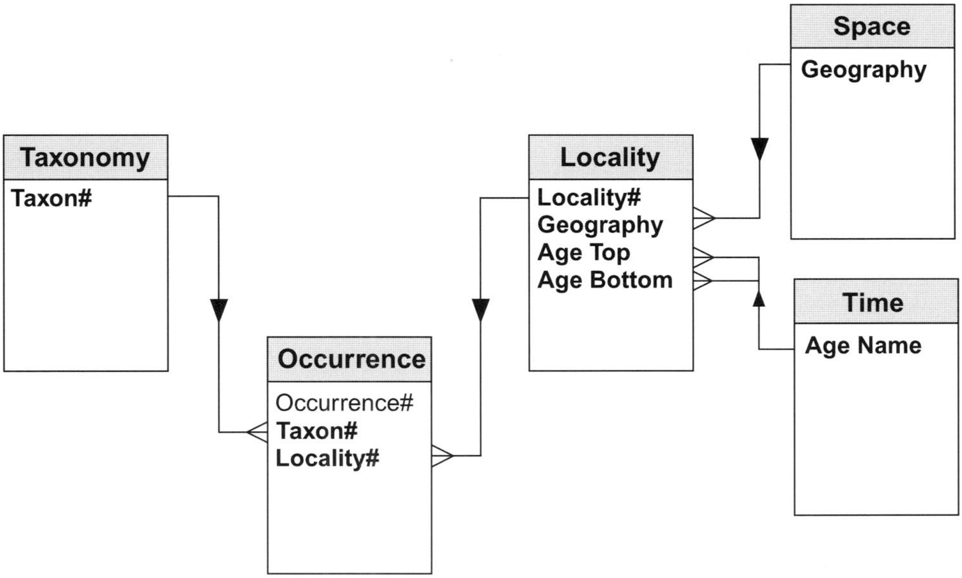

Fig. 3. Basic database structure with the locality data linked to two additional tables that store the spatial and temporal information separately.

palynological slide or geochemical sample. Additional tables can then be added as necessary for data provenance (references), higher taxonomy, timescales, etc.

Although this chapter is not concerned with a detailed description of database design (see Peuquet (1988) for a more detailed discussion), three logistical points are worthy of mention because they directly affect the qualification of the data. First, it is advisable that all records in each database table should have a 'unique identifier'. This is a field (variable) that has a unique value for each record and should have no other meaning (i.e. should not include an age code or taxon name that could potentially change in the future). These identifiers can then be used to link tables (e.g. in Fig. 2 linking locality and taxon records to an individual occurrence). Links can of course be made on any other fields in a table, but care must be taken in knowing the relationships of the data (one-to-one, many-to-one, one-to-many)

The second logistical consideration is data provenance. In order that the data in the database can be used with confidence it is essential to ensure that all data are referenced and audited. The provenance of information is critical to ensuring the integrity of the data, such that the issues of precision and error can be traced back to source. A distinction also should be drawn between raw data (observations) that are more or less immutable, and interpretations based on those data. If data are to be compiled from the published literature, it is also advisable to design the database to record data as it was written in the source or to record explicitly changes made to the data (e.g. correction of obvious misspellings or selected age assignment among disputed alternatives) as it is entered. For example, an author might misspell a taxon's name and this error may be amended immediately, but because some spelling variants are truly different taxonomic entities (e.g. *Cicatricosisporites*, a trilete spore, and *Cicatricososporites*, a monolete spore), any change should be noted in a comment field in case examination should verify the 'error'. In the end, original data represent facts that can be accepted or disputed (and perhaps modified) by different users of the database according to their scientific opinion. Making corrections or changes at the time of entry without annotation precludes verification without returning to the original publication.

The final logistical point is the treatment of error (inaccuracy). Errors in a database can be of three types: errors due to mistakes in data entry; errors due to mistakes in the original data; and errors due to subsequent changes to that data (e.g. new phylogenetic hypotheses or age reassignments). In general, the first of these is easily remedied by systematic checking of the data. The second and third require that the database be designed to be dynamic and allow updates as necessary.

Scale

Scale is a critical issue in ecology (Levin 1992) and palaeoecology (Kidwell & Behrensmeyer 1993), but frequently obfuscated by ambiguous terminology. In the ecological literature, scale refers to the spatial and/or temporal dimensions that describe an object (e.g. 2 cm tooth or 4 ha plot), event (e.g. 4 month rainy season) or observation (e.g. 2 year study of a 4 ha plot) (O'Neill & King 1998). This has the opposite meaning to scale in the cartographic sense, which refers to the level of detail; thus 'large-scale' to an ecologist refers to a large area or duration, but a 'large-scale map' is usually of great detail but small area. This can lead to confusion when using GIS for examining ecology and palaeoecology. To combat this we have adopted two terms from landscape ecology: grain, which is the minimum resolution/scale of an observation (the smallest spatial or temporal interval of observation); and extent, which is the total amount of space or time observed, usually defined as the maximum size of the study area (O'Neill & King 1998). Therefore, a 'large-scale map' is fine-grained but of limited extent. The important issue is to specify explicitly what the grain and extent are for each study.

In studies of the fossil record, scale can be treated in the same manner. The grain of an observation is equivalent to, for example, a rock sample, or locality, or basin (and the amount of time and space that they represent) and is determined by the size – thickness, area or volume – measured. Which grain is used depends on the questions asked of the data. A global study (global extent) might only require a summary of the fossil fauna or flora for each sedimentary basin in the world, and therefore the grain is defined by the size of each basin. Conversely, a study of a specific basin (basin extent) might require a grain based on localities, or sites, or samples within that basin. The term 'resolution' can be taken as a synonym of grain, thus 'time resolution' refers to the interval of elapsed time represented by an assemblage (see Kidwell & Behrensmeyer 1993, table 1).

Precision, strictly defined, is the ability to repeat a result, or the degree of consistency among several results, whereas accuracy is the ability to achieve the real or true value. Here we may loosen the definition of precision to refer to how easily we could return to (literally revisit) a site given the information provided in the database. To record that a site is located in 'Yorkshire' may be accurate, that is, true, but it does not get us easily to the actual site at the base of a specific cliff. Likewise, a site might really be of Eocene age, but this would not be helpful to track down the actual site. Thus precision can be construed as uncertainty in the grain or extent of a sample/analysis of the fossil record.

Table 1. *Geographic precision*

Code	Explanation
1	Precise location, within 1 km (equivalent to 'site/locality')
2	Within 10 km (equivalent to 'nearest town')
3	Within 100 km (equivalent to 'US county')
4	Within 500 km (equivalent to 'US state')
5	Very imprecise, not know to within 500 km (equivalent to 'country')

Grain (resolution)

Scaling issues are compounded in palaeoecology by taphonomic (i.e. preservational) processes that affect the apparent grain and extent of analyses through combining elements of assemblages that did not co-occur in space (spatial averaging; e.g. wind-blown pollen from outside the depositional area) or in time (time-averaging; e.g. reworking of shells from different depositional events), and by inaccuracies in the data. In terms of a grain represented by 'localities' these issues can be summarized into two principal questions: Where is the locality? How much 'space' and therefore time, is represented by the locality?

Accuracy. A fossil comes from a definite location, but it is not always possible to know the locality with precision, either because the details are not reported in the literature, or the location could not be known at the time due to poor maps or difficult terrain. The advent of global positioning systems (GPS) has mitigated many of these problems, but in the older literature, localities were often described with respect to a local geographic feature, e.g. a town, river confluence, etc. By using GIS to plot detailed geographic datasets (topographies, roads, rivers) at various scales, these localities usually can be placed in latitude–longitude space. Nonetheless, a simple qualifier can allow for imprecisely known localities to be distinguished from well resolved sites, if that is important in analyses (geographic precision (Markwick 1996); Table 1). It needs to be remembered that given plate motions (and the uncertainty therein), absolute spatial resolution will deteriorate the further back in geological time that the interval under investigation occurred (Fig.4).

Locations can also be misplaced. This can be mitigated by checking locations against the coordinates given in published gazetteers and atlases, but can be performed most effectively using GIS. Again, detailed map datasets of rivers, roads, political boundaries, topography, outcrop geology, etc. can be superimposed digitally in latitude–longitude (or x–y) space with the datapoints to be checked. This provides an immediate visual indication of error. Intentionally misplaced localities (for political or site conservation reasons) can be dealt with similarly.

Age assignments can be made incorrectly, based on incorrect radiometric ages or fossil sparcity, or subject to change based on later analysis (different timescales). Ziegler *et al.* (1985) tried to qualify confidence in age assignments by recording the provenance of the age date (Table 2). Such a scheme may be refined by distinguishing between different dating techniques within a particular category (e.g. Ar/Ar or K/Ar age dating). By keeping the absolute age data as a separate table, updates, and multiple timescales can be accommodated readily.

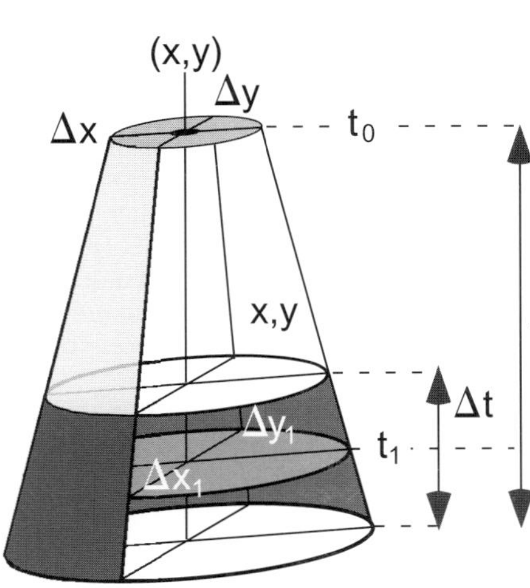

Fig. 4. A representation of the uncertainty in spatial and temporal position of a locality at point (x,y) at the present day (t_0), with present uncertainty in spatial location Δx and Δy. The past position at time t_1 is more uncertain both spatially (Δx_1, Δy_1) and temporally (Δt) due to uncertainties in the plate reconstruction.

Table 2. *Stratigraphic reliability codes (Paleogeographic Atlas Project, Chicago)*

Code	Explanation
A	Complete biostratigraphic control
B	Some biostratigraphic control
C	Stratigraphic interpolation (= dating of rocks above or below)
D	Geological inference (= correlation with other site[s])
E	Radiometric dating
F	Secondary information (= methods or source unspecified)
G	Guesswork (= no age provided, or dated to Period only)

Mixing and averaging. Behrensmeyer *et al.* (2001) provide an up-to-date summary of the field and implications of taphonomic studies for palaeoecological interpretations. In short, taphonomic processes mix assemblages and the amount of space and/or time encompassed by a sample is the spatial or temporal resolution of that sample. A single 'locality' may comprise many taxa and vary spatially from a few centimetres (such as a palynological preparation) to a few tens of centimetres or metres (e.g. a bed of rock) to hundreds of kilometres (e.g. a formation within a basin). The larger the area or volume of rock encompassed, the greater the amount of time that might be represented ('analytical time averaging'); (Behrensmeyer & Hook 1992). However, biological and taphonomic processes specific to a particular group of organisms reduce generality. A series of palynological samples through a core, each very small and representing depositional instants, implies a tight temporal grain, but mixing and transport of pollen in wind and water might imply coarse spatial grain for the same samples. Furthermore, the temporal duration of a single palynological preparation from a well core may present a depositional instant if made parallel to bedding, or a few years or tens of years if made perpendicular to bedding. This will also be reflected in the interpretation of the contemporary environment, including climate.

Separate biological and taphonomic processes produce a distinctly different grain implied by most vertebrate localities. Because of the relative sparsity of specimens in most cases, a vertebrate locality might include an area that is on the order of kilometres, or even tens of kilometres, in size, and which may encompass a thickness of hundreds of metres of sediment. As such, it might represent hundreds (or thousands) of years of deposition, depending on the tectonic setting (Behrensmeyer 1982; Behrensmeyer & Chapman 1993; Rogers 1993), but if the animals are migratory, it would be necessary to obtain a sample that adequately reflects the local fauna.

The physical mixing of earlier faunas within contemporary faunas ('taphonomic time averaging'; Behrensmeyer & Hook 1992; Behrensmeyer & Chapman 1993) further degrades resolution. The consequence of these problems is that as a palaeontological event (such as an extinction or a response to climate change) or environmental interpretation is examined over broader areas, so the temporal resolution with which it can be defined decreases. Conversely, the more finely events are resolved in time, the more difficult it is to know how large a region is affected. This is referred to as the 'paleontological uncertainty principle', analogous to the 'uncertainty principle' in quantum physics (S. Wing, pers. comm. 1991).

One solution is to use only data of a specified grain (resolution), but this can lead to loss of data, including information that, although poorly resolved, is nonetheless important. For example, if the location of a fossil is given as 'India' this may be considered spatially poorly resolved and therefore ignored, but if it is the only report of that fossil from India, then it is still useful information. However, this requires that the precision can be qualified; descriptors such as 'sample', 'composite locality', 'quarry', 'site' can be used, but each of these terms has numerous definitions, and so must be defined for every database. Landscape ecologists, faced with a similar problem, have derived numerous (mostly hierarchical) classification schemes for describing different scales of landscape system based on the areal extent considered (Huggett 1995). A similar approach might be appropriate for palaeontological databases.

An alternative solution is to include all data at the resolution at which it is reported (using a qualifier), and then to coarsen all 'localities' to some standard spatial or temporal scale by concatenating faunal and floral lists, in order to eliminate local variability. This has been used by

Markwick (2002) for present-day faunas and floras in order to examine the relationship of climate, biogeography and diversity (see also Markwick 1996). The selection of the smallest sampling unit determines the highest resolution (finest grain) possible in analysis based on information in the database. It is relatively easy to coarsen the resolution of data at a later date. It is impossible to refine it.

Taxonomy

Taxonomy influences grain, because different organisms scale with the environment differently, but this is a matter to consider when analysing the data. The major problem to be qualified in database design is taxonomic error (inaccuracy). Errors in taxonomic assignments can be due to several causes, among them the following: (i) incomplete preservation (absence of diagnostic characters); (ii) morphological uniformity (e.g. pollen of grasses); (iii) form taxa (e.g. separate genera for leaves, seed, pollen, etc.); and (iv) unreported taxonomy. Classification schemes for all biological entities are subject to change and disagreement. This is particularly true for fossil taxa, which may have no extant representatives, and which might be represented by incomplete and/or limited numbers of specimens. Different workers may adopt different taxonomic schemes depending on their own experience and opinions, and the relevant literature may incorporate a long history of taxonomic changes. The solution is partly an issue of accommodating uncertainty because assignments at a low taxonomic level may be poorly supported and disputed widely, whereas the higher level assignments can be made with considerable confidence and general agreement among professionals. Potential errors can be minimized by coarsening the data to a more 'confident' taxonomic level, and/or by recording specimen information as a guide to the characters used in the taxonomic assignment. This will vary according to the group studied, such that this method may create problems when assemblages are compared (the question of which taxonomic level to use, and whether the same level should be applied to all groups in the analysis). A species assignment based on an isolated fossil tooth will probably be of low confidence for a lizard, but significant for a mammal.

Another potential solution is to adopt a 'standard', preferably published, taxonomy and use this throughout the database. This ensures that the higher level taxonomy is at least consistent, although consistency is no guarantee of truth. Multiple standards can be made available as separate relations in the database structure.

Synonymy

Synonymization is the method of transferring a specimen or species to its appropriate taxonomic unit (e.g. species or genus) for any of several reasons, but usually because it is identical to a previously designated taxon. This can be dealt with by adding a 'synonym table' to the database structure that is used as a look-up library for all taxon names entered into the taxonomy table. The links can be structured such that if the entered taxon is found to belong within another species according to the synonym table, the most recent synonymized form replaces it. Again, the issue of data provenance is emphasized as species nomenclature is particularly fluid and contentious.

The rules of biological nomenclature state that no two animal or plant species may have the same name, and the rules establish how to designate and name a new species. Yet different species are often encountered in the literature that have the same name given informally during a study. This is particularly so in palynology and occurs primarily in the stratigraphic literature where interest focuses on distinguishing rock units from one another by segregation of pollen types. The frequent expression of this is the designation of many species named by combining informally a genus name with 'sp. A' or 'sp.1', as in Agasie (1969) and Ravn (1995) who record '*Tricolpites* sp. 1' from their sites in Arizona and Wyoming. However, sharing the same name does not imply that these pollen types represent the same biological entity, which is implied when formally named species share the specific epithet. Indeed, '*Tricolpites* sp. 1' in the paper by Agasie (1969) does not appear similar to '*Tricolpites* sp. 1' of Ravn (1969). The simplest method to overcome this problem is to treat 'sp. 1' etc. of every author as a distinct taxonomic unit, distinguished by a unique name, for example '*Tricolpites* sp. 1 of Agasie (1969)'.

Discussion

With the ready availability of desktop computerized relational database and GIS software, the logistics of building databases to cope with the large volumes of palaeontological data is no longer a major issue. While it is useful to remember certain guidelines as to database structure (Fig. 2) and the physical amount of data to be included (a database should be 'simple enough that it can be used, but comprehensive enough that it will be useful,'; Markwick 1996, p. 921), the principal problem facing designers of palaeontological databases is how to accommodate and qualify heterogeneities within the record, specifically of scale. We have suggested here that it is always better to collect information at the finest grain (resolution) possible and to append the appropriate confidence estimate (as a qualifier that can be queried on), since higher resolution data can always be degraded to lower resolution, but the reverse is impossible. The question of how observations made at different scales can be compared has been discussed by numerous authors for both modern and fossil settings (see Signor 1978; Hatfield 1985; Levin 1992; Anderson & Marcus 1993; Brown 1995; Rosenzweig 1995). But it is important to understand why scale is so important, especially for researchers integrating datasets from different fields, which has been made so much easier through GIS.

We have already noted how the apparent grain of a fossil assemblage can be affected by physical mixing and averaging in time and space, and that this problem worsens as the extent of the study increases. Consequently, this problem is greatest for global studies. For example, Markwick (1998), using the global distribution of fossil crocodilians to reconstruct palaeoclimate, calculated that the probability that 100 Eocene fossil crocodilian localities represented the identical 30 year timespan within the Eocene (21 000 000 years) and therefore the same 'climate', was $1/700\,000^{99}$. The problem of correlating age-equivalent samples is further exacerbated when multiple lines of evidence are used (e.g. palynology, floras and vertebrates to reconstruct palaeoclimate), each subject to different taphonomic processes. Failure to recognize the mixture of biological and environmental phenomena operating at different scales can produce spurious and misleading results. Even within the same biological group, mixing data of different resolutions can have strong effects on derived interpretations, especially in quantitative analyses. Lupia *et al.* (1999) analysed palynological samples from North America to investigate the possible replacement of conifers and free-sporing plants by angiosperms. They chose to restrict analyses to individual palynological samples, from a single site and stratigraphic horizon, rather than including samples created by combining multiple samples from several sites or stratigraphic horizons. Lupia *et al.* (1999) found nearly constant within-flora diversity through the Cretaceous compared to previous results from Lidgard & Crane (1990) that showed increasing within-flora diversity from Early to Late Cretaceous. By examining Lidgard and Crane's (1990) dataset, Lupia *et al.* (1999) concluded that the difference was attributable to the former's inclusion of combined samples, preferentially of Late Cretaceous age, in their analyses.

Likewise, the scale of biotic processes responding to abiotic conditions combined with resolution may decrease methodological power. For example, published data on using the foliar physiognomic method for reconstructing palaeoclimate suggest that the method, which seems to work well over large geographic gradients (Wolfe 1971, 1993), may break down at smaller scales probably due to the bias of local effects (Dolph & Dilcher 1979). Such problems are exacerbated when palaeontological data are compared with global climate model results, which can be of coarse spatial resolution, on the order of 4–5° of latitude and longitude (McGuffie & Henderson-Sellers 1997). Such coarseness may hide the finer scale variations in the real contemporary climate system, as experienced by the fossil organisms (climate proxies) themselves (Markwick 1998). Precipitation, for example, is very sensitive to local orography and moisture sources, and has been found to vary by 30% over a matter of a few kilometres (Linacre 1992). This may be particularly important in areas of rapid relief changes, such as the Eocene of the western United States (Sloan 1994).

The effect of error (inaccuracy) in databases also depends on the question being addressed. For North American Cambrian trilobites, Westrop & Adrain (2001) found that despite 70% of the generic records in the Sepkoski generic database being inaccurate (compiled from the published literature), when compared to their own field-based compilation, both datasets showed the same large-scale (coarse grain) patterns in Phanerozoic biodiversity (Adrain & Westrop 2000; Westrop & Adrain 2001). With finer grain, such errors become more important (Westrop & Adrain 2001).

The consequences of scale (grain) and error depend on the fossil group or assemblage investigated, the extent of the study and the questions to be asked. Palaeontological databases must therefore be designed to accommodate these issues.

Conclusions

The fossil record is the only direct evidence about the biological evolution of life on Earth. This represents a huge volume of data, and computerized databases provide the most efficient means of storing and examining the records for large-scale patterns and processes. The quantity and quality of these data are always less than ideal. But, we suggest that it is the heterogeneities in scale (grain) in the fossil record that are potentially the major cause of problems in palaeoecology especially when different and varied datasets are mixed, as increasingly they are in studies examining the interplay of biotic and abiotic phenomena using GIS. Scale can, and should, be qualified in a database, and to facilitate this we offer the following suggestions.

(i) Data should be collected at the finest available grain (resolution) since the finest scale of data stored in the database defines the finest grain possible in any analysis using that data. The grain in each case needs to be recorded (e.g. by a qualifier in a separate data field). In this way the database can be queried for data of a specified grain, or data of mixed resolutions can be coarsened to some common grain.
(ii) All data in the database should have an audit trail, recording all sources and changes made to that data. To this end a distinction should be made between raw data (observations) and interpretation.
(iii) Database structure is dictated by the nature of the fossil record, such that the most efficient structure is based around separate tables of spatial and temporal location, taxonomy and occurrence, which can also include taphonomic information (Fig. 3). The database structure should also facilitate the inclusion of multiple standards (e.g. alternative timescales and taxonomies).

We would like to thank A. Crame for the opportunity to publish this work, and the reviewers, S. Lidgard and R. Livermore, whose suggestions greatly improved this paper. This work draws on our experience of designing and compiling palaeontological databases at The University of Chicago, where we benefited from the advice of P. Crane, S. Lidgard, the late J. Sepkoski, and especially the members of the Palaeogeographic Atlas Project: A. Ziegler, D. Rowley, A. Lottes and M. Hulver. Finally, we wish to thank the libraries of the University of Chicago, especially the staff of the John Crerar Science Library – at the heart of all great universities is a great library.

References

ADRAIN, J. M. & WESTROP, S. R. 2000. An empirical assessment of taxic paleobiology. *Science*, **289**, 110–112.

AGASIE, J. M. 1969. Late Cretaceous palynomorphs from northeastern Arizona. *Micropaleontology*, **15**, 13–30.

ALROY, J. 1995. *Quantitative mammalian biochronology, biogeography, and diversity history of North America*. PhD thesis, University of Chicago.

ALROY, J., MARSHALL, C. R., BAMBACH, R. K., *ET AL.* 2001. Effects of sampling standardization on estimates of Phanerozoic marine diversification. *Proceedings of the National Academy of Sciences, U.S.A.*, **98**, 6261–6266.

ANDERSON, S. & MARCUS, L. F. 1993. Effect of quadrat size on measurement of species density. *Journal of Biogeography*, **20**, 421–428.

BEHRENSMEYER, A. K. 1982. Time resolution in fluvial vertebrate assemblages. *Paleobiology*, **8**, 211–227.

BEHRENSMEYER, A. K. & CHAPMAN, R. E. 1993. Models and simulations of time-averaging in terrestrial vertebrate accumulations. *In*: KIDWELL, S. M. & BEHRENSMEYER, A. K. (eds) *Taphonomic Approaches to Time Resolution in Fossil Assemblages*. University of Tennessee, Knoxville, Short Courses in Paleontology, **6**, 125–149.

BEHRENSMEYER, A. K. & HOOK, R. W. 1992. Paleoenvironmental contexts and taphonomic modes. *In*: BEHRENSMEYER, A. K., DAMUTH, J. D., DIMICHELE, W. A., POTTS, R., SUES, H.-D. & WING, S. L. (eds) *Terrestrial Ecosystems Through Time. Evolutionary Paleoecology of Terrestrial Plants and Animals*. University of Chicago Press, Chicago, 15–136.

BEHRENSMEYER, A. K., KIDWELL, S. M. & GASTALDO, R. A. 2001. Taphonomy and paleobiology. *In*: ERWIN, D. H. & WING, S. L. (eds) *Deep Time: Paleobiology's Perspectives*. *Paleobiology*, Supplement, **26** (4), 103–147.

BENTON, M. J. (ed.) 1993. *Fossil Record 2*. Kluwer Academic Publishers, Dordrecht.

BOULTER, M. C., CHALONER, W. G. & HOLMES, P. L. 1991. The IOP plant fossil record: are fossil plants a special case? *In*: HAWKSWORTH, D. L. (ed.) *Improving the Stability of Names: Needs and Options*. Koeltz Scientific Books, Königstein, Regnum Vegetabile **123**, 231–242.

BROWN, J. H. 1995. *Macroecology*. University of Chicago Press, Chicago.

DAMUTH, J. 1993. *ETE Database Manual*. Evolution of Terrestrial Ecosystems Consortium, Washington, DC.

DOLPH, G. E. & DILCHER, D. L. 1979. Foliar physionomy as an aid in determining palaeoclimate. *Palaeontographica B*, **170**, 151–172.

HATFIELD, C. B. 1985. Fossil vertebrate taxonomic diversity correlated with outcrop area. *Ohio Journal of Science*, **85**, 64–67.

HUGGETT, R. J. 1995. *Geoecology. An Evolutionary Approach*. Routledge, London.

KIDWELL, S. M. & BEHRENSMEYER, A. K. (eds) 1993. *Taphonomic Approaches to Time Resolution in Fossil Assemblages*. Univeristy of Tennessee, Knoxville, Short Courses in Paleontology, **6**.

KREBS, J. W., KAESLER, R. L., BROSIUS, E. A., MILLER, D. L. & CHANG, Y.-M. 1996. Paleobank, a relational database for invertebrate paleontology: the data model. *The University of Kansas Paleontological Contributions New Series*, **8**, 1–7.

LABANDEIRA, C. C. & SEPKOSKI, J. J., JR. 1993. Insect diversity in the fossil record. *Science*, **261**, 310–315.

LEVIN, S. A. 1992. The problem of pattern and scale in ecology. *Ecology*, **73**, 1943–1967.

LIDGARD, S. & CRANE, P. R. 1990. Angiosperm diversification and Cretaceous floristic trends: a comparison of palynofloras and leaf macrofloras. *Paleobiology*, **16**, 77–93.

LINACRE, E. 1992. *Climate Data and Resources. A Reference Guide*. Routledge, London.

LUPIA, R. 1999. Discordant morphological disparity and taxonomic diversity during the Cretaceous angiosperm radiation: North American pollen record. *Paleobiology*, **25**, 1–28.

LUPIA, R., LIDGARD, S. & CRANE, P. R. 1999. Comparing palynological abundance and diversity: implications for biotic replacement during the Cretaceous angiosperm radiation. *Paleobiology*, **25**, 305–340.

MCGUFFIE, K. & HENDERSON-SELLERS, A. 1997. *A Climate Modelling Primer*. Wiley, Chichester.

MARKWICK, P. J. 1996. *Late Cretaceous to Pleistocene climates: nature of the transition from a 'hot-house' to an 'ice-house' world*. PhD Thesis, The University of Chicago.

MARKWICK, P. J. 1998. Fossil crocodilians as indicators of Late Cretaceous and Cenozoic climates: implications for using palaeontological data for global change. *Palaeogeography, Palaeoclimatology, Palaeoecology*, **137**, 205–271.

MARKWICK, P. J. 2002. Integrating the present and past records of climate, biodiversity and biogeography: implications for palaeoecology and palaeoclimatology. *In*: CRAME, J. A. & OWEN, A. W. (eds) *Palaeobiogeography and Biodiversity Change: the Ordovician and Mesozoic–Cenozoic Radiations*. Geological society, London, Special Publications, **194**, 179–199.

O'NEILL, R. V. & KING, A. W. 1998. Homage to St. Michael or why are there so many books on scale? *In*: PETERSON, D. L. & PARKER, V. T. (eds) *Ecological Scale, Theory and Applications*. Columbia University Press, New York, 3–15.

PEUQUET, D. J. 1988. Issues involved in selecting appropriate data models for global databases. *In*: MOUNSEY, H. (ed) *Building Databases for Global Science*. Taylor and Francis, London, 66–78.

RAUP, D. M. & SEPKOSKI, J. J. 1986. Periodic extinction of families and genera. *Science*, **231**, 833–836.

RAVN, R. L. 1995. Miospores from the Muddy Sandstone (upper Albian), Wind River basin, Wyoming, USA. *Palaeontographica, Abteilung B*, **234**, 41–91.

ROGERS, R. R. 1993. Systematic patterns of time-averaging in the terrestrial vertebrate record: a Cretaceous case study. *In*: KIDWELL, S. M. & BEHRENSMEYER, A. K. (eds) *Taphonomic Approaches to Time Resolution in Fossil Assemblages*. University of Tennessee, Knoxville, Short Courses in Paleontology, 228–249.

ROSENWEIG, M. L. 1995. *Species Diversity in Space and Time*. Cambridge University Press, Cambridge.

SEPKOSKI, J. J. JR. 1982. *A Compendium of Fossil Marine Families*. Milwaukee, Public Museum, Contributions in Biology and Geology, **51**.

SIGNOR, P. W. III 1978. Species richness in the Phanerozoic: an investigation of sampling effects. *Paleobiology*, **4**, 394–406.

SLOAN, L. C. 1994. Equable climates during the early Eocene: significance of regional paleogeography for North American climate. *Geology*, **22**, 881–884.

WESTROP, S. R. & ADRAIN, J. M. 2001. Sampling at the species level: impact of spatial biases on diversity gradients. *Geology*, **29**, 903–906.

WOLFE, J. A. 1971. Tertiary climatic fluctuations and methods of analysis of Tertiary floras. *Palaeogeography, Palaeoclimatology, Palaeoecology*, **9**, 27–57.

WOLFE, J. A. 1993. A method of obtaining climatic parameters from leaf assemblages. *U.S. Geological Survey Bulletin*, **2040**, 1–71.

ZIEGLER, A. M., ROWLEY, D. B., LOTTES, A. L., SAHAGIAN, D. L., HULVER, M. L. & GIERLOWSKI, T. C. 1985. Paleogeographic interpretation: with an example from the Mid-Cretaceous. *Annual Review of Earth and Planetary Sciences*, **13**, 385–425.

Integrating the present and past records of climate, biodiversity and biogeography: implications for palaeoecology and palaeoclimatology

PAUL J. MARKWICK

Robertson Research International Limited, Llandudno, Conwy, LL30 1SA, UK

(e-mail: paul.markwick@which.net*)*

Abstract: A geographic information system (GIS) based, integrated dataset of Recent North American, European, southern African and Australian non-avian tetrapod faunas is used to examine the macroscale relationship between climate, biogeography and terrestrial taxonomic and functional species diversity (richness). The results support a modified form of the species–energy hypothesis, with the pattern of terrestrial biodiversity reflecting the manner in which species procure energy, rather than only the absolute amount of 'available energy'. Area and history are also found to be important. Ectotherms show the simplest relationship with environmental variables (and strongest latitudinal diversity gradients), and endotherms the most complex. A strong linear relationship is found between the proportion of each fauna represented by ectotherms and temperature (mean annual temperature and coldest month mean temperature). This relationship is used in an experiment to retrodict the palaeotemperature for the Middle Eocene lagerstätten fauna from Messel, Germany. Results compare well with interpretations based on other climate proxies.

Climate interpretations based on fossil data depend heavily on analogy with recent species and to this end an understanding of living groups and their distribution is crucial. Terrestrial organisms are neither randomly nor evenly distributed on the globe, and natural historians have long postulated that this is due, at least in part, to climate and other environmental factors (Humboldt & Bonpland 1807; Wallace 1876; Matthew 1915; Darlington 1948). Some taxa, such as crocodilians, are demonstrably limited by temperature and have been used by geologists since the early nineteenth century as tools for reconstructing palaeoclimate (Lyell 1830; Colbert *et al.* 1946; Hibbard 1960; Markwick 1994, 1998*a*). A climate origin has also been postulated to explain observed 'latitudinal' species diversity (or richness) gradients in many Recent terrestrial organic groups (Pianka 1966; Stevens 1989; Currie 1991; Rohde 1992; Wright *et al.* 1993). Ostrom (1970) has suggested that such gradients might provide a better tool for retrodicting palaeoclimate than comparison of an individual fossil with the climate of its living relative (see also Fischer 1961, p. 50).

Many theories have been proposed to account for observed latitudinal diversity gradients in the Recent (see Rohde (1992) for review), of which area–history and species–energy have perhaps received the most attention. Arguments based on area (Rosenzweig 1995) derive from the observation that island habitats tend to have fewer species than non-isolated habitats of the same area. This is embodied in the island biogeography theory of MacArthur & Wilson (1967), in which standing diversity reflects the balance between immigration and extinction rates, as dictated by island area (smaller areas hold smaller populations that are therefore assumed to be more susceptible to extinction) and the proximity in time (history) and space to the dispersal source. Historical changes in insularity, for example the repeated fragmentation and coalescence of 'islands' during Pleistocene interglacial–glacial changes, are envisaged to increase the opportunity for allopatric speciation, and have been used to account for the high species diversities in SE Asia (Qian & Ricklefs 2000). Similar arguments have been used to explain patterns in the pre-Pleistocene marine invertebrate fossil record (Flessa 1975; Sepkoski 1976; Flessa & Sepkoski 1978; Crame 2001). But the relationship between area and diversity is not unequivocal, even for well documented examples such as SE Asia (Harrison *et al.* 2001), and the use of area *per se* must be viewed with caution. As Rohde (1997) has pointed out in regard to latitudinal diversity gradients, the low latitudes today do not necessarily contain the largest areas, despite being the location of the greatest species richness. However, islands need not be geographic entities (ocean islands), but may include other isolated physiological features (mountains, lakes) or distinct habitats

From: CRAME, J. A. & OWEN, A. W. (eds) 2002. *Palaeobiogeography and Biodiversity Change: the Ordovician and Mesozoic–Cenozoic Radiations*. Geological Society, London, Special Publications, **194**, 179–199.
0305-8719/02/$15.00

(habitat 'islands'), although few habitat 'islands' are as absolute as ocean islands (Rosenzweig 1995).

The association of areas of high species diversity with low latitudes, high temperatures and high incident solar energy fluxes, has given rise to the widely promoted species–energy theory (Pianka 1966; Currie 1991; Rohde 1992; Stevens 1992; Currie & Fritz 1993; Latham & Ricklefs 1993; Wright *et al.* 1993; Brown 1995; Rosenzweig 1995). Currie (1991), comparing North American terrestrial species diversity against 21 environmental variables, found that species diversity correlated best with potential evapotranspiration (PET), considered to be an appropriate measure of ambient energy. However, Kerr & Packer (1997) found that this was only true for mammals in North America where PET < 1000 mm a^{-1} (Canada and Alaska). Stevens (1989) has suggested that climate (energy) variability is the most important determinant of species numbers, with few taxa being able to survive large seasonal variations. Such taxa should therefore have the largest latitudinal ranges (Rapoport's rule), but this pattern has not been found in all groups (Rohde *et al.* 1993; Roy *et al.* 1994). For many this theory remains equivocal (Pianka 1966; Rohde 1992; Roy *et al.* 1994; Brown 1995; Rosenzweig 1995; Kerr & Packer 1997) because not all groups show clear diversity gradients and the mechanism by which energy can dictate the number of species is uncertain.

The problem is that individual species respond to different environmental factors (and combinations of factors) and to different extents depending on their physiology and ecology (Root & Schneider 1993). Measuring only the number of species (taxonomic diversity) rather than the distribution of differences between organisms (functional and ecological diversity) may obfuscate the processes dictating diversity patterns (Gaston 2000). It is therefore essential to be able to examine the macroscale spatial structure of diversity in the context of physiology and behaviour (what an organism does), as well as taxonomy (what an organism is called), although for traditional classification methods there is often considerable overlap between the two. What is more, species richness does not vary only with latitude (Brown 1995), nor is it independent of history (evolution and palaeobiogeography). To understand global patterns therefore requires large, intercontinental datasets. Unfortunately, there have been few studies at this scale, and these are restricted to analyses at relatively coarse taxonomic levels and resolutions/grain (e.g. Gaston *et al.* 1995).

This paper presents a new, digital, geographic information system-based dataset with which the relationships between present-day terrestrial biodiversity, biogeography and climate are examined. Examples of the observed macroscale modern patterns are illustrated (both as maps and bivariate plots) in order to facilitate comparison with previous studies of North America, especially that of Currie (1991). The consequences of derived relationships for interpretations of palaeoclimate and palaeoecology are discussed. As a test of Ostrom's (1970) suggestion of using diversity gradients to retrodict palaeoclimate, an experiment is presented in which the Middle Eocene palaeotemperature of Messel, Germany, is reconstructed using modern-day regressions between observed taxonomic diversity and temperature, and the results compared with values obtained from other methods.

Methods

The dataset used in this study is part of a large computer-based ecological database of fossil and modern faunal and floral localities compiled by Markwick (1996). The database is designed to facilitate analysis at any specified taxonomic level, such that differences between the response of families, genera and species can be systematically analysed. Data can also be queried for any combination of parameters included in the database. Since this study was begun in 1990, the database has been integrated into a geographic information system (ArcView GIS and ArcInfo). The fossil data include detailed specimen, environmental and stratigraphic information on about 6000 Cretaceous and Cenozoic fossil vertebrate localities, with taxonomic and ecological data for 22 000 extant and fossil vertebrate and floral taxa (including habitat, size and diet). The modern data analysed here draw on the climate information from 1060 climate stations (Fig. 1) taken from Müller's (1982) compilation for vegetation studies. Each station contains monthly data for 14 climate variables including mean daily temperature, mean precipitation, radiation and potential evapotranspiration (PET). A large number of additional parameters, including annual metrics and combinations of variables, have been calculated using these data. A list of variables mentioned in this paper, their abbreviations and explanations are given in Table 1. The propensity for stations to occur in lowland sites reflects Müller's (1982) original requirements: acceptable stations must contain data for a large array of climate parameters representing time

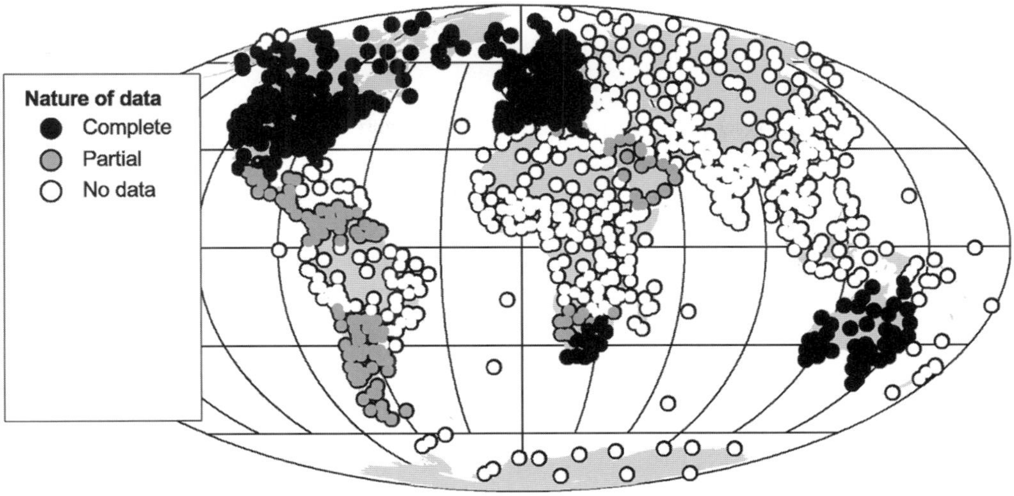

Fig. 1. The distribution of climate stations from Müller (1982) used in this study. Black circles represent stations with complete non-avian tetrapod lists; grey circles represent those stations with only partial data.

series on the order of 30 years (typical of 'climate'). Consequently most stations are located in large towns or cities that historically occupy lowland sites. This bias closely mimics the biases in the geological record that tend to over-represent low-elevation environments.

Complete non-avian tetrapod faunal lists have been assigned to 408 of these stations: Europe (204); North America (143); southern Africa (21); and Australia (40). Seventy-two lists from South America and 18 lists from Arabia contain mammal data only. Floral data (Little 1977) have been assigned to the North American stations, and the database also includes global coverage for all palms (Uhl & Dransfield 1987), turtles (Iverson 1992) and crocodilians (see references in Markwick 1998*a*).

Faunal lists are assigned to each station by overlaying climate station distributions onto published species distribution maps (Hall & Kelson 1959; Van der Brink 1967; Little 1977; Arnold & Burton 1978; Cook 1984; Stebbins 1985; Uhl & Dransfield 1987; Branch 1988; Bouchardy & Moutou 1989; Eisenberg 1989;

Table 1. *Explanation of environmental variables used in this paper*

Abbreviation	Explanation
MAT	Mean annual temperature
MART	Mean annual range of temperature
CMM	Coldest month mean temperature
WMM	Warmest month mean temperature
Radiation	Incident solar radiation measured at surface
Cumulative $T \geq 0$	Sum of all degrees above 0°C ('degree months')
Cumulative $T \geq 5$	Sum of all degrees above 5°C ('degree months')
Anuual precipitation	Mean annual precipitation
P range	Mean annual range in precipitation
Months $T \geq 10$ $P \geq 40$	Number of months with mean temperatures greater than 10°C and Precipitation greater than 40 mm
PET	Potential evapotranspiration
Mean annual NDVI	Mean annual negative difference vegetation index: a satellite-based metric used as a proxy for primary productivity
NDVI 1SD	One standard deviation of mean monthly NDVI

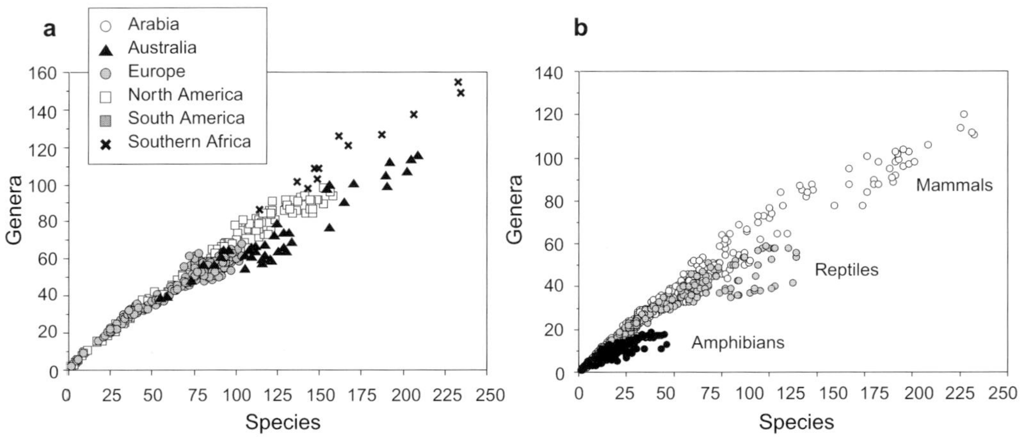

Fig. 2. The relationship between the numbers of species and genera at each site used in the analysis. (**a**) Non-avian tetrapod species and genera by continent. (**b**) Species and genera by taxonomic group.

Kingdon 1990; Conant & Collins 1991; Grenard 1991; Cogger 1992; Iverson 1992; Redford & Eisenberg 1992; Strahan 1992). A 50 km radius circle is drawn around each station, and an occurrence registered where the taxon's distribution intersects this circle. A radial limit of 50 km was chosen as it represents a typical decorrelation distance for precipitation, which is the most sensitive climate parameter to spatial heterogeneity. This means that the faunas and floras can be confidently assumed to have experienced the climate assigned to them, except in areas with rapid relief changes (such as the Alps), where this methodology mixes high- and low-elevation faunas. These points are found to fall off the derived regressions but do not significantly affect results. Ecologically, this approach removes local, small-scale faunal and floral heterogeneities, and thereby emulates the spatial and temporal time-averaging in the fossil record, with which this modern dataset can thereby be directly compared. The 50 km radius also approximately equates with the scale of regional general circulation models (0.5° × 0.5°).

This approach differs from existing diversity gradient studies (e.g. Currie 1991) in two ways. Firstly, the use of point (station) rather than gridded data reduces the area effects implicit in quadrat techniques (Anderson & Marcus 1993), and allows a direct comparison with climate (the use of latitudinal zones is to be avoided because it ignores longitudinal effects such as 'continentality'). Secondly, the method integrates data from more than one continent, thus reducing the potential effects of regional biogeographical or historical artifacts.

The diversity data have been plotted on present-day maps using ArcView GIS for each taxonomic group in the database, as well as for each habitat and diet category. This provides a qualitative indication of spatial similarities between biogeography, taxonomic diversity and environmental factors. The Spearman rank test (using the Statview software; Haycock *et al.* 1992–1993) was used to investigate correlations between environmental variables and global taxonomic diversity (Table 2), and regional diversity (Tables 3–7). The relationship of taxonomic assemblage composition to environmental factors was examined with correspondence analysis using the CANOCO software (Ter Braak 1987–1992). The use of generic rather than species-level presence–absence information for this analysis is due to computational limitations. However, analysis of the dataset has shown the close relationship between the numbers of species and genera (Fig. 2), such that results derived from either taxonomic level are comparable. Generic level assignments are probably more robust for fossil taxa.

Results

General patterns

Plots of diversity against absolute latitude are shown in Figure 3. For comparison, a selection of environmental variables is plotted against absolute latitude in Figure 4 (see Table 1 for an explanation of each variable). Total non-avian tetrapod species diversity (amphibians + reptiles

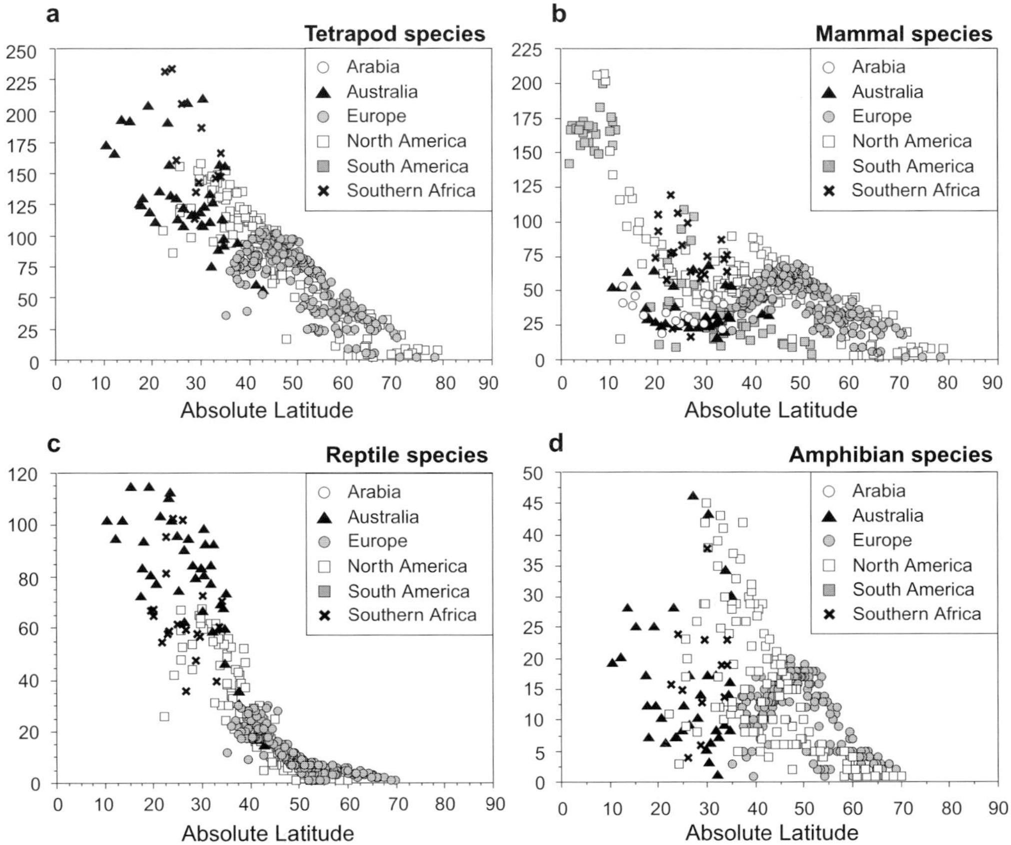

Fig. 3. Species diversity plotted against absolute latitude for (**a**) total non-avian tetrapods, (**b**) mammals, (**c**) reptiles and (**d**) amphibians.

+ mammals) shows a strong linear (or slightly curvilinear) trend, with the data from all regions falling along this latitudinal gradient (Fig. 3a). There is no apparent hemispheric asymmetry, as observed in other faunal groups such as marine bivalves (Crame 2000), and the spatial distribution shows few longitudinal heterogeneities (Fig. 5). This global trend correlates strongly with temperature and productivity metrics (Table 2), consistent with the species–energy hypothesis. However, regional correlations are not significant for either Australia or southern Africa (Table 3), which may reflect the smaller sample sizes representing these regions and/or the lack of strong environmental gradients within their borders (especially within southern Africa). However, this apparently simple global relationship belies great variation in the trends of each taxonomic and physiological component, and it is these differences that may be critical in understanding potential causes.

Reptile species diversity (Fig. 3c) follows a strongly curvilinear trend, which increases monotonically from high to low latitudes. There is no apparent asymmetry between northern and southern hemisphere groups, with a marked change in gradient at about 45° latitude. This pattern most closely approximates to the global distribution of incident energy (temperature; Fig. 4a–e), specifically cumulative energy (cumulative T; Fig. 4d, Table 2). The baseline of either 0° or 5°C for cumulative T is related to the typical critical minimum body temperatures observed for modern reptiles (for crocodilians this is 4–5°C; Brisbin *et al.* 1982). The geographic distribution of reptile species diversity (Fig. 6) shows a similar simplicity, and regionally, reptile diversity correlates strongly with temperature for all areas except southern Africa (Table 4).

Amphibian species diversity also shows a monotonic increase in taxonomic diversity from high to low latitudes (Fig. 3d). Nevertheless, the

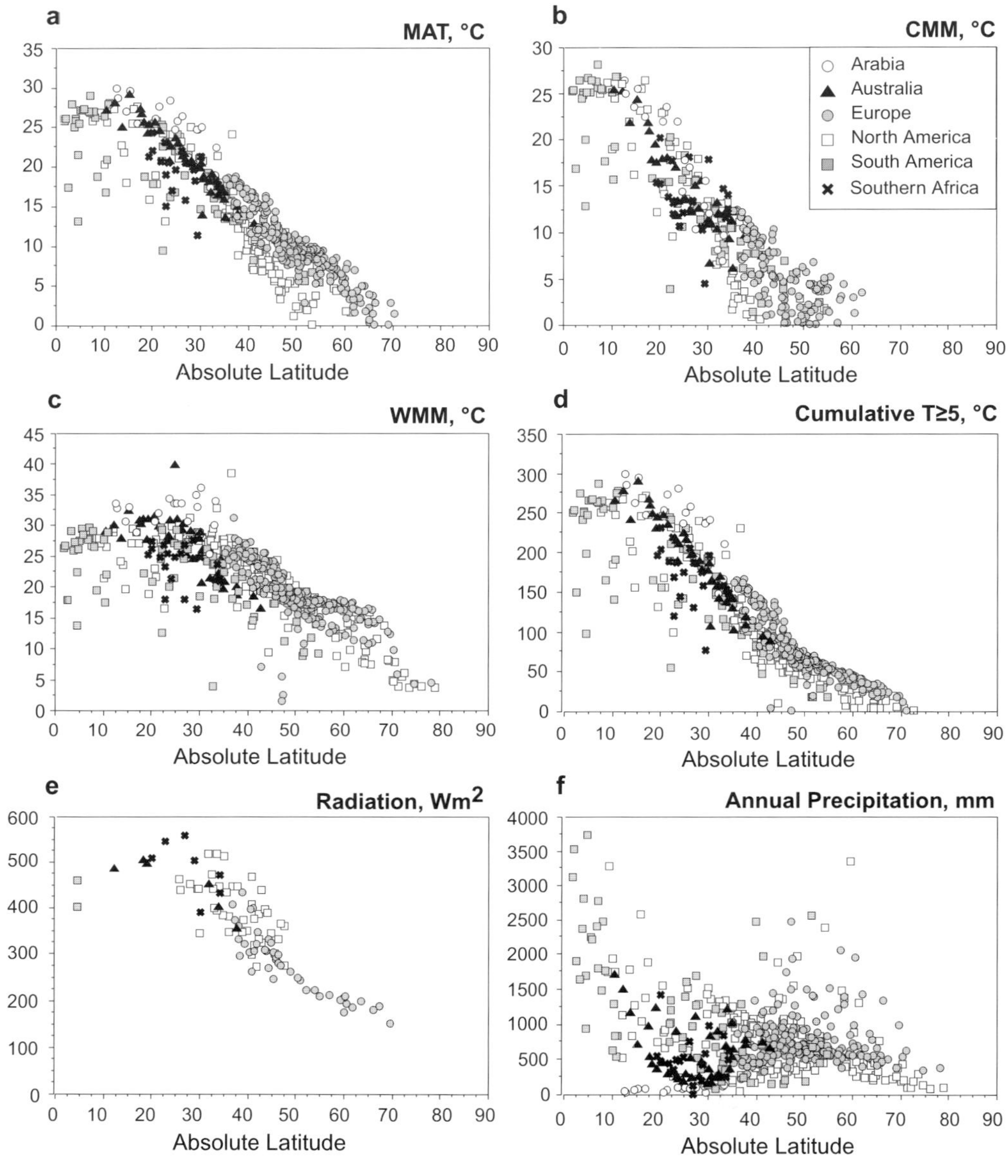

Fig. 4. Environmental parameters plotted against absolute latitude for (**a**) MAT, (**b**) CMM, (**c**) WMM, (**d**) cumulative T ≥ 5°C, (**e**) radiation, (**f**) annual precipitation, (**g**) months with T ≥ 10°C and P ≥ 40 mm (see Lottes & Ziegler 1994), (**h**) PET, (**i**) mean annual NDVI, (**j**) NDVI, 1SD. See Table 1 for an explanation of each parameter.

spread of data away from this trend is far greater than that observed for reptiles. Spatially this shows strong longitudinal gradients superimposed on the overall latitudinal trend (Fig. 7), and a general pattern that is similar to the modern distribution of precipitation around the globe (i.e. with highs in the wettest regions of North America (NW and SE USA) and NE Australia). The low-diversity, low-latitude data points in Figure 3d all reflect areas of reduced water availability (the deserts of southern Africa and Australia; Table 5). Like reptiles, amphibians are also precluded from the coldest parts of the world (Fig. 7). This distribution

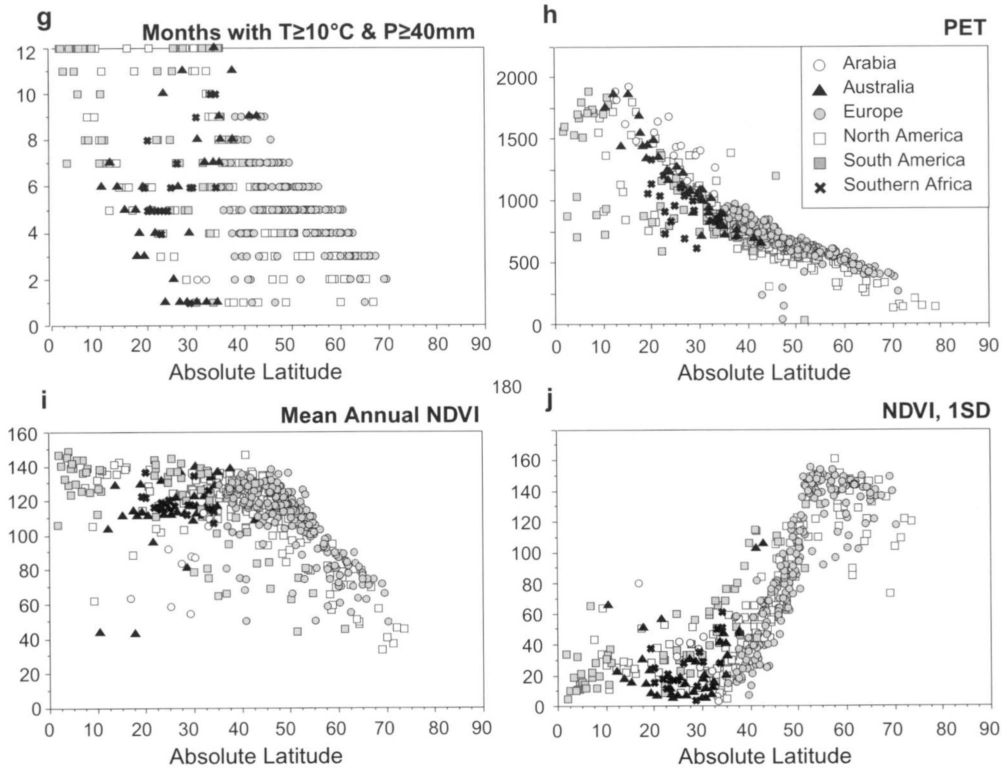

Fig. 4. continued

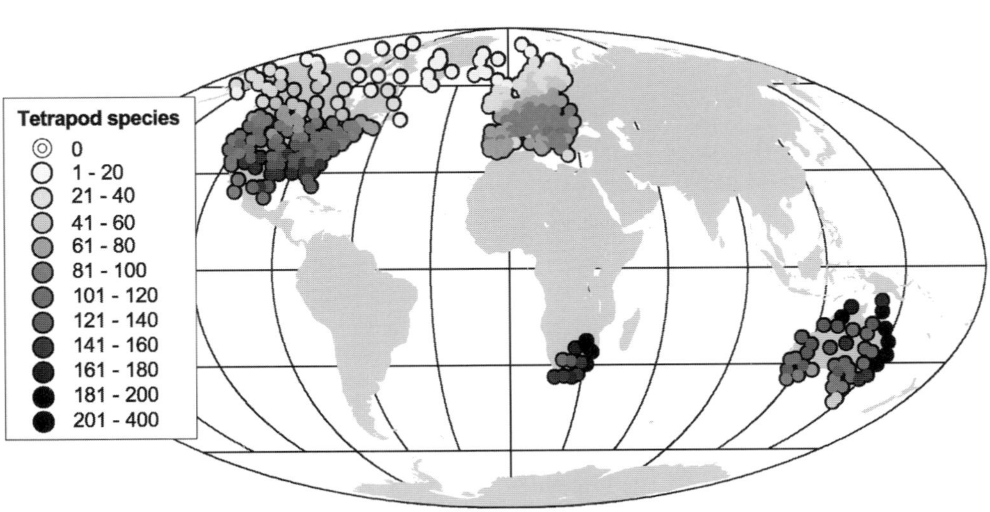

Fig. 5. The distribution of total non-avian tetrapod species diversity for North America, Europe, southern Africa and Australia.

Table 2. *Spearman rank test rho values for environmental variables and taxonomic species diversity for each group: all regions*

	Amphibians	Reptiles	Herptiles	Mammals	Non-bat mammals	Non-avian tetrapods
n	400	411	411	524	524	400
Absolute latitude	–0.576	–0.956	–0.931	–0.462	–0.355	–0.869
Elevation	0.256	0.268	0.259	0.406	0.448	0.362
MAT	0.503	0.892	0.863	0.363	0.219	0.754
MART	NS	–0.358	–0.339	NS	NS	NS
CMM	0.396	0.768	0.739	0.275	NS	0.591
WMM	0.515	0.879	0.852	0.388	0.294	0.803
Radiation	NS	0.825	0.741	0.445	0.450	0.733
Cumulative T ≥ 0	0.509	0.903	0.873	0.380	0.244	0.824
Cumulative T ≥ 5	0.531	0.926	0.893	0.410	0.281	0.803
Annual precipitation	0.484	NS	NS	0.327	0.253	NS
P range	0.227	0.231	0.230	0.377	0.285	0.193
Months T ≥ 10 P ≥ 40	0.670	0.402	0.462	0.475	0.388	0.450
PET	0.485	0.895	0.869	0.327	0.180	0.765
Mean annual NDVI	0.694	0.598	0.638	0.562	0.472	0.628
NDVI 1SD	–0.450	–0.882	–0.851	–0.398	–0.288	–0.768

($p < 0.0001$; NS, not significant, $p > 0.0001$)

Table 3. *Spearman rank test rho values for environmental variables and non-avian tetrapod species diversity by region*

	South America	North America	Europe	Arabia	Southern Africa	Australia
n	–	144	204	–	12	40
Absolute latitude	–	–0.919	–0.688	–	NS	–0.617
Elevation	–	0.345	0.478	–	NS	NS
MAT	–	0.891	0.484	–	NS	NS
MART	–	–0.451	NS	–	NS	NS
CMM	–	0.805	NS	–	NS	NS
WMM	–	0.899	0.639	–	NS	NS
Radiation	–	NS	NS	–	NS	NS
Cumulative T ≥ 0	–	0.906	0.504	–	NS	NS
Cumulative T ≥ 5	–	0.910	0.586	–	NS	NS
Annual precipitation	–	0.376	NS	–	NS	NS
P range	–	NS	NS	–	NS	NS
Months T ≥ 10 P ≥ 40	–	0.688	0.462	–	NS	NS
PET	–	0.895	0.539	–	NS	NS
Mean annual NDVI	–	0.780	0.662	–	NS	NS
NDVI 1SD	–	–0.858	–0.622	–	NS	NS

($p < 0.0001$; NS, not significant, $p > 0.0001$)

pattern is consistent with their dependence on both available water and energy (as temperature). Such dependence is reflected in the highest correlations in Table 2 being with combinations of temperature and precipitation (compare with Fig. 4g). Negative difference vegetation index; (NDVI, Fig. 4i,j) is a satellite-derived proxy for net primary productivity (Goward *et al.* 1985; Goward & Dye 1987; Cihlar *et al.* 1991), which Lottes & Ziegler (1994) have shown closely correlates with the number of months with mean temperatures greater than 10°C and mean precipitation greater than 40 mm (as a proxy for effective 'growing season'). If this is the case, then plants, which also depend on incident energy and water, should show a similar diversity pattern to amphibians, and this is what is observed (at least for tree species; Fig. 8). O'Brien (1993) has shown that climate provides the first-order explanation for plant species

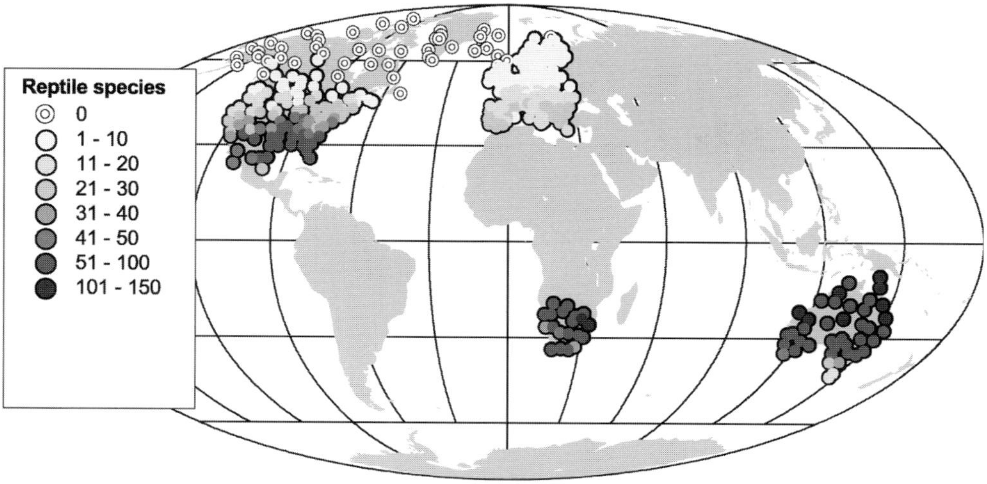

Fig. 6. The distribution of total reptile species diversity for North America, Europe, southern Africa and Australia.

diversity gradients in southern Africa. Regional correlations for amphibian diversity show similar relationships (Table 5).

Mammal species diversity shows greater complexity than either amphibian or reptile diversity (Fig. 3b), and the greatest hemispheric asymmetry. Although the highest diversities are also in the lowest latitudes, there is a secondary peak in mid-latitudes (about 45–50°) for North America and Europe, but not for South America. Geographically, the highest diversities in North America are in the western interior, while in Australia, diversity is highest along the NE coast, as it is for amphibians (Fig. 9). This complexity is apparent from the correlations in Table 2 (see also Table 6), which are consistently lower for mammals than for the other major groups. The highest correlation is with mean annual NDVI, although note the similarity between mammal species diversity and the

Table 4. *Spearman rank test rho values for environmental variables and reptile species diversity by region*

	South America	North America	Europe	Arabia	Southern Africa	Australia
n	–	144	204	–	21	40
Absolute latitude	–	–0.936	–0.928	–	NS	–0.726
Elevation	–	NS	NS	–	NS	NS
MAT	–	0.932	0.784	–	NS	0.638
MART	–	–0.490	NS	–	NS	NS
CMM	–	0.847	0.524	–	NS	NS
WMM	–	0.931	0.887	–	NS	NS
Radiation	–	NS	0.901	–	NS	NS
Cumulative T ≥ 0	–	0.939	0.801	–	NS	0.637
Cumulative T ≥ 5	–	0.946	0.860	–	NS	0.637
Annual precipitation	–	0.360	NS	–	NS	NS
P range	–	NS	NS	–	NS	NS
Months T ≥ 10 P ≥ 40	–	0.690	0.399	–	NS	NS
PET	–	0.940	0.817	–	NS	0.640
Mean annual NDVI	–	0.746	0.638	–	NS	NS
NDVI 1SD	–	–0.890	–0.861	–	NS	NS

($p < 0.0001$; NS, not significant, $p > 0.0001$)

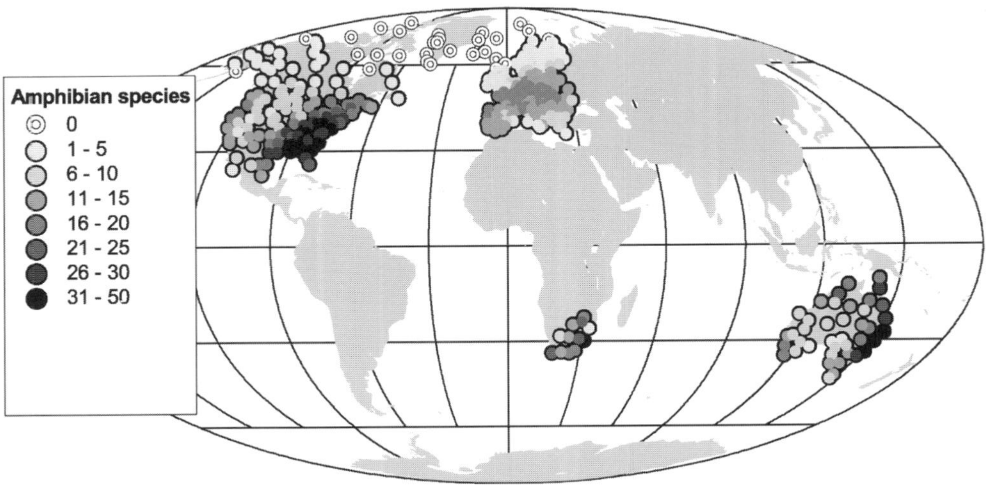

Fig. 7. The distribution of total amphibian species diversity for North America, Europe, southern Africa and Australia.

latitudinal distribution of annual precipitation (Fig. 4f). This complexity is further revealed when mammalian diversity is divided into its constituent taxonomic parts, as shown in Figure 10, in which the most speciose orders of mammals are considered as a proportion of the total mammalian fauna. The proportion of bats increases equatorward to dominate the high-diversity faunas of northern South America (perhaps following the increases in insect diversity upon which most feed). Nevertheless, the proportion of both carnivores and rodents decrease, and it is largely the high rodent diversity of the western interior USA that generates the mammal species diversity pattern seen in the mid-latitudes in Figures 3b and 9.

The physiological and ecological composition of each fauna is therefore important in understanding the relationship between diversity and

Table 5. *Spearman rank test rho values for environmental variables and amphibian species diversity by region*

	South America	North America	Europe	Arabia	Southern Africa	Australia
n	–	144	204	–	12	40
Absolute latitude	–	–0.791	–0.457	–	NS	NS
Elevation	–	NS	0.467	–	NS	NS
MAT	–	0.779	0.286	–	NS	NS
MART	–	–0.425	NS	–	NS	NS
CMM	–	0.711	NS	–	NS	NS
WMM	–	0.761	0.372	–	NS	NS
Radiation	–	NS	NS	–	NS	NS
Cumulative T ≥ 0	–	0.793	0.295	–	NS	NS
Cumulative T ≥ 5	–	0.792	0.367	–	NS	NS
Annual precipitation	–	0.644	NS	–	NS	0.799
P range	–	0.339	NS	–	NS	NS
Months T ≥ 10 P ≥ 40	–	0.844	0.458	–	NS	0.677
PET	–	0.745	0.325	–	NS	NS
Mean annual NDVI	–	0.864	0.601	–	NS	NS
NDVI 1SD	–	–0.683	–0.407	–	NS	NS

($p < 0.0001$; NS, not significant, $p > 0.0001$)

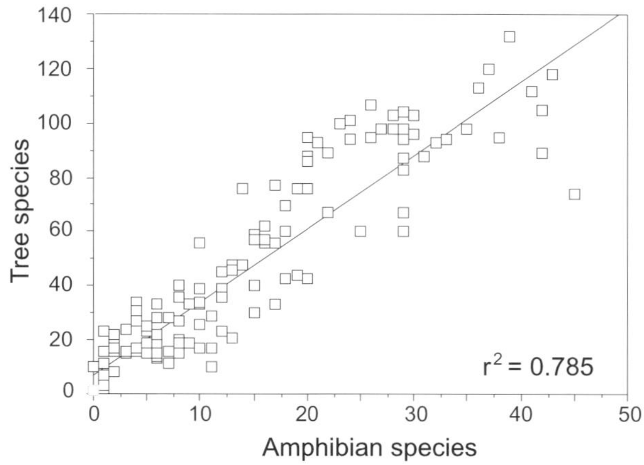

Fig. 8. The relationship between North American amphibian and tree species diversity, reflecting the dependence of both on water availability and temperature.

environment. At the coarsest physiological scale this resolves itself into the distinction between the ectothermic and endothermic components of non-avian tetrapod diversity, which is shown plotted against mean annual temperature (MAT) in Figure 11. This reveals a strong linear relationship, with no hemispheric asymmetries or other regional heterogeneities, and no apparent signal in the residuals (Fig. 12). This is supported by the high rho values in Table 7.

An additional question is the extent to which taxonomy masks the physiological diversity signal. Figure 13 shows the results of a correspondence analysis of modern non-avian tetrapod genera in North America and Europe and these results are presented geographically in Figure 14. The correlations between each axis and the environmental variables are assessed using the Spearman rank test and are listed in Table 8. These results show how the first axis, which accounts for 37.2% of the variance, is dominated by the historical biogeographical differences between North America and Europe (this is not seen when Europe and North

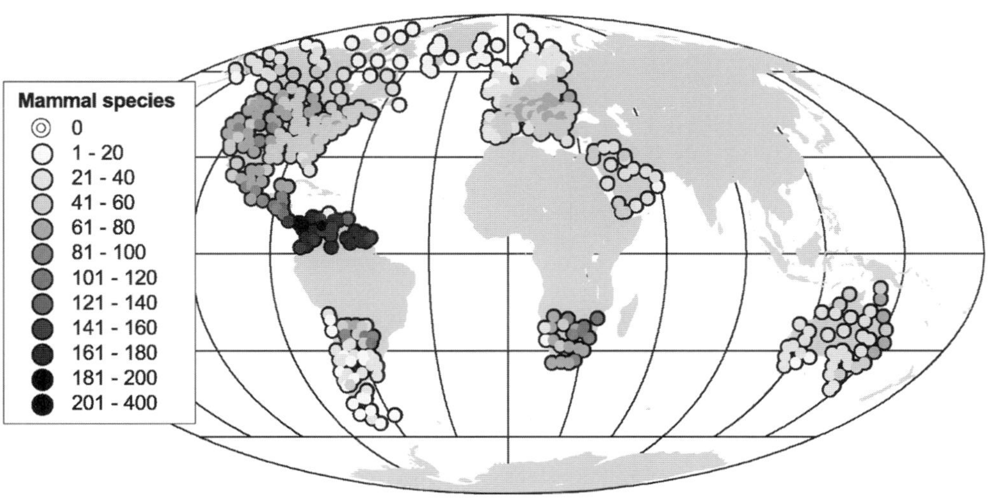

Fig. 9. The distribution of mammal species diversity for North America, Europe, southern Africa, Arabia, tropical South America and Australia.

Table 6. *Spearman rank test rho values for environmental variables and mammal species diversity by region*

	South America	North America	Europe	Arabia	Southern Africa	Australia
n	72	165	204	18	21	40
Absolute latitude	–0.831	–0.658	–0.552	NS	NS	NS
Elevation	NS	0.511	0.517	NS	NS	NS
MAT	0.715	0.615	0.323	NS	NS	NS
MART	–0.670	–0.400	–0.321	NS	NS	NS
CMM	0.786	0.583	NS	NS	NS	NS
WMM	0.513	0.529	0.513	NS	NS	NS
Radiation	NS	NS	NS	NS	NS	NS
Cumulative T ≥ 0	0.714	0.628	0.347	NS	NS	NS
Cumulative T ≥ 5	0.715	0.627	0.444	NS	NS	NS
Annual precipitation	0.565	NS	NS	NS	NS	0.728
P range	0.712	0.323	NS	NS	NS	NS
Months T ≥ 10 P ≥ 40	0.713	0.337	0.411	NS	NS	0.643
PET	0.633	0.526	0.385	NS	NS	NS
Mean annual NDVI	0.636	0.442	0.616	NS	NS	NS
NDVI 1SD	–0.765	–0.511	–0.478	NS	NS	NS

($p < 0.0001$; NS, not significant, $p > 0.0001$)

America are treated separately, where axis 1 approximates axis 2 in this analysis). The greatest differences in the faunas are between southern Europe and southern North America, consistent with the historical faunal interchanges being via the North Atlantic land bridges that linked the two continents during the Early Cenozoic. The second axis (Fig. 14) accounts for 28% of the variance and correlates most strongly with temperature and therefore incident energy (Table 8). Axis 3 (13.4% of variance) is dominated by the precipitation pattern.

Application to the fossil record

Ostrom (1970) postulated the use of diversity gradients as a palaeoclimate tool. In order to examine the viability of this suggestion, an experiment is made in which the linear relationship between the proportion of the fauna represented by ectotherms and MAT (Fig. 11) is used to reconstruct the palaeotemperature of the Middle Eocene Messel Shale, Germany, based on its fossil non-avian tetrapod fauna. Messel, as a lagerstätten, is used in order to minimize the effects of compositional biases (taphonomy, taxonomy and collection). Species, genus and family diversities for this fauna are derived from the faunal lists given in Schaal & Ziegler (1992). Additional linear relationships between the proportion of ectotherms and coldest month mean temperature (CMM) and latitude have also been made. Errors from these calculations represent one standard deviation of the residuals of MAT, CMM and absolute latitude (Fig. 12, Table 9), reflecting potential effects of history, biogeography and taxonomy on present-day data.

The results for this analysis of the Messel Shale fauna are given in Table 9 and compared with estimates from other climate proxies in Table 10. The values for MAT and CMM calculated using the proportion of reptiles or ectotherms agree with the climate interpretations based on the presence of fossil crocodilians (Markwick 1998*a*) and palms (Markwick 1996), although they are somewhat lower than the MAT estimated by Wilde (1989) based on plant physiognomy. However, estimates of absolute latitude are generally lower than the 44° palaeolatitude calculated for this site using palaeomagnetic data (D.B. Rowley pers comm. 1995). Regressions based on the total number of non-avian tetrapods give consistently lower MATs and CMMs than those using the proportion of reptiles or ectotherms; they also give higher absolute latitudes (Table 9). In both cases the results are similar to what would be expected at the palaeolatitude of Messel in the Eocene (44°N) if the Earth's latitudinal thermal gradient was the same as today's, which would imply that the absolute number of species of non-avian tetrapods is indeed a function of latitude (and thus of incident solar energy flux). However, such an interpretation may be premature, since it is based only on one fossil fauna. Variations in predicted temperatures between taxonomic

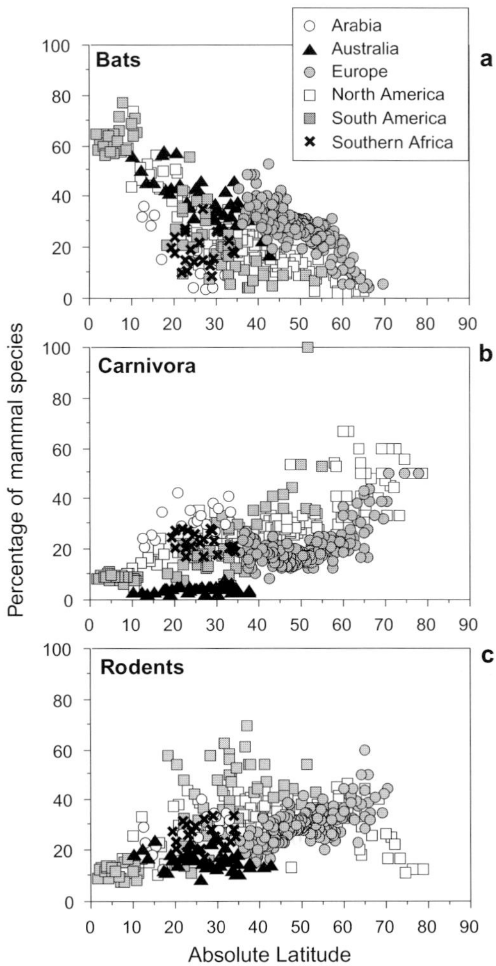

Fig. 10. The proportion of mammal species represented by (**a**) bats, (**b**) Carnivora and (**c**) rodents, as a function of absolute latitude.

levels may, at least in part, reflect the degree to which each is completely sampled by the fossil record.

Discussion

In general, the results presented here (Figs 3, 5, 6, 7 and 9) are consistent with the continental-scale patterns and trends found by Currie (1991) for North America, and by Pianka (1981) for Australia, and appear to support the species–energy hypothesis. However, Currie's conclusion, that PET (as a proxy for energy) correlates best with species diversity patterns, is not found in this study to apply to all continents, nor to all taxonomic groups (Table 2, Fig. 15). The problem, as stated at the beginning of this paper, is that different organisms are limited by different environmental factors (and combinations of factors) and to different extents. It is therefore difficult to imagine why any one environmental parameter should account for all of the diversity patterns around the globe (Gaston 2000). It is also important to distinguish between the pattern of species diversity, which requires a causal explanation, and latitudinal gradients, which are a graphical abstraction of the data. Latitude *per se* cannot be a determinant of species richness (Gaston 2000).

The links between climate, biogeography and diversity are obviously complex and, given the results presented here, must be considered in terms of not only absolute numbers of taxa, but also the physiology and ecology of those taxa. Reptiles show the simplest and steepest diversity gradients, which appear to be independent of regional and hemispheric biases (Fig. 3c). As ectotherms, reptile survival depends, primarily, upon absorption of energy from the environment, above some critical minimum energy (temperature) level (Fig. 4d), and this is supported by the correlations found in this study (Tables 2 and 4). The pattern of amphibian diversity (Fig. 3d) reflects amphibians' physiological and ecological dependence on both temperature and water (Tables 2 and 5; Fig. 7). This is a requirement shared by plants, which for North America at least, show a similar diversity distribution to amphibians (Fig. 8).

The pattern of mammalian species diversity is far more complex (Fig. 3b). There is no simple monotonic gradient from high to low latitudes and they do show a hemispheric asymmetry. As endotherms, mammals' dependence on primary energy sources is indirect through the filter of their various feeding strategies. Their diversity patterns may still ultimately reflect a climate signal (Frey 1992; Janis 1993; Janis *et al.* 2000), and consequently the physiological and ecological structure of mammalian faunas can be used to reconstruct past habitats and thereby climate (Andrews *et al.* 1979), but an understanding of mammal diversity in terms of energy requires a detailed understanding of mammalian ecology. Endothermic herbivores, for example, should show their highest diversity in regions of high plant productivity, which is itself a function of temperature and precipitation (Lottes & Ziegler 1994). Such a conclusion is supported by the highest rho values for mammal diversity in Table 2 correlating with mean annual NDVI. Interestingly, Janis *et al.* (2000) have postulated that the decline in North American ungulate

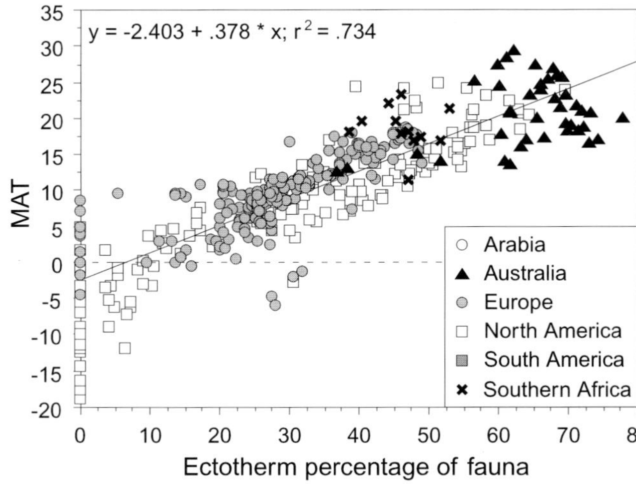

Fig. 11. Mean annual temperature (MAT) as a function of the percentage of each fauna represented by ectotherm (amphibian and reptile) species. The least squares regression through these data is used to retrodict MATs for specified fossil faunal compositions.

species diversity during the Neogene may reflect a corresponding decrease in primary productivity.

These physiologically based differences in the pattern of taxonomic diversity suggest that the species–energy hypothesis should be modified to account for how organisms procure energy, rather than describing diversity as simply a function of the amount of available energy. But diversity must also be considered in the context of history. Organisms rarely occupy the full geographic area that physiologically they could survive in. Alligatorids are not found in Australasia or Africa (Markwick 1998*a*, *b*); salamanders are absent from sub-Saharan Africa, SE Asia and Australasia, which partly contributes to the lower amphibian diversities observed for these regions in Figure 3d. These absences reflect the

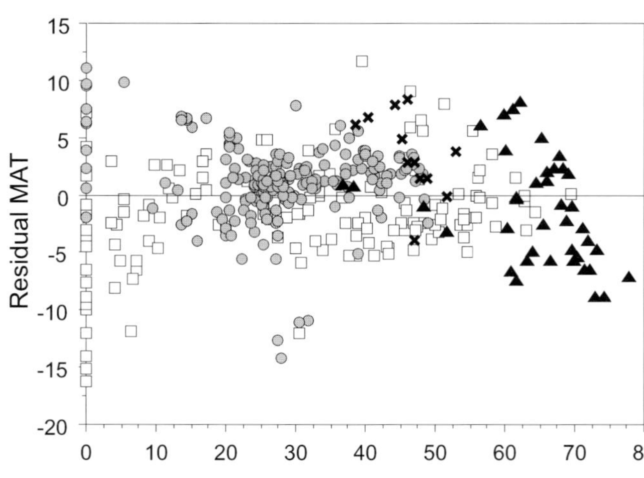

Fig. 12. The residuals for MAT from Figure 11, as a function of percentage ectotherm species. No trends are observed.

Table 7. *Spearman rank test rho values for environmental variables and the percentage of fauna that are represented by herptile (amphibian + reptile) species, by region*

	South America	North America	Europe	Arabia	Southern Africa	Australia
n	–	144	204	–	21	40
Absolute latitude	–	–0.923	–0.861	–	NS	NS
Elevation	–	NS	NS	–	NS	NS
MAT	–	0.902	0.769	–	NS	NS
MART	–	–0.489	NS	–	NS	NS
CMM	–	0.820	0.602	–	NS	NS
WMM	–	0.897	0.792	–	NS	NS
Radiation	–	NS	0.830	–	NS	NS
Cumulative T ≥ 0	–	0.913	0.772	–	NS	NS
Cumulative T ≥ 5	–	0.918	0.807	–	NS	NS
Annual precipitation	–	0.520	NS	–	NS	–0.731
P range	–	0.333	NS	–	NS	NS
Months T ≥ 10 P ≥ 40	–	0.799	0.380	–	NS	–0.763
PET	–	0.901	0.781	–	NS	NS
Mean annual NDVI	–	0.816	0.588	–	NS	NS
NDVI 1SD	–	–0.832	–0.817	–	NS	NS

($p < 0.0001$; NS, not significant, $p > 0.0001$)

consequences of barriers to dispersal during the Cenozoic. Similarly, the diversity of most organic groups follows distinct patterns of diversification through time (Sepkoski 1996), which means that the overall diversity of any group at a specific moment will depend on where it is placed in its diversification history.

These historical factors appear to affect the taxonomic composition of the fauna (names), rather than absolute numbers or the relationship between species diversity and environmental variables. The correspondence analysis of North American and European non-avian tetrapod generic names (Figs 13 and 14, Table 8), supports this. Axis 1 (Fig. 14) reflects historical compositional differences between the two continents, but the remaining variance comprises the physiological structure of the diversity (temperature, axis 2, and water, axis 3; Figs 13 and 14), which dominate when the continents are considered individually. This is consistent with studies of Pleistocene beetle assemblages that provide a sensitive measure of palaeoclimate throughout the Quaternary (Atkinson *et al.* 1986). Despite the fact that with each interglacial–glacial fluctuation the composition of the

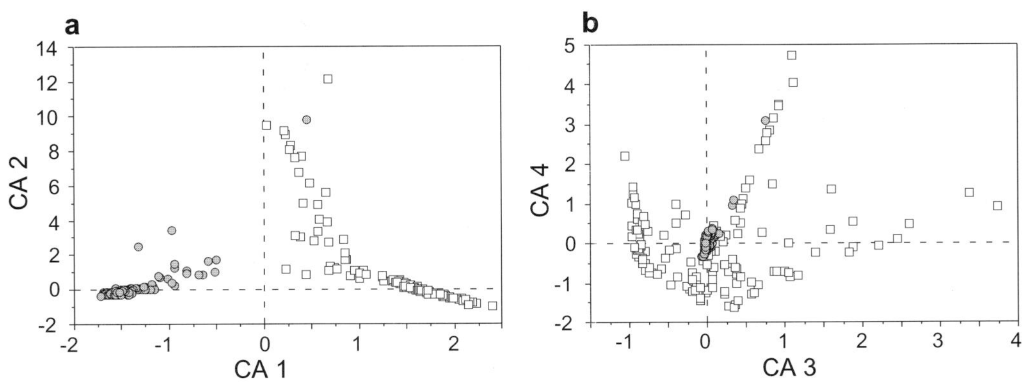

Fig. 13. The results of a correspondence analysis (CA) of North American and European genera. (**a**) Axes 1 and 2, which comprise 65.2% of the total variance. (**b**) Axes 3 and 4.

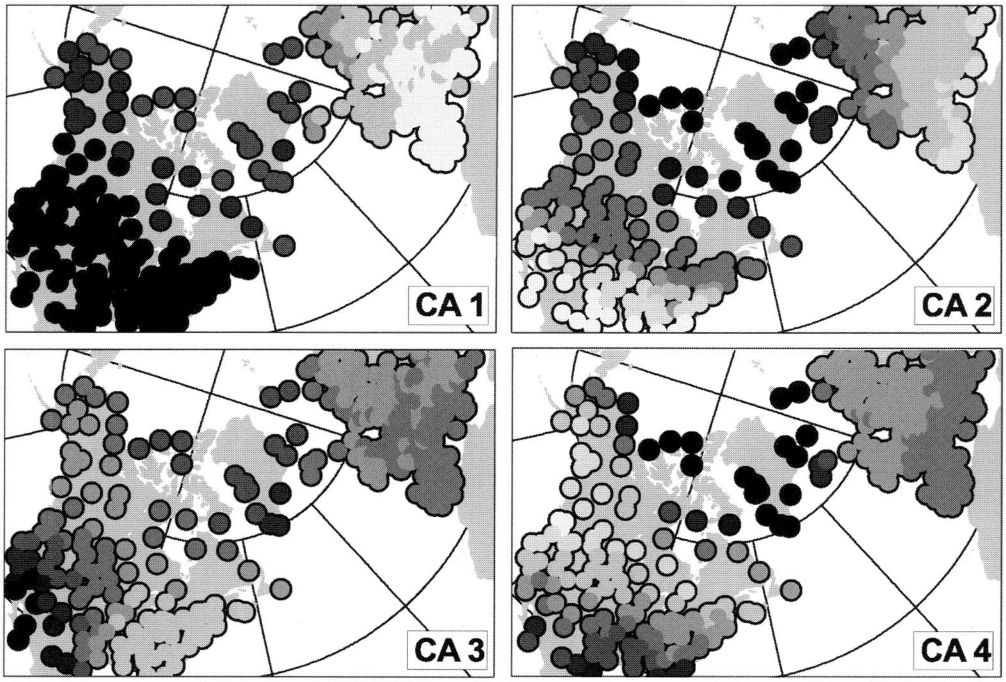

Fig. 14. The geographic distribution of the scores from Figure 13. The scores for each axis are represented by shades of grey, from white (the most negative extreme in each case), through grey to black (the most positive score for each axis).

Table 8. *Spearman rank test rho values for the correspondence analysis of North American and European genera*

	CA 1	CA 2	CA 3	CA 4
Cumulative percentage variance	37.2	65.2	78.6	86.5
n	347	347	347	347
Absolute latitude	NS	0.823	NS	–0.275
Longitude	–0.673	NS	NS	NS
Elevation	NS	NS	NS	NS
MAT	NS	–0.886	NS	0.403
MART	0.355	0.374	NS	–0.294
CMM	NS	–0.799	NS	0.410
WMM	NS	–0.792	NS	0.270
Radiation	0.550	–0.614	NS	NS
Cumulative T ≥ 0	NS	–0.873	NS	0.365
Cumulative T ≥ 5	NS	–0.865	NS	0.342
Annual precipitation	NS	–0.216	–0.499	NS
P range	NS	–0.261	NS	NS
Months T ≥ 10 P ≥ 40	NS	–0.571	–0.410	NS
PET	NS	–0.868	NS	0.522
Mean annual NDVI	NS	–0.626	NS	0.288
NDVI 1SD	NS	–0.775	NS	–0.490

($p < 0.0001$; NS, not significant, $p > 0.0001$)

Table 9. *Predicted palaeotemperature and palaeolatitude values for the Middle Eocene Messel Shale fauna, Germany*

	Composition	MAT (°C)	CMM (°C)	Absolute latitude (deg.)
Species				
Reptiles	39.7%	16.2 ± 3.6	7.6 ± 5.7	35.8 ± 5.6
Ectotherms	46.0%	13.6 ± 3.8	4.4 ± 6.2	40.4 ± 5.5
Tetrapods	n = 63	7.5 ± 5.4	–2.8 ± 8.5	50.8 ± 6.0
Genera				
Reptiles	43.4%	17.9 ± 3.5	9.5 ± 5.6	32.9 ± 5.3
Ectotherms	50.9%	16.5 ± 3.6	8.1 ± 6.0	35.5 ± 5.0
Tetrapods	n = 53	10.0 ± 5.5	0.0 ± 8.7	46.2 ± 6.3
Families				
Reptiles	40.0%	20.0 ± 3.6	12.5 ± 5.4	30.6 ± 6.8
Ectotherms	47.1%	13.9 ± 4.3	4.8 ± 6.5	40.5 ± 7.7
Tetrapods	n = 34	15.7 ± 6.1	6.1 ± 9.3	3.4 ± 8.5

See text for description of the method

Table 10. *Palaeotemperature and latitude estimates for the Middle Eocene of Messel, based on other data (see text for details)*

	MAT (°C)	CMM (°C)	Absolute latitude (deg.)
Modern values	8.6	0	49.8
Fossil floras	25 – 30	> 10	
Presence of fossil crocodilians	> 14.2	> 5.5	
Presence of fossil palms	> 13	> 0	
Palaeogeography			44.0 ± 0.9

faunas changed, such that the assemblage of one interglacial will be different from the next in the same place, they still represent the same climate interpretation (e.g. Elias 1994): different taxonomic participants, but the same physiologically dependent diversity structure.

Although the results of this study strongly imply an underlying fundamental relationship between environmental energy procurement and terrestrial taxonomic diversity, the exact mechanism by which this proceeds has not been addressed. It is possible that it could be through affecting evolutionary rates (i.e. generation, mutation and selection) as postulated by various authors (Stevens 1989; Currie 1991; Rohde 1992). Area is also important, and has been shown to be a fundamental determinant of global diversity patterns (MacArthur & Wilson 1963; Sepkoski 1976; Rosenzweig 1995), and it is interesting to note that not only do areas of high orographic relief fall off derived regressions using the methodology in this study, but so too do islands (Fig. 16).

These relationships between diversity and energy procurement have some important consequences. Firstly, if the relationship between the proportion of ectotherms and endotherms is a function of climate (Fig. 11) then this may indeed provide a tool for palaeoclimate as suggested by Ostrom (1970), and supported by the experiment using the Messel Fauna presented in his paper. However, at this stage such an application should be viewed with caution, since it is unclear why the proportion of ectotherms to endotherms should follow the strong linear trend observed in Figure 11. Secondly, if energy, as temperature, is the limiting factor for ectotherms, then as temperatures increase towards low latitudes not only should ecotherm diversity increase, but the proportion of ectotherms comprising faunas should also increase. This is indeed the case (Fig. 11), and this would seem to imply that the limit on reptile diversity in the modern world is contemporary climate, which has major implications for the consequences of future climate

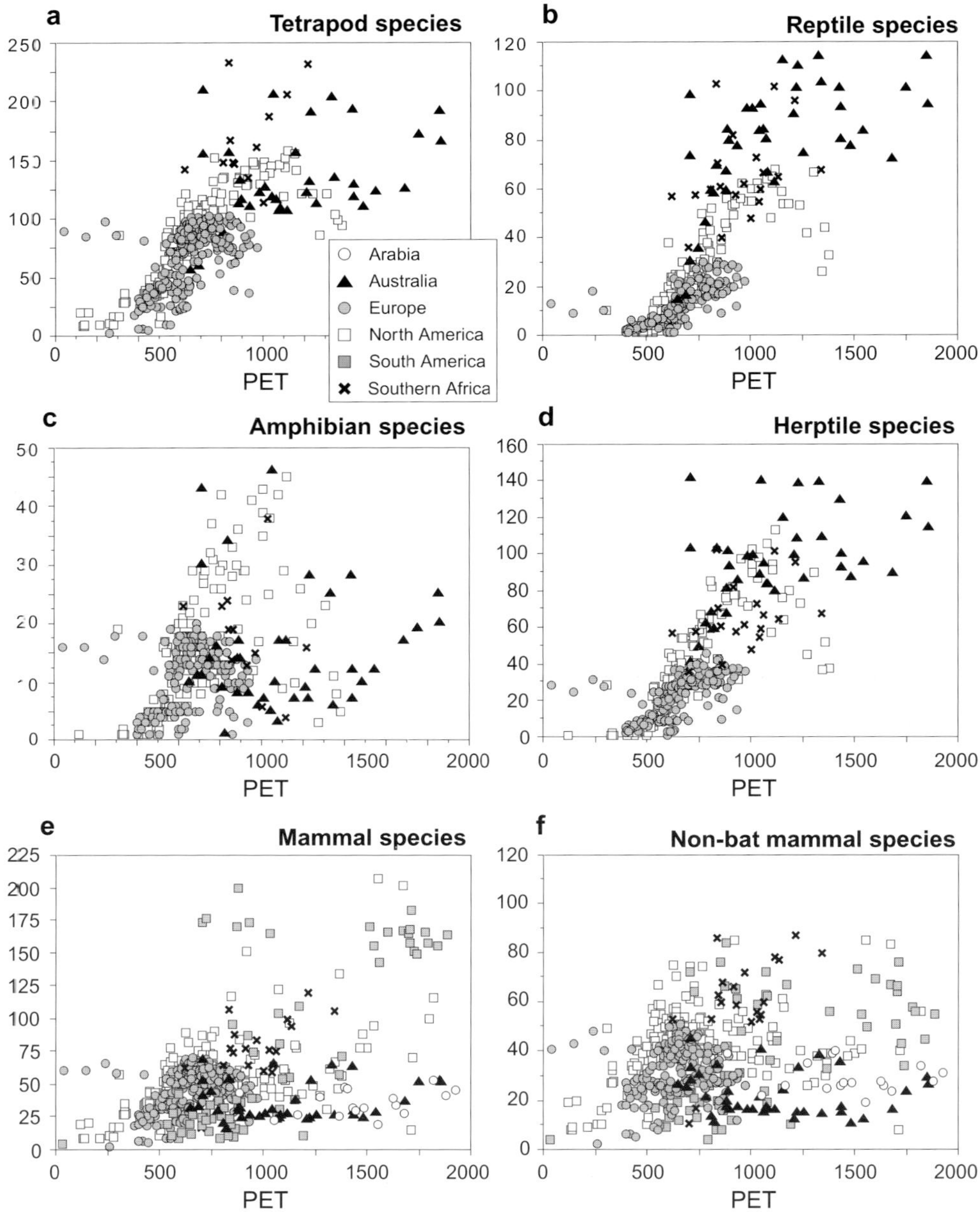

Fig. 15. Diversity plotted with respect to PET. The close correlation between PET and the diversity of all groups in North America is not seen in other parts of the world.

change. It also suggests that the critical limit on species numbers is survival (extinction rate). Ultimately, an organism is limited by its ability to survive, and that depends on its adaptive success in procuring energy and its physiological ability to exist in a particular environmental space.

Conclusions

A new GIS-based dataset of Recent taxonomic diversity has been used to investigate the link between modern terrestrial biodiversity patterns and environmental variables. The results suggest the following.

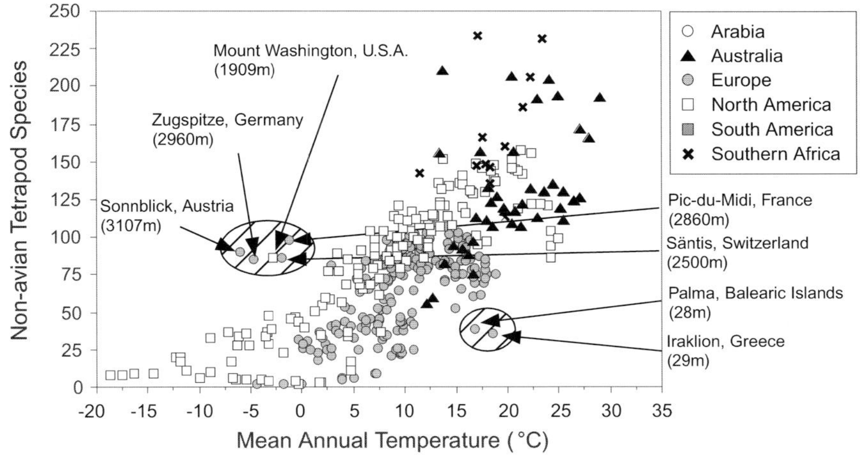

Fig. 16. The relationship of mean annual temperature (MAT) and the diversity of non-avian tetrapod species, showing how points of high orographic relief (in which diversity is overestimated by mixing low- and high-elevation faunas for climate stations located on the peaks rather than in the valleys) and islands (in which diversities are lower than would be expected) fall off the general trend observed.

(i) Terrestrial non-avian tetrapod species diversity is a function of differing adaptive strategies for procuring energy, either directly via solar energy (e.g. reptiles) or indirectly via food (e.g. mammals), as indicated by the different patterns of diversity shown by ectotherms and endotherms. The link between diversity and environmental parameters therefore depends on the physiology and ecology of the organisms themselves.

(ii) History influences the taxonomic composition of terrestrial diversity, but much less the physiological structure.

(iii) The proportion of ectotherm species, genera and families in faunas is found to increase linearly with increasing temperature. This is used to retrodict the palaeotemperature of Middle Eocene Messel, Germany, with results that agree well with estimates of temperature from other climate proxies.

(iv) Given an understanding of animal physiology, the link between biogeography, biodiversity and environmental variables can be demonstrated. This has major implications for predicting the response of organisms to future climate and environmental changes.

I would like to thank A. Crame for giving me the opportunity of presenting this material and for organizing the Lyell 2001 meeting. This study was begun in 1990 at The University of Chicago where funding was provided through a Teaching Assistantship. I would like to thank A. Ziegler, D. Jablonski, D. Rowley, R. Lupia and P. Wagner for reading earlier drafts and for helpful criticisms. The final version greatly benefited from the reviews of A. Crame and an anonymous reviewer.

References

ANDERSON, S. & MARCUS, L. F. 1993. Effect of quadrat size on measurement of species density. *Journal of Biogeography*, **20**, 421–428.

ANDREWS, P., LORD, J. M. & EVANS, E. M. N. 1979. Patterns of ecological diversity in fossil and modern mammalian faunas. *Biological Journal of the Linnean Society*, **11**, 177–205.

ARNOLD, E. N. & BURTON, J. A. 1978. *A Field Guide to the Reptiles and Amphibians of Britain and Europe*. Collins, London.

ATKINSON, T. C., BRIFFA, K. R., COOPE, G. R., JOACHIM, M. J. & PERRY, D. W. 1986. Climatic calibration of coleopteran data. *In*: BERGLUND, B. E. (ed.) *Handbook of Holocene Palaeoecology and Palaeohydrology*. Wiley, Chichester, 851–858.

BOUCHARDY, C. & MOUTOU, F. 1989. *Observing British and European Mammals*. British Museum (Natural History), London.

BRANCH, B. 1988. *Field Guide to the Snakes and Other Reptiles of Southern Africa*. New Holland, London.

BRISBIN, I. L., JR., STANDORA, E. A. & VARGO, M. J. 1982. Body temperatures and behavior of American alligators during cold winter weather. *The American Midland Naturalist*, **107**, 209–218.

BROWN, J. H. 1995. *Macroecology*. University of Chicago Press, Chicago.

CIHLAR, J., ST.-LAURENT, L. & DYER, J. A. 1991. Relation between the Normalized Difference Vegetation Index and ecological variables. *Remote Sensing of Environment*, **35**, 279–298.

COGGER, H. G. 1992. *Reptiles and Amphibians of Australia*. Comstock, Cornell.

COLBERT, E. H., COWLES, R. B. & BOGERT, C. M. 1946. Temperature tolerances in the American alligator and their bearing on the habits, evolution, and extinction of the Dinosaurs. *Bulletin of the American Museum of Natural History*, **86**, 327–374.

CONANT, R. & COLLINS, J. T. 1991. *A Field Guide to Reptiles and Amphibians. Eastern/Central North America*. Houghton Mifflin, Boston.

COOK, F. R. 1984. *Introduction to Canadian Amphibians and Reptiles*. National Museums of Canada, Ottawa.

CRAME, J. A. 2000. Evolution of taxonomic diversity gradients in the marine realm: evidence from the composition of Recent bivalve faunas. *Paleobiology*, **26**, 188–214.

CRAME, J. A. 2001. Taxonomic diversity gradients through geological time. *Diversity & Distributions*, **7**, 175–189.

CURRIE, D. J. 1991. Energy and large-scale patterns of animal- and plant- species richness. *The American Naturalist*, **137**, 27–49.

CURRIE, D. J. & FRITZ, J. T. 1993. Global patterns of animal abundance and species energy use. *OIKOS*, **67**, 56–68.

DARLINGTON, P. J. 1948. The geographical distribution of cold-blooded vertebrates. *The Quarterly Review of Biology*, **23**, 1–26, 105–123.

EISENBERG, J. F. 1989. *Mammals of the Neotropics. The Northern Neotropics. Volume 1. Panama, Colombia, Venezuela, Guyana, Suriname, French Guiana*. University of Chicago Press, Chicago.

ELIAS, S. A. 1994. *Quaternary Insects and their Environments*. Smithsonian Institution Press, Washington, DC.

FISCHER, A. G. 1961. Latitudinal variations in organic diversity. *American Scientist*, **49**, 50–74.

FLESSA, K. W. 1975. Area, continental drift and mammalian diversity. *Paleobiology*, **1**, 189–194.

FLESSA, K. W. & SEPKOSKI, J. J. JR 1978. On the relationship between Phanerozoic diversity and changes in habitable area. *Paleobiology*, **4**, 359–366.

FREY, J. K. 1992. Response of a mammalian faunal element to climatic changes. *Journal of Mammalogy*, **73**, 43–50.

GASTON, K. J. 2000. Global patterns in biodiversity. *Nature*, **405**, 220–227.

GASTON, K. J., WILLIAMS, P. H., EGGLETON, P. & HUMPHRIES, C. 1995. Large scale patterns of biodiversity: spatial variation in family richness. *Proceedings of the Royal Society, London, B*, **260**, 149–154.

GOWARD, S. N. & DYE, D. G. 1987. Evaluating North American net primary productivity with satellite observations. *Advances in Space Research*, **7**, 165–174.

GOWARD, S. N., TUCKER, C. J. & DYE, D. G. 1985. North American vegetation patterns observed with the NOAA-7 advanced very high resolution radiometer. *Vegetatio*, **64**, 3–14.

GRENARD, S. 1991. *Handbook of Alligators and Crocodiles*. Krierger Publishing Company, Malabar, Florida.

HALL, E. R. & KELSON, K. R. 1959. *The Mammals of North America*. Ronald Press, New York.

HARRISON, S. P., YU, G., TAKAHARA, H. & PRENTICE, I. C. 2001. Plant diversity and palaeovegetation in East Asia. *Nature*, **413**, 129–130.

HAYCOCK, K., ROTH, J. & GAGNON, J. 1992–1993. *Statview*. Abacus Concepts, Inc., Berkeley, California.

HIBBARD, C. W. 1960. An interpretation of Pliocene and Pleistocene climates in North America. *Annual Report of the Michigan Academy of Science, Arts, and Letters*, **62**, 5–30.

HUMBOLDT, A. V. & BONPLAND, A. 1807. *Essai sur la géographie des plantes; accompagne d'un tableau physique des régions équinoxiales*. Fr. Schoell, Paris.

IVERSON, J. B. 1992. *A Revised Checklist with Distribution Maps of the Turtles of the World*. Privately published, Richmond, Indiana.

JANIS, C. M. 1993. Tertiary mammal evolution in the context of changing climates, vegetation, and tectonic events. *Annual Review Ecological Systems*, **24**, 467–500.

JANIS, C. M., DAMUTH, J. & THEODOR, J. M. 2000. Miocene ungulates and terrestrial primary productivity: where have all the browsers gone? *Proceedings of the National Academy of Science USA*, **97**, 7899–7904.

KERR, J. T. & PACKER, L. 1997. Habitat heterogeneity as a determinant of mammal species richness in high-energy regions. *Nature*, **385**, 252–254.

KINGDON, J. 1990. *Arabian Mammals. A Natural History*. Academic Press, London.

LATHAM, R. E. & RICKLEFS, R. E. 1993. Global patterns of tree species richness in moist forests: energy-diversity theory does not account for variation in species richness. *Oikos*, **67**, 325–333.

LITTLE, E. L. 1977. *Atlas of United States Trees*. United States Department of Agriculture Forest Service, United States Government Printing Office, Washington, DC.

LOTTES, A. L. & ZIEGLER, A. M. 1994. World peat occurrence and the seasonality of climate and vegetation. *Palaeogeography, Palaeoclimatology, Palaeoecology*, **106**, 23–37.

LYELL, C. 1830. *Principles of Geology, being an attempt to explain the former changes of the Earth's surface, by reference to causes now in operation*, vol. 1. John Murray, London.

MACARTHUR, R. H. & WILSON, E. O. 1963. An equilibrium theory of insular zoogeography. *Evolution*, **17**, 373–387.

MACARTHUR, R. H. 1967. *The Theory of Island Biogeography*. Princeton University Press, Princeton, NJ.

MARKWICK, P. J. 1994. 'Equability', continentality and Tertiary 'climate': the crocodilian perspective. *Geology*, **22**, 613–616.

MARKWICK, P. J. 1996. *Late Cretaceous to Pleistocene climates: nature of the transition from a 'hot-house' to an 'ice-house' world*. PhD Thesis, University of Chicago.

MARKWICK, P. J. 1998*a*. Fossil crocodilians as indicators of Late Cretaceous and Cenozoic climates: implications for using palaeontological data for global change. *Palaeogeography, Palaeoclimatology, Palaeoecology*, **137**, 205–271.

MARKWICK, P. J. 1998*b*. Crocodilian diversity in space and time: the role of climate in paleoecology and its implication for understanding K/T extinctions. *Paleobiology*, **24**, 470–497.

MATTHEW, W. D. 1915. Climate and evolution. *Annals of the New York Academy of Science*, **24**, 171–318.

MÜLLER, M. J. 1982. *Selected climatic data for a global set of standard stations for vegetation science. Tasks for vegetation science*. Dr W. Junk Publishers, The Hague.

O'BRIEN, E. M. 1993. Climatic gradients in woody plant species richness: towards an explanation based on an analysis of southerm Africa's woody flora. *Journal of Biogeography*, **20**, 181–198.

OSTROM, J. H. 1970. Terrestrial vertebrates as indicators of Mesozoic climates. *Proceedings of the North American Paleontological Convention*, 347–376.

PIANKA, E. R. 1966. Latitudinal gradients in species diversity: a review of concepts. *The American Naturalist*, **100**, 33–46.

PIANKA, E. R. 1981. Diversity and adaptive radiations of Australian desert lizards. *In*: KEAST, A. (ed.) *Ecological Biogeography of Australia*. Dr W. Junk Publishers, The Hague, 1377–1392.

QIAN, H. & RICKLEFS, R. E. 2000. Large-scale processes and the Asian bias in species diversity of temperate plants. *Nature*, **407**, 180.

REDFORD, K. H. & EISENBERG, J. F. 1992. *Mammals of the Neotropics. The Southern Cone. Volume 2. Chile, Argentina, Uruguay, Paraguay*. University of Chicago Press, Chicago.

ROHDE, K. 1992. Latitudinal gradients in species diversity: the search for the primary cause. *Oikos*, **65**, 514–527.

ROHDE, K. 1997. The larger area of the tropics does not explain latitudinal gradients in species diversity. *Oikos*, **79**, 169–172.

ROHDE, K., HEAP, M. & HEAP, D. 1993. Rapoport's Rule does not apply to marine teleosts and cannot explain latitudinal gradients in species richness. *The American Naturalist*, **142**, 1–16.

ROOT, T. L. & SCHNEIDER, S. H. 1993. Can large-scale climatic models be linked with multiscale ecological studies? *Conservation Biology*, **7**, 256–270.

ROSENZWEIG, M. L. 1995. *Species Diversity in Space and Time*. Cambridge University Press, Cambridge.

ROY, K., JABLONSKI, D. & VALENTINE, J. W. 1994. Eastern Pacific molluscan provinces and latitudinal diversity gradient: no evidence for 'Rapoport's rule'. *Proceedings of the National Academy of Science U.S.A.*, **91**, 8871–8874.

SCHAAL, S. & ZIEGLER, W. 1992. *Messel. An Insight into the History of Life and of the Earth*. Clarendon Press, Oxford.

SEPKOSKI, J. J. JR. 1976. Species diversity in the Phanerozoic: species-area effects. *Paleobiology*, **2**, 298–303.

SEPKOSKI, J. J. JR. 1996. Competition in macroevolution: the double wedge revisited. *In*: JABLONSKI, D. , ERWIN, D. H. & LIPPS, J. H. (eds) *Evolutionary Paleobiology: Essays in Honor of James W. Valentine*. University of Chicago Press, Chicago, 211–255.

STEBBINS, R. C. 1985. *A Field Guide to Western Reptiles and Amphibians*. Houghton Mifflin Company, Boston.

STEVENS, G. C. 1989. The latitudinal gradient in geographical range: how so many species coexist in the tropics. *The American Naturalist*, **133**, 240–256.

STEVENS, G. C. 1992. The elevational gradient in altitudinal range: an extension of Rapoport's latitudinal rule to altitude. *The American Naturalist*, **140**, 893–911.

STRAHAN, R. 1992. *Encylopoedia of Australian Animals. Mammals*. Angus and Robertson, Pymble, NSW.

TER BRAAK, C. J. F. 1987–1992. *CANOCO – a FORTRAN program for Canonical Community Ordination*. Microcomputer Power, Ithaca, NY.

UHL, N. W. & DRANSFIELD, J. 1987. *Genera Palmarum. A Classification of Palms Based on the Work of Harold E. Moore, Jr.* Allen Press, Lawrence, Kansas.

VAN DER BRINK, F. H. 1967. *A Field Guide to the Mammals of Britain and Europe*. Collins, London.

WALLACE, A. R. 1876. *The Geographical Distribution of Animals*, vol. 1. MacMillan, London.

WILDE, V. 1989. Untersuchungen zur Systematik der Blattreste aus dem Mitteleozän der Grube Messel bei Darmstadt (Hessen, Bundesrepublik Deutschland). *Courier Forschungsinstitut Senckenberg*, **115**.

WRIGHT, D. H., CURRIE, D. J. & MAURER, B. A. 1993. Energy supply and patterns of species richness on local and regional scales. *In*: RICKLEFS, R. E. & SCHLUTER, D. (eds) *Species Diversity in Ecological Communities. Historical and Geographical Perspectives*. University of Chicago Press, Chicago, 66–74.

Index

Page numbers in *italics* refer to Figures and page numbers in **bold** refer to Tables.